Engineering Materials

This series provides topical information on innovative, structural and functional materials and composites with applications in optical, electrical, mechanical, civil, aeronautical, medical, bio- and nano-engineering. The individual volumes are complete, comprehensive monographs covering the structure, properties, manufacturing process and applications of these materials. This multidisciplinary series is devoted to professionals, students and all those interested in the latest developments in the Materials Science field, that look for a carefully selected collection of high quality review articles on their respective field of expertise.

Indexed at Compendex (2021) and Scopus (2022)

Kavindra Kumar Kesari · Chander Prakash · Mohammad Khalid · Arvind Negi
Editors

Functionalized Cellulose Materials

Sustainable Manufacturing and Applications

Springer

Editors
Kavindra Kumar Kesari
Chandigarh University
University Center for Research
and Development
Mohali, Punjab, India

Department of Applied Physics
Aalto University
Espoo, Finland

Mohammad Khalid
Sunway Centre for Electrochemical Energy
and Sustainable Technology (SCEEST)
School of Engineering and Technology
Sunway University
Subang Jaya, Selangor, Malaysia

Chander Prakash
Chandigarh University
University Center for Research
and Development
Mohali, Punjab, India

Arvind Negi
Department of Bioproducts and Biosystems
Aalto University
Espoo, Finland

ISSN 1612-1317 ISSN 1868-1212 (electronic)
Engineering Materials
ISBN 978-3-031-76952-8 ISBN 978-3-031-76953-5 (eBook)
https://doi.org/10.1007/978-3-031-76953-5

This Springer imprint is published by the registered company Springer Nature Switzerland AG
The registered company address is: Gewerbestrasse 11, 6330 Cham, Switzerland

Contents

Unveiling the Transformative Power of Smart Cellulosic Nanomaterials: Revisiting Potential Promises to Sustainable Future

Abhijeet Singh, Simrandeep Kaur, Harshita Thakur, Rashi, Sakshi Kashyap, Asyakina Lyudmila, and Gaurav Mudgal

Abstract This chapter explores the transformative potential of cellulosic nanomaterials in addressing global challenges and advancing sustainable development. As our planet faces interconnected crises of climate change, resource depletion, pollution, and public health, traditional materials and technologies are proving inadequate. Cellulosic nanomaterials, derived from the most abundant biopolymer on Earth, offer a promising solution. The chapter delves into the unique properties of cellulosic nanomaterials, including cellulose nanocrystals (CNCs) and cellulose nanofibrils (CNFs), highlighting their exceptional strength, biocompatibility, and renewability. It examines how manipulation at the nanoscale unlocks extraordinary functionalities, enabling innovations across various sectors such as advanced materials, energy storage, sustainable chemical processes, healthcare, and environmental monitoring. The text emphasizes the role of cellulosic nanomaterials in creating a more sustainable future, offering alternatives to petroleum-based products, and addressing critical global issues. While acknowledging the challenges in production, integration, and environmental impact assessment, the chapter underscores the rapid advancements in research and development aimed at overcoming these hurdles. By providing an in-depth exploration of cellulosic nanomaterials' properties, production methods, and applications, this chapter aims to accelerate their development and deployment. It presents a vision of a greener, more sustainable future enabled by these innovative materials, positioning them as key players in the transition towards a more resilient

A. Singh · S. Kaur · H. Thakur · Rashi · S. Kashyap
University Institute of Biotechnology, Chandigarh University, Mohali 140413, Punjab, India

G. Mudgal (✉)
Center for Waste Management and Renewable Energy, Saveetha Dental College and Hospitals, Saveetha Institute of Medical and Technical Sciences, Saveetha University, Chennai 600077, India
e-mail: gauravmdgl@gmail.com

Research and Development, Vanproz Agrovet LLP, Bagroda Industrial Area, Bhopal 462026, Madhya Pradesh, India

A. Lyudmila
Laboratory for Phytoremediation of Technogenically Disturbed Ecosystems, Kemerovo State University, Krasnaya Street, 6, Kemerovo, Russia 650000

K. K. Kesari et al. (eds.), *Functionalized Cellulose Materials*, Engineering Materials,
https://doi.org/10.1007/978-3-031-76953-5_1

and environmentally conscious society. At the outset, this compilation demonstrates how cellulosic nanomaterials contribute to achieving UN's SDGs 3, 6, 7, 9, 12, 13, and 15, offering innovative solutions for health, clean water, renewable energy, sustainable industry, responsible consumption, climate action, and environmental protection.

Keywords Nano · Cellulose · Purification · Light-weight · Durable · Ecofriendly · Sustainability · SDG

1 Introduction

Products, services, and commodities are a gift of human intellect-derived innovations that fuel the industrial processes for growth and development and upliftment from resource-poor, technologically deprived, and underdeveloped societies to more able solution mindsets. Our planet, however, stands at a critical juncture, grappling with myriad interconnected challenges. Climate change, resource depletion, pollution, and public health crises converge to form a complex and urgent global predicament. Traditional materials and technologies are often inadequate in addressing these multifaceted issues. At the heart of these challenges lies a material crisis. Conventional materials are reaching their limits regarding performance, sustainability, and resource efficiency. The demand for innovative materials that can deliver superior properties while minimizing environmental impact has never been greater. Nanomaterials, engineered at the atomic and molecular scale, offer a groundbreaking solution. These materials possess extraordinary properties not found in their bulk counterparts, enabling breakthroughs in various fields. From energy generation and storage to environmental remediation and healthcare, nanomaterials are poised to revolutionize how we address global challenges. The manipulation of matter at the atomic and molecular scale is poised to revolutionize our world. By engineering materials at such a minuscule level, we unlock extraordinary properties and functionalities that can address some of humanity's most pressing challenges. From energy generation and storage to healthcare, environmental remediation, and agriculture, nanotechnology offers innovative solutions. For instance, nanomaterials can enhance solar cell efficiency, create high-capacity batteries, and purify water contaminated with harmful pollutants. In the realm of healthcare, nanomedicine is transforming diagnostics and therapeutics, with applications ranging from targeted drug delivery to tissue engineering. Moreover, nanotechnology is instrumental in creating a more sustainable future. By developing materials with enhanced properties, we can reduce resource consumption, minimize waste, and improve energy efficiency. For example, nanomaterials can be used to create lightweight and durable materials for transportation, reducing fuel consumption and emissions. The potential of nanotechnology extends beyond technological advancements. It has the power to improve quality of life, address global health disparities, and foster economic growth. As we continue to explore the possibilities of this emerging field, it is essential to prioritize ethical considerations and responsible development to maximize its benefits while mitigating potential risks.

Cellulose, the most abundant biopolymer on Earth, has emerged as a promising feedstock for the development of sustainable nanomaterials. These materials, including cellulose nanocrystals (CNCs) and cellulose nanofibrils (CNFs), possess exceptional properties such as high strength, biocompatibility, and renewability [1, 2]. By manipulating matter at the nanoscale, exists a plethora of possibilities, to list a few such as those which can help to (i) develop advanced materials with enhanced strength, durability, and lightweight properties [3–5]; (ii) create efficient energy storage and conversion systems [6, 7]; (iii) design innovative catalysts for sustainable chemical processes [8–10]; (iv) develop targeted drug delivery systems for improved healthcare [11–15] and (v) create advanced sensors for detecting health issues, environmental monitoring and pollution control [16–18].

2 Cellulosic Nanomaterials Stand as a Beacon of Hope in Our Pursuit of a Sustainable Future

Derived from abundant and renewable cellulose, these materials offer a compelling alternative to traditional, often petroleum-based counterparts. Their exceptional properties, including high strength, biodegradability, and versatility, position them as key players in addressing global challenges such as climate change, resource depletion, and environmental pollution. By integrating cellulosic nanomaterials into various sectors, we can create a more sustainable and resilient society. From renewable energy and advanced materials to healthcare and environmental remediation, the potential applications are vast. While the potential of cellulosic nanomaterials is immense, several challenges need to be addressed for their widespread adoption. These include the development of cost-effective production methods, improving the dispersion of nanocellulose in polymer matrices, and understanding the long-term environmental impact of these materials. Despite these challenges, the field of cellulosic nanomaterials is rapidly advancing. Ongoing research and development efforts are focused on overcoming these hurdles and unlocking the full potential of this sustainable resource. By addressing these challenges and capitalizing on the opportunities, we can create a more sustainable future.

This chapter peeps deep into the exciting world of cellulosic nanomaterials, exploring their unique properties, production methods, and promising applications. By understanding the challenges and opportunities associated with these materials, we can accelerate their development and deployment for a greener and more sustainable future.

3 Smart Nanomaterials: How is It a Paradigm Shift?

Nanomaterials represent a paradigm shift in materials science, offering unprecedented control over properties at the atomic and molecular levels. Unlike traditional materials, which are limited by bulk properties, nanomaterials exhibit exceptional characteristics, including enhanced strength, conductivity, and reactivity, this versatility positions nanomaterials as a cornerstone for addressing global challenges such as energy, environment, and healthcare. For instance, graphene's unique structure enables its application in energy storage, water filtration, and electronics. Similarly, titanium dioxide nanoparticles demonstrate promise in self-cleaning surfaces and solar energy harvesting [19]. By tailoring the size, shape, and composition of nanomaterials, scientists can design materials with specific functionalities to meet diverse needs. This ability to engineer materials at the nanoscale opens new avenues for innovation and sustainable development. From clean energy technologies to advanced medical treatments, nanomaterials have the potential to revolutionize multiple industries and improve quality of life [20]. Table 1 summarises the key attributes of the nanocellulosic variants of technological interventions.

4 Advanced Nanocellulosic Materials with Enhanced Strength, Durability, and Lightweight Properties

The advent of nanotechnology has ushered in a new era of materials science, where properties can be engineered at the atomic and molecular levels. This unprecedented control has led to the development of nanocellulosic materials with extraordinary strength, durability, and lightweight characteristics. Some of the nuanced approaches are discussed ahead.

4.1 Enhancing Cement Composites with Nano-Cellulose and Nano-Silica

A recent study investigated the potential of combining nano-cellulose derived from wood sawdust with nano-silica to create a more sustainable and high-performance cement composite [87]. The research focused on the impact of these nanomaterials on the mechanical properties and microstructure of the cement matrix. Results demonstrated that the incorporation of even small amounts of nano-cellulose significantly improved the composite's strength and crack resistance. This enhancement was attributed to the ability of nano-cellulose fibers to bridge cracks and reinforce the cement matrix. Simultaneously, nano-silica contributed to improved pozzolanic reactivity, reducing the formation of calcium hydroxide and consequently lowering the material's carbon footprint. The study highlights the potential of utilizing waste

Table 1 An overview of key nanocellulosic materials, potential counterparts, and primary applications

Nanocellulosic material	Counterpart	Applications	Sectors/solutions	References
Cellulose nanofibrils (CNF)	Wood pulp, plant fibers	Reinforcement in composites; Packaging materials; Paper and cardboard enhancement; Biomedical scaffolds; Water purification membranes; 3D printing materials; Aerogels for insulation; Conductive materials (when combined with conductive polymers); Oil spill cleanup; Wound dressings	Automotive; Aerospace; Construction; Healthcare; Environmental; Electronics; Energy; Packaging	[21–28]
Cellulose nanocrystals (CNC)	Cotton, tunicate cellulose	Optical films and coatings; Drug delivery systems; Antimicrobial surfaces; Rheology modifiers; Emulsion stabilizers; Reinforcement in nanocomposites; Photonic materials; Sensors and biosensors; Energy storage devices; Security paper	Optics; Pharmaceuticals; Food industry; Materials science; Energy storage; Security	[22, 27, 29–34]
Bacterial nanocellulose (BNC)	Bacterial synthesis	Artificial skin and wound dressings; Blood vessel replacements; Food additives and packaging; High-fidelity acoustic diaphragms; Flexible displays; Tissue engineering scaffolds; Cosmetic face masks; Dental implants; Protective clothing; Water treatment filters	Healthcare; Food industry; Electronics; Cosmetics; Textiles; Environmental	[35–42]
TEMPO-oxidized cellulose nanofibrils	Chemically modified wood pulp	Transparent films; Gas barrier coatings; Flame-retardant materials; Hydrogels for drug delivery; Conductive paper; Supercapacitors; Nanopaper for printed electronics; Thickeners in cosmetics; Emulsion stabilizers in foods; Cement reinforcement	Packaging; Electronics; Energy storage; Construction; Cosmetics; Food industry	[39, 43–48]

(continued)

Table 1 (continued)

Nanocellulosic material	Counterpart	Applications	Sectors/solutions	References
Cellulose nanoyarns	Twisted nanofibrils	High-strength textiles; Artificial muscles; Wearable electronics; Sutures for surgery; Filtration membranes; Reinforcement in composites; Smart textiles; Energy harvesting devices; Protective clothing; Electrodes for batteries	Textiles; Healthcare; Electronics; Energy; Defense	[49–57]
Nanocellulose aerogels	Freeze-dried nanocellulose	Thermal insulation; Oil spill cleanup; Catalyst support; Drug delivery systems; Acoustic insulation; Tissue engineering scaffolds; Supercapacitor electrodes; Water purification; Gas sensors; Impact protection materials	Construction; Environmental; Chemical industry; Healthcare; Energy; Electronics	[10, 58–73]
Nanocellulose-polymer composites	Various polymer matrices	Lightweight structural materials; Barrier films for food packaging; Biodegradable plastics; Flame-retardant materials; Electromagnetic shielding; 3D printing filaments; Automotive interior parts; Sporting goods; Aerospace components; Marine applications	Automotive; Aerospace; Packaging; Sports; Marine; Electronics	[22, 30, 37, 41, 51, 56, 63, 74]
Functionalized nanocellulose	Surface-modified nanocellulose	Controlled drug release; Heavy metal removal from water; Self-cleaning surfaces; Antimicrobial packaging; Smart paper with sensing capabilities; Stimuli-responsive materials; Catalysis support; Nanocellulose-based inks; Biomedical implants; Nanocomposite dental fillings	Healthcare; Environmental; Packaging; Electronics; Chemical industry; Dentistry	[9, 63, 75–86]

products, such as wood sawdust, to create value-added materials with significant environmental benefits. By combining nano-cellulose and nano-silica, researchers have developed a promising approach to producing sustainable and durable cement-based composites.

In another study, the researchers explored the potential of bacterial nanocellulose (BNC) as a reinforcement for cement mortar [88]. By incorporating BNC in various forms—powder, gel, and as a coating on polypropylene fibers—researchers investigated its impact on mechanical properties. Results indicated that BNC significantly improved both flexural and compressive strength of cement mortar. BNC powder proved more effective than BNC gel in enhancing these properties. Additionally, the integration of BNC with polypropylene fibers created a synergistic effect, improving the overall performance of the composite material. This enhancement is attributed to the increased roughness and surface area of the fibers, facilitating better bonding with the cement matrix. Furthermore, the study demonstrated that BNC contributed to reduced water absorption in cement mortar, enhancing its durability and resistance to degradation [88]. These findings highlight the potential of BNC as a valuable additive for producing high-performance cement-based materials. By harnessing the unique properties of this bio-based nanomaterial, it is possible to develop more sustainable and durable construction materials [88, 89].

4.2 The Nanocellulosic Realm of Aerospace and Automotive Engineering

By incorporating nanomaterials into composites, engineers have created structures that are both lighter and stronger than traditional materials. This translates to significant fuel savings, reduced emissions, and enhanced vehicle performance. For instance, carbon nanotubes, when embedded in polymer matrices, produce composites with exceptional strength-to-weight ratios [90], making them ideal for aircraft/satellite components [91]. Beyond this, MIT engineers have revolutionized the aerospace industry by recently developing a groundbreaking method for producing high-performance composites without the need for massive autoclaves [92]. Traditionally, creating aircraft fuselages and wings involves stacking layers of composite materials and curing them under immense pressure and heat within enormous autoclaves. This process is time-consuming, energy-intensive, and requires substantial infrastructure. The MIT team has devised a novel approach that utilizes carbon nanotubes as a heating element and a pressure-generating mechanism. By sandwiching a layer of carbon nanotubes between composite materials and applying an electric current, the researchers can effectively fuse the materials without relying on external pressure vessels. This innovative technique significantly reduces manufacturing time and energy consumption while maintaining the exceptional strength and durability of traditionally produced composites. This breakthrough has the potential to transform the aerospace industry, enabling faster production of aircraft and other composite structures. Additionally, it opens up new possibilities for the manufacturing of other composite products, such as wind turbine blades and automotive components.

4.3 Supercar Crafted from Cellulose Nanofibers

Another prime example is the development of a supercar crafted from cellulose nanofibers (CNFs) [93]. This innovative material, derived from wood, offers exceptional strength and durability while being significantly lighter than traditional materials like steel. Japan's Ministry of Environment spearheaded this groundbreaking project, aiming to reduce carbon emissions associated with vehicle manufacturing. The resulting supercar showcases the potential of CNFs to revolutionize the automotive industry. By replacing conventional materials with this plant-based alternative, the car achieves a remarkable 50% weight reduction, leading to improved fuel efficiency and reduced environmental impact [93]. Beyond weight reduction, CNFs offer additional advantages. Their inherent properties contribute to enhanced vehicle performance, including improved impact resistance and durability. Moreover, the use of a renewable resource like cellulose aligns with the growing emphasis on sustainable manufacturing practices. This pioneering development underscores the potential of nanotechnology in addressing global challenges. As research and development in this field continue to advance, we can anticipate even more groundbreaking applications of cellulose nanofibers in the automotive industry and beyond.

4.4 Natural Filtration Capabilities of Xylem Tissue

Researchers have successfully harnessed the filtration capabilities of plant xylem to create sustainable and efficient water purification systems [94]. By replicating the xylem's structure and function, scientists have developed cellulose-based filters capable of removing harmful contaminants, including bacteria and viruses, from water sources. This approach offers a promising alternative to traditional water treatment methods, which often rely on energy-intensive and chemically intensive processes. The inherent properties of cellulose, such as biodegradability, renewability, and high adsorption capacity, make it an ideal material for water purification. As research in this field progresses, we can anticipate the development of even more advanced cellulose-based filtration technologies, addressing a wide range of water quality issues. For example, researchers at MIT have harnessed the natural filtration capabilities of trees to develop a sustainable and effective water purification system [95]. By utilizing xylem tissue, the plant's water transport system, they have created filters capable of removing harmful bacteria and viruses from contaminated water. The innovative approach involves transforming cross-sections of sapwood into functional filters. These filters mimic the natural process by which trees purify water, effectively trapping contaminants within the xylem's intricate structure. Laboratory tests have demonstrated the filters' ability to remove over 99% of bacteria and viruses, surpassing World Health Organization standards for drinking water. To address challenges such as filter clogging and durability, the research team developed a simple treatment process involving hot water and ethanol. This treatment significantly extends the filter's lifespan without compromising its filtration efficiency. Furthermore, the researchers have successfully implemented their technology in India, collaborating with local communities to develop a user-friendly filtration system [95]. By leveraging the abundance of trees and simple fabrication processes, this approach holds the potential to provide clean drinking water to millions of people worldwide. Abreast this, other nanocellulosic matrices from other species are being trialled, such as from bamboo [96], and honeycombs [97].

4.5 A Bacterial Nanocellulose Makes the Brand of Bags

Another fabulous eco-friendly and innovative context is a Bou bag being developed with bacterial nanocellulose by researchers at Modern Synthesis, a biotech company, in collaboration with a Danish fashion brand Ganni [98]. The bacterial biofilm can be converted into a superfine thread which can be combined with textile fabric to make green composites devoid of any synthetic material or processes with toxic chemicals. This potentially is more ethical compared to animal killing for leather products and even generates lower carbon emissions relatively.

4.6 Computer Chips Using Cellulose Nanofibrils

The burgeoning e-waste crisis, characterized by the rapid obsolescence of electronic devices and their harmful environmental impact, demands innovative solutions. Researchers at the University of Wisconsin-Madison have introduced a groundbreaking approach by developing computer chips using cellulose nanofibrils (CNFs), a sustainable material derived from wood [99]. Unlike traditional silicon-based chips, which require energy-intensive manufacturing processes and pose environmental risks upon disposal, these wood-based chips offer an eco-friendlier alternative. By replacing the rigid silicon wafer with a flexible and biodegradable CNF substrate, the researchers have created chips that are not only functional but also environmentally benign. The potential implications of this breakthrough are far-reaching. The ability to decompose electronic waste safely and efficiently can significantly mitigate the environmental impact of the electronics industry. Moreover, the use of renewable and abundant resources like wood aligns with the growing emphasis on sustainability. While challenges such as scaling up production and ensuring long-term performance need to be addressed, the development of wood-based computer chips represents a significant step forward in creating a more sustainable and circular electronics industry.

5 Reimagining Clean Energy

Another area where nanomaterials excel is in the energy sector. Their unique properties are being harnessed to develop advanced batteries, solar cells, and fuel cells. For example, silicon-based nanomaterials are improving the efficiency and lifespan of lithium-ion batteries, paving the way for electric vehicles with extended ranges. Additionally, nanostructured materials are enhancing the performance of solar cells, making solar energy a more viable and cost-effective option.

1. Energy Storage:

   - Supercapacitors: Nanocellulose-based electrodes are being developed for supercapacitors due to their high surface area and ability to form porous structures.
   - Batteries: Nanocellulose is being explored as a binder and separator material in lithium-ion batteries, potentially improving performance and safety.
   - Solid electrolytes: Nanocellulose-based solid electrolytes are being researched for next-generation batteries.

2. Solar Energy:

   - Transparent substrates: Nanocellulose films can be used as transparent substrates for solar cells.
   - Dye-sensitized solar cells: Nanocellulose is being investigated as a component in dye-sensitized solar cells, potentially improving efficiency and reducing costs.

3. Fuel Cells:

   - Proton exchange membranes: Nanocellulose-based membranes are being developed for use in fuel cells, offering potential improvements in proton conductivity and mechanical stability.

4. Biofuels:

   - Enzyme immobilization: Nanocellulose can be used as a support for enzyme immobilization in biofuel production processes, potentially improving efficiency and reusability.
   - Biomass conversion: Nanocellulose-based catalysts are being researched for more efficient conversion of biomass to biofuels.

5. Thermal Energy Management:

   - Insulation: Nanocellulose aerogels are excellent thermal insulators and can be used in buildings for energy efficiency.

- Phase change materials: Nanocellulose is being incorporated into phase change materials for thermal energy storage and management.

6. Energy Harvesting:

- Piezoelectric materials: Some researchers are exploring the potential of functionalized nanocellulose in piezoelectric energy harvesting devices.
- Triboelectric nanogenerators: Nanocellulose-based materials are being investigated for use in triboelectric nanogenerators for mechanical energy harvesting.

7. Oil and Gas Industry:

- Enhanced oil recovery: Nanocellulose-based materials are being researched for use in enhanced oil recovery processes.
- Drilling fluids: Nanocellulose can be used as an additive in drilling fluids to improve their properties.

8. Hydrogen Storage:

- Nanocellulose-based materials are being explored for their potential in hydrogen storage applications.

9. CO_2 Capture:

- Nanocellulose-based materials and membranes are being researched for CO_2 capture and separation, which is crucial for clean energy technologies.

10. Smart Grids:

- Sensors and actuators: Nanocellulose-based sensors and actuators are being developed, which could find applications in smart grid technologies.

These applications are at various stages of development, from early research to near-commercialization. The use of nanocellulosic materials in the energy sector is particularly attractive due to their renewable nature, the potential for high performance, and the possibility of creating more sustainable energy technologies.

5.1 Clean Energy: Beyond Efficiency Frontiers

In addition to Efficiency Frontiers Facing the necessity of the shift towards sustainable energy sources, and seeing that the business-as-usual approach is no longer acceptable, it becomes obvious that only radical changes will suffice. We are saying that the basic concepts around power generation, storage, and supply have to be redesigned. Smart nanomaterials provide the prospect for disruptive innovations, which go beyond the current conventional practices, become new paradigms, and make way for the creation of a considerable amount of green energy.

5.2 Solar Revolution 2.0

Generally, the conventional strategy for solar power has been directed mainly at enhancing the efficiency of the silicon photovoltaic cell. Thus, it is evident that, although these efforts have produced considerable advances, such improvements are nevertheless constrained by the nature of the silicon medium and the range of the light spectrum that can be collected at the core of the technology [100, 101]. To go further in changing solar energy we must look at new ways and new materials, varied silicon rays can also collect light from the infrared and ultraviolet regions to improve the efficiency of energy collection. For example, the quantum dot solar cells utilize semiconductor nanocrystals which can be designed for selective absorption of a specific range of wavelengths. Such materials enhance the efficiency of photovoltaic cells and allow for the creation of thin-film and lightweight solar modules that can be applied to different surfaces [102]. Figure 1 gives a schematic overview of the potential of a solar revolution.

Furthermore, biomimetic systems based on natural photosynthesis are one of the promising directions for the efficient conversion of solar energy. Several artificial photosynthetic systems are gradually being created because of replicating the process

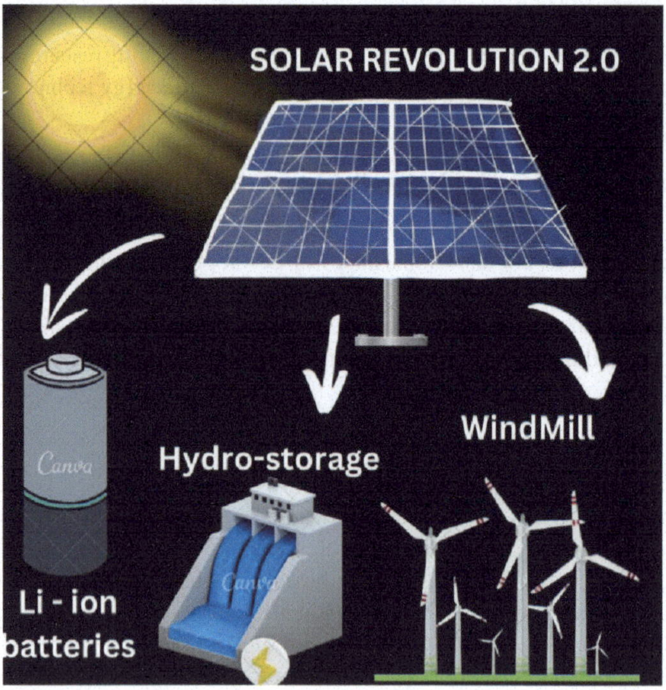

Fig. 1 Solar Revolution 2.0: integrated systems for li-ion batteries, hydro storage, and windmill applications

that plants undergo in their conversion of solar energy into chemical energy. Most such systems use nanostructured catalysts to mediate the water-splitting procedures, which are essential for preparing hydrogen from sunlight and water, which is a renewable and clean energy [103]. Another prospect of using perovskite material is in solar cells. These materials have demonstrated an increase in efficiency in a short time and they can be manufactured employing low cost. A combination of nanomaterials with perovskites can also increase their stability and effectiveness making them a good potential for the development of the next generation of solar products [104].

5.3 *Energy Storage Metamorphosis*

Storing energy is very vital in managing the supply and demand of energy in renewable energy systems. Currently, lithium-ion batteries are widely used in portable electronics and electric vehicles but they present problems at the level of energy density, charge rate, and recyclability. Smart nanomaterials are therefore a way that can unlock new directions in energy storage that do not fit into the above classifications [105]. One of them is supercapacitors which have the ability of ultrafast charging and high power density. Supercapacitors differ from conventional batteries in terms of the mechanism of storage of the energy where instead of the chemical reactions that are used to store energy in batteries, physical separation of charges is used in the supercapacitors [106]. The addition of such nanomaterials as graphene and carbon nanotubes into the supercapacitors' electrodes can notably improve the characteristics that can make them useful for grid-scale energy storage and fast charging [107]. The two areas that are particularly well known for investment include lithium-ion batteries as well as the wide category of solid-state batteries. These batteries eliminate the flammable liquid electrolyte which is used in normal batteries and use a solid one making them safer and having more energy density. A higher ionic conductivity of solid electrolytes results in more efficient and longer-life batteries through Nanostructured materials [108]. In addition, smart nanomaterials enable a boost in the functions of the energy conversion and distribution systems. Thermoelectric materials which are used to directly convert temperature differences into electrical energy can have their efficiency enhanced with the help of operation at the nanoscale. They can be applied in waste heat to power systems, which harness and transform the waste heat from any manufacturing processes, electrical power stations, etc. into usable power [109]. Besides, smart nanomaterials can facilitate the provision of new grid structures that do not exist in the present systems. For instance, for local renewable generation and storage devices that are integrated to create microgrids, this would be of great benefit to nanomaterial-based technologies. These Systems can be off-grid or on-grid systems, which makes the system flexible at the same time quite robust [110]. Revolutionizing clean energy with smart nanomaterials is about far more than enhancing current methods to achieve maximum outcomes mostly regarding completely re-architecting the approach to energy generation, storage, and distribution. Nanotechnology becomes a workbench for a new generation of solar panels with tailored light absorption coefficients, high-performance batteries with new attributes not seen in current Li-ion batteries, and a smart grid with self-healing characteristics. Smart nanomaterials have excellent possibilities in various fields, however, to achieve all these possibilities, further development, discovery, and cooperation are needed. The sight of new frontiers being opened should therefore go hand in hand with the creation of technologies that are done right, considering the impact on the environment as well as possible social implications. The road map to attain sustainable and clean energy is not easy but with the help of smart nanomaterials, we can make significant strides that we have never imagined possible. Thus, implementing these innovations will successfully build a world based on clean, efficient, and reliable energy for everyone that will improve future generations' lives in a prosperous manner.

6 Conversion of Smart Nanomaterials to Achieve Sustainable Goals

6.1 A Nanoscale Renaissance Beyond Desalination

A shift initiated through the notion of the Nanoscale Renaissance revealed that the population is increasing day by day and industries are also being set up, the provision of safe and hygienic water is one of the major responsibilities of society. Despite the several advantages and versatility of use in several sectors of the conventional water purification techniques, attaining the set goals and objectives is quite difficult due to the following limitations of the techniques: Therefore, concerning the smart nanomaterials the 'Water purification sub-sector' has begun experiencing a new dawn of additional technologies that cannot be offered even by physical standard methods otherwise. The chapter describes the revolution in water purification with the help of smart nanomaterials, and attention is paid to the new aspects of the concept and the comparison with other types of water purification. Aside from desalination Conventional water treatment also referred to as desalination has for a long time been a way of tackling the issues of water scarcity in society, mainly in the areas of water deficit. However, technologies, that are direct in applications such as reverse osmosis which is a conventional way of desalination, require a lot of energy and do not eliminate all the impurities in water [111]. The new smart nanofiltration membranes are another giant leap in this field for the same reason that it is beyond the simple process of desalination of water. A high level of selectivity is attained as a result of indicating peculiarities of nanomaterials in the smart nanofiltration membranes [112]. Figure 2 exemplifies various merits of nanoparticles in the remediation of contaminants in public waters.

These membranes can also be designed to have certain dimensions of the holes to act as a filter for pollutants such as microplastics, pharmaceutical wastes, and other

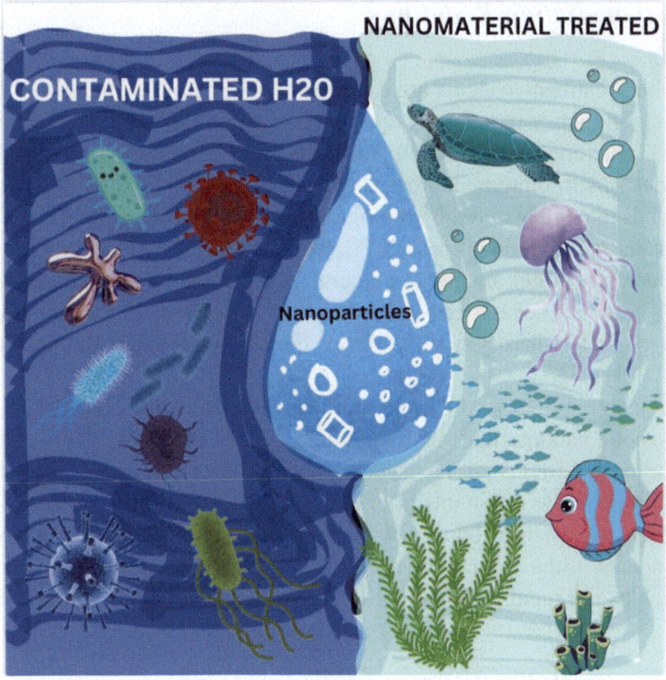

Fig. 2 Nanoparticle-based remediation of contaminated water

unnamed pollutants. Such a level of filtration is hardly possible with the help of standard methods and can be considered as a potential solution to some of the key issues of water pollution. For example, MOF-derived graphene oxide membranes used in the water purification process are outstanding. These are the membranes that can easily be designed to have very thin channels that can only allow the water molecules to pass through while preventing any other particles [113]. This not only improves the different procedures of filtration but also the energy spent is comparatively less than the other types. These smart membranes' benefits are not only for such centralized water treatment plants but are also for the following. That is why they can be applied in decentralized water purification systems that are being introduced closer to the points of demand. This is especially so with the areas that are hard to reach or areas in which only a few players in the clean water and water provision industries exist. Smart nanofiltration systems even when put in operation at lower levels make a point to deliver water of good quality and this could be done with little emphasis put on the top structures of facilities.

6.2 Bioremediation Redefined

Conventional treatment systems are useful for handling large quantities of wastewater but lack efficiency in handling specific pollutants and need a lot of energy and operational structures. Smart nanomaterials present a revolution in bioremediation that provides in-situ bioremediation of the environment naturally. Advanced smart nanomaterials that are inspired by biological enzymes introduce new possibilities to effective water purification processes. For example, nano-enzymes can eat up pollutants in sources of water as they disintegrate dangerous components into nontoxic products. These nano-enzymes can be designed to treat specific pollutants, a level of selectivity that is hardly achievable by the conventional pollutants' treatment processes [114] (Fig. 3).

According to previous studies, one promising direction in this field is the application of magnetic nanoparticles with the outer layer of the catalytically active materials. These nanoparticles are envisaged to be incorporated in contaminated water and thus eliminate the pollutants after which the remains are adequately removed using magnetic fields; the product is water that has been purified without any residual treatment agents. It thus spares the need for developing very extensive structures as well as enables the rapid purification of water at camp [115]. Another outstanding

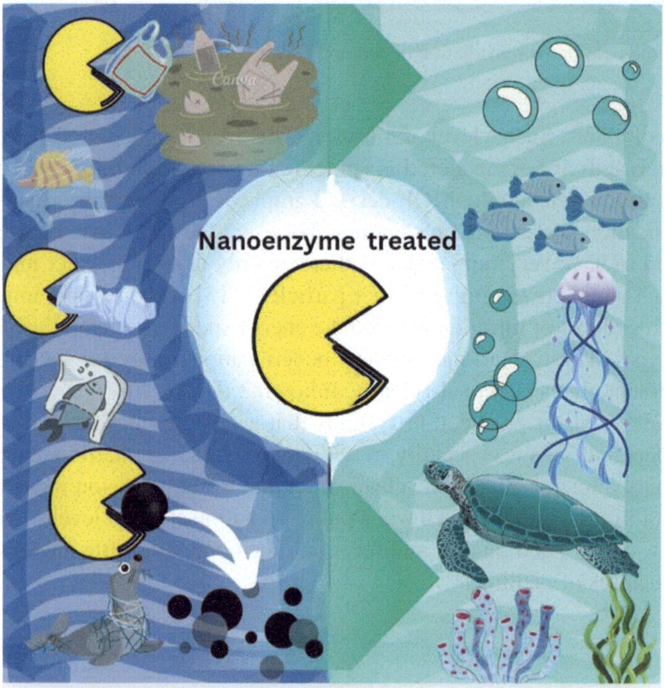

Fig. 3 Visualizing the impact of nanoenzyme treatment

area of the smart nanomaterials application in bioremediation is a nanosponge which is a nanostructured material which is intended for the adsorption and immobilization of definite pollutants. These nanosponges that can be used in water sources can effectively pull out pollutants such as heavy metals and organic materials. The nanosponges can be reused several times for continued water treatment once the pores become filled with contaminants, they can be recycled [116].

6.3 Smart Water Management

The conservation of water should be done efficiently for the general Water Resource sustainability. Conventional methods of water distribution and storage are consequently less effective because they are always leaking hence resulting in a lot of loss of water. Smart nanomaterials change the approaches to leak identification and elimination for better management of water systems. One exotic mixture is in the use of self-sealing technologies for pipes. These coatings based on smart nanomaterials are also designed to feel and stop tiny holes on their own. The essence of nanotechnology is that when the crack or the leak is discovered the nanomaterials within the coating swell and block the water leakage. Besides cutting numerous expenses in maintenance, it also increases the durability of water installations, proving beneficial for municipalities and industries [117]. Real-time monitoring of water flow and pressure within pipes, the other inventive concept and product of smart nanomaterials, are the embedded sensors. It is possible to identify derivate sensors, which can reveal signs of a leak or a blockage before the process develops into a catastrophic one. For instance, piezoelectric carbon nanotube sensors offer tiny pressure differential and flow meter data resolution for water management infrastructures. Thus, by incorporating such sensors into water systems it becomes much easier to control the supply of water and at the same time prevent major breakdowns in the system [118]. Smart nanomaterials have been adopted into water management systems to participate in the creation of channelized smart networks or smart grids for water. These networks also employ collected data from sensors to control and regulate water intake and distribution as per the requirements and circumstances. Such mobility of water entails efficiency in the use of water hence conservation of water to make the usage of water as sustainable as possible.

Smart nanomaterials in water purification can be described as a renaissance in the ways to maintain clean and safe water for the population. Thus, it is possible to overcome many of the existing problems of water pollution and enhance the effectiveness and efficiency of water supply management using interdisciplinary approaches based on nanotechnology capabilities. Highly selective and energy-efficient integrated nanofiltration membranes will provide efficient solutions for the depuration of various contaminants, suggesting its proper application for depuration at the point of use. The use of smart nanomaterials offers novel strategies for in-situ bioremediation where smart nano-systems replicate nature's detoxification mechanism for efficient pollutant removal. At the same time, smarter techniques in water management such as self-healing coatings and sensor integration improve leakage detection and reduce its occurrence resulting in a flexible and sustainable water network. While we further expand our research and innovations in these areas of advanced technologies it is critical to worry about sustainability and ethical approaches. It is inherent hence that when innovativeness is wed to responsibility, we will indeed get the absolute most out of intelligent nanomaterials to foster greater availability of clean water and sustainable use of Water Resources. This nanoscale renaissance in water purification is therefore not only a technological move but a mile towards the realization of sustainable standards within the entire humanity and the generations to come.

7 Circular Economy: From Waste to Resource with Nanotech Muscle

The Renaissance of waste has been for some time a fashionable issue stemming from human greed for wealth, power, comfort, luxury, and material possessions. While demand for products grows more worldwide, the apparent and customary model "take, make, dispose" is unsustainable. Subsequently, a circular economy concept, which can be described as a reuse economy or a closed-loop system, is a widely supported vision of the future. Smart nanomaterials are on the precipitate of this revolution due to enabling measures for transforming waste into high-potential resources, enhancing product durability, and enabling intelligent disassembly. The chapter discusses the enlarged scopes that nanotechnology introduces in transforming a circular economy, which outlines its new directions and values it [119].

7.1 Upcycling Revolution

Common recycling techniques lead to the production of down-cycle goods that are intrinsically of a lower standard. Yes, recycling is good but upcycling will be the best thing to do when you want to fully make the best out of waste. These applications of smart nanomaterials with their characteristics and possibilities are at the helm of this upcycling shift. Among the most promising trends in this area, it is possible to distinguish the application of nanocatalysts concerning the chemical recycling of plastics. Chemical upcycling is different from mechanical recovery in a way that the latter may end up lowering the quality of the plastic by breaking down its polymers into smaller parts while the former transforms the polymers into their monomers or other useful chemicals. These reactions may recover fuels, chemical feedstocks, or even new high-performance materials from the piles of plastic wastes and, the efficiency and selectivity of these reactions can be further improved through the use of nanocatalysts [120].

For instance, there are nanocatalysts from transition metals that have been designed to depolymerize polyethylene and polypropylene- the two most extensively used polymers and transform them into liquid fuels or monomers for other polymers. These nanocatalysts work at lower temperatures with improved selectivity, which indicates that chemical upcycling is the best solution for plastic waste [121].

7.2 *Self-Healing Materials for Extended Lifespan*

It is important to note that the mentioned nanomaterials can be applied not only to upcycle the type of plastics described above but also to use other types of waste. For instance, nanocatalyst and enhanced depolymerization of agricultural waste products makes them very effective in producing useful products such as bio-fuels and bio-plastics. Likewise, circuit cables that could comprise several metals can in a similar manner as described above be elevated for useful recycled compositions which can in return be useful in making fresh circuit cables [122]. One of the possible categories of material that can be incorporated into structures is the self-healing material for the longer life span of the structure. Since the aim is the circular economy, many factors affect it and one of the most significant is the fast product turnovers and thus, short product lives. Substituting a new concept of Self-healing materiality or materials that can heal themselves and hence can help to be useful for a longer time, Conductive polymers came into existence [123]. It can be described regarding the nanocapsules with healing agents where the part of the component has been damaged to release the agents. These nanocapsules could be designed such that they release their contents when there is a crack or micro-damage—this will in effect trigger, the healing of the material. It can be incorporated into several material products that include polymers and that would in turn increase the life span of the product among other gains. For example, self-healing concrete that is used in construction has nanocapsules embedded in it either containing calcium carbonate or epoxy resin, in the event of cracking or any form of damage, the capsule will rupture and release the chemical that helps to fill the crack or damage and seals it in a way that it cannot be opened again and is very expensive to repair [124]. In the same regard, self-healing polymers as employed in consumer electronics would enhance the gadget's durability through healing abrasion lines that may negatively impact the gadget's operation [74].

The advantages of articles that possess SEL awareness are not confined to the life cycles of unique products. These materials can therefore help cut down the use of other consumables as well as reduce the instances of having to dispose of replacement items. This conforms with the circular economy concept where the aim is to use a product and its components for as long as possible, thereby getting the most benefit from it in the process avoid harming the environment.

7.3 Intelligent Deconstruction

In the case of cyclic use, efficient and rapid dismantling and recovery of the end product at a stage of the life cycle when it no longer fulfills its purpose is desirable. Smart nanomaterials can facilitate sui generis deconstruction in that they have characteristics that can be run in parallel such that when they wish to deconstruct them in other to get their materials back, this can easily be done. Other techniques are in assembly, and planning for building elements that contain weaknesses or substructures that can be dissolved by given stimuli. For example, mobile phones can be designed with nanomaterial contact points that readily fuse when in contact with a specific solvent or when exposed to a certain temperature at which solvents are applied for recycling. This wise dismantling is applied in the process of recycling in a manner that tends to enhance the quality aspect of the recycled materials [125].

Another possible application might be intelligent adhesives that might get untangled with the help of an external signal. These adhesives can therefore set to a rigid state at the point of application of the product they are applied on but can be easily made to debond if subjected to some conditions such as ultraviolet light or an electric field. They play an important role in deconstructing the products without causing any destruction to the point that components and parts can be reused in a good state [126].

For instance, the best intelligent deconstruction strategies in the automobile industries can improve the recycling of parts including batteries, and electronics systems. Therefore, by making cars with components that are easily removable, car producers added to the list of recyclates the aforementioned metals, including lithium, cobalt, and rare Earth elements, thus reducing the harm of mining [127]. The shift towards the circular economy is crucial for attaining sustainable outcomes; furthermore, smart nanomaterials contribute significantly to this process. In doing so recycling material and upcycling elevate the idea of recycling past the conventional and making use of waste in creating high-value-added products improves resource efficiency. Self-healing products due to smart nanomaterials are an innovation that helps to increase the durability of products and, therefore, minimize the use of resources. Smart disassembly methods make the products easier to dismantle and help in the recycling of important materials, to complete a cycle within the circular economy.

In this chapter, the innovations presented reveal the explicative character of smart nanomaterials as a system changer in the context of the circular economy (Fig. 4). It is also important to point out that as the development of these technological advances, along with their application into practical uses, their sustainability and ethical aspects should be taken into consideration to make sure that nanotechnology acts mainly for the benefit of people, rather than the opposite.

With smart nanomaterials fully deployed, it is possible to develop a world that may have less waste while resources are recycled on a cyclical basis, thus lowering the adverse effects of human beings on the environment. This nanoscale revolution in material science is not simply a technological progression but one of civilization's most significant developments toward a green, abundant era.

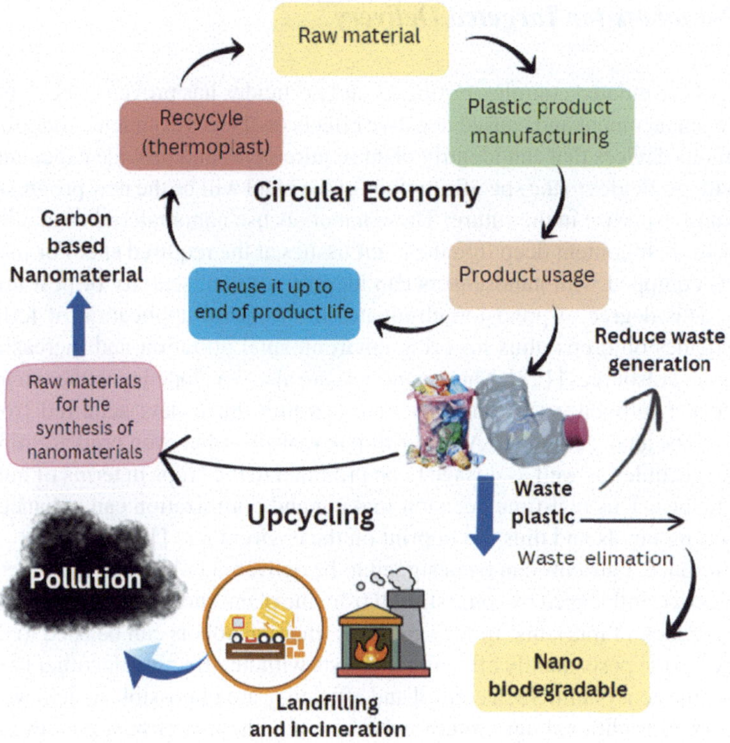

Fig. 4 Model of the circular economy showing nanotechnology contribution, empowering environment sustainability

7.4 Sustainable Agriculture: Precision at the Nanoscale

Sustainable agriculture is defined by its practices and values as the process of feeding the entire world's population in a manner that will not deplete the Earth's resources. Old farming practices though useful in the past are today offering minimal solutions to new challenges in farming such as productivity, utilization of resources, control of pests and diseases among others as well as the health of the soil. The application of smart nanomaterials in the field of agriculture holds the promise of improvement at the nanoscale that can improve yields, optimize the use of resources, and reduce the negative effects of mankind on the environment. This chapter focuses on new uses of smart nanomaterials in providing smarter solutions for sustainable agriculture and the maximal advantages of its usage.

7.5 Nanobots for Targeted Delivery

The use of formulated complex fertilizers and pesticides has proven to be inefficient resource management and caused negative effects on the environment. Nanorobotics autonomous devices that can identify plant requirements and provide exact nutrients along with pesticides to the specific regions of the plant will be the new progression in agricultural expertise in the future. These nanobots use nanomaterials and allow the delivery of their content deep into the plant tissues at the required site. For instance, nanobots equipped with nanosensors can identify nutrient scarcity or pest invasion in cells. This degree of precision eliminates the frequent application of fertilizers and pesticides on crops, thus lowering environmental pollution and increasing the utilization of resources [128]. These nanobots are also developed with the integration of artificial intelligence (AI) and machine learning. Such data acquired from the ground can be analyzed by the AI algorithm in real-time that even predetermines the delivery schedules as well as dosage to be provided to the crops in terms of nutrients and pesticides. This real-time decision-making and optimization can enhance yield while cutting inputs and thus the imprint on the environment [129].

For instance, nutrients can be designed to be delivered only when the function of the nanobots is influenced by some stimuli from the plant, such as chemical indicators of stress or lack of nutrients. In the same way, these nanobots can be used to deliver pesticides to the pests mainly affecting the pests without affecting the other harmless insects or the ecosystem. Such control and efficiency are impossible to achieve in the framework of traditional agriculture, which is why the use of nanobots in modern farming is revolutionary [130].

7.6 Biological Imitation for Disease and Pest-Resistant Houses

Negaters on the other hand are conventional pesticides that although they show efficiency in pest and disease control in most cases are environmentally unfriendly and affect other organisms. Employing smart nanomaterials related to the defensive structures of plants provides a new opportunity for plant protection. Nanomaterials can thus be synthesized in such a way that they replicate the characteristics of the plant defense mechanisms. For instance, nano-coatings of plant cuticular structures can be used to spray crops, thus jamming the signals of pests (Fig. 5). These coatings disrupt the sorts of chemical messages that pests employ to find their hosts or partners, thus controlling pest populations without the usage of deadly toxins [131].

The other unique strategy that can be implemented includes the application of smart nanomaterials that boost the plants' pro-active immune system functions. These nanomaterials replicate the actions of natural elicitors to release defense reactions in the plant for instance producing antibiotics or enhancing the rigidity of the cell walls. This biomimetic strategy also benefits farmers by safeguarding their crops

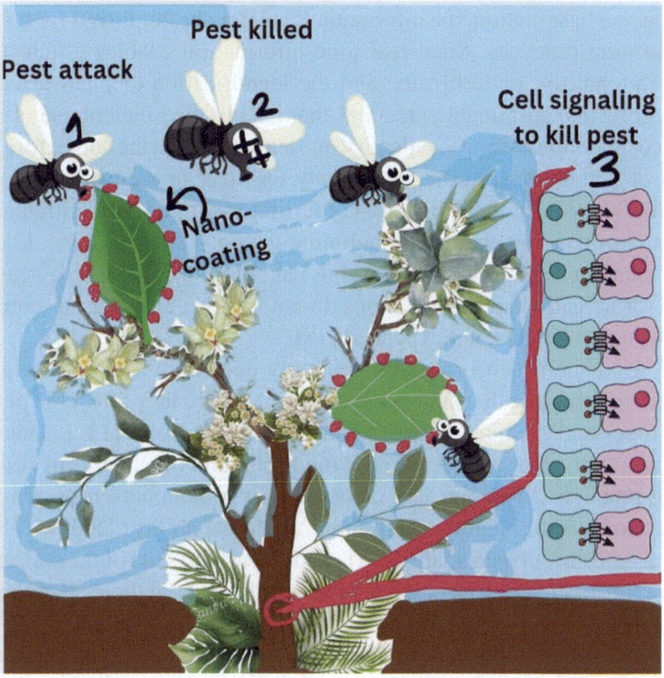

Fig. 5 Nanocoating and cell signaling are used to detect and repel pests in this innovative pest control method

against diseases but spares the use of synthetic pesticides that harm the environment by reducing their use and thus enhancing biodiversity [132]. For example, instead of overall biocides, silver nanoparticles, which have established antimicrobial activity, can be designed to be used against the root of germs causing damage to the usual resident beneficial microorganisms in soil [133]. In the same way, it is possible to use nanoparticles to deliver plant hormones to strengthen the immunity of plants against diseases. These two biomimetic strategies are eco-friendly compared to conventional pesticide usage, which improves crop governance [134]. Therefore, the monitoring and optimization of soil health is an important prerequisite for obtaining high yields [135].

Most noteworthy, soil status is at the center of sustainable agriculture mostly because of an impact on productivity, and moisture conservation. Keeping track of the soil health using conventional practices can be quite a task and, at the same time, ineffective due to the mere provision of few details. Thus, investing in smart nanobiosensors placed in the ground could be a great solution that provides for the constant control and adjustment of the conditions in the ground. Smart nanobiosensors with well-developed nanomaterials can sense the diverse aspects of soil quality such as nutrients, moisture, and pH as well as the pathogens [16]. These sensors relay information to a base station, the information that may be employed for the improvement of farming practices. Such real-time information enables efficient watering practices, rational use of fertilizers, and the identification of plant diseases all of which are good and sustainable practices that lead to the efficient use of resources. For instance, nanobiosensors can identify nutrient deficits in the soil before it impacts the health of plants; thus, farmers can apply fertilizer in the exact area and at the correct time. This meant the application of fertilizers was done in a strategic manner that would eliminate the chances of polluting the environment [136]. Likewise, by explicitly tracking the humidity of the soil, nanobiosensors can help determine the right time for irrigation so that the crops do not lack water but at the same time the available water resource is not expended [18].

Finally, the nanobiosensors enable the early detection of plant diseases. These sensors help in detecting the pathogens in the soil before they affect the plants hence reducing losses and the need to use chemicals in the process [137]. This is correct because disease prevention is central to crop health and the long-term sustainability of farming practices. However, the incorporation of nanobiosensors into artificial intelligence and machine learning helps improve the capability of SHM [138]. Thanks to historical and current data analysis, AI can forecast the state of the soil and suggest how to care for the crops in the best possible way throughout their development stage [18].

Smart nanomaterials in sustainable farming involve the use of smart structures and innovations in the way farming is done. Given that nanomaterials possess special characteristics, it is possible to control the delivery of micro rations of nutrients to the plants and optimize the use of insecticides and fungicides, as well as monitor the condition of soil. The former advances add to the productivity and effectiveness of usage of the resources and the latter reduces the effects on the environment thus making the farming system more sustainable and more resilient.

In targeted delivery, nanobots in the case of plants would enable efficient delivery of nutrients and pesticides to plants when they are required. Smart nanomaterials based on the copies of biological structures in animals and plants can be used in disease and pest control without the use of toxic chemicals that harm biodiversity. Smart nanobiosensors assist in real-time, precise measurement of soil health with the help of which efficient agriculture and sustainable farming can be achieved [139].

Thus, it becomes important to lay down key principles of sustainability and ethics as we progress in the deployment of enabled advanced technologies. With equal focus on innovation and sustainability, the area of smart nanomaterials can be optimally developed to achieve a prosperous future of agriculture that can sustain the growing global population as well as adapt to change. This revolution in shading the agriculture industry at the nanoscale is not only marked by the development in technology but it is also a fundamental stride in getting to the formulated sustainable future and food security for the generations to come.

7.7 *Biocompatible Nanomaterials*

There is a highly sought requisite for a new break in the healthcare systems and how to make them sustainable for the future. While the numbers of people living into their seventh and eighth decades of life continue to increase, and the percentage of the population suffering from chronic diseases, doctors and pharmaceuticals are under tremendous pressure to cultivate new, efficient, and cost-appropriate therapeutic options [140]. In this revolution, Biocompatible nanomaterials play a Lead role since they provide out-of-the-box solutions to diagnostics, therapeutics, and Regenerative Medicine. This chapter focuses on the new uses of biocompatible nanomaterials to develop health care and discusses how innovations can change the healthcare sector and be sustainable [17].

7.8 *Personalized Nanomedicine*

Conventional delivery processes are characterized by their ability to distribute the therapeutics throughout the body and thus pose limited effects on the desired target organs or tissues and relatively large side effects. Nanomedicine based on the usage of smart nanocarrier systems is a rather innovative approach to targeted drug delivery that is extolled for its high specificity and individualized approach [3]. The smart nanocarriers can be designed in such a way that they only release their drugs at the affected biological site in the patient's body. One of the most generic and promising trends, that can be observed in this sphere, is the creation of stimuli-responsive nanocarriers. These nanocarriers are built to unload the therapeutic cure at the site of action only where is there a change in pH, or the presence of enzymes or proteins related to diseased tissues. For example, pH-sensitive nanoparticles can release anticancer drugs due to the lower pH level in tumor tissues sparing healthy cells and increasing the side effects. Nanocarriers can be responsive to peculiar enzymes in the human body and thus possess the ability to deliver the therapeutic agents only upon the presence of these enzymes which are usually over-expressed in diseased tissues [141]. These accuracies help in lowering the drug quantities to be given to the patient and at the same time increasing the effectiveness of the treatment [142]. These potentials have been depicted in Fig. 6.

In addition, the combination of AI and machine learning along with smart nanocarriers would help in making treatments even more personalized. The integration of the patient's genetic and proteomic profiles allows for avoiding futile attempts, identifying the most effective drug delivery strategies, and optimizing the design of nanocarriers depending on the patient's characteristics [143]. The level of personalization thus constitutes the potential of realizing therapeutic benefits and therapeutic risks that are as high as possible while the risks are as low as possible, which is a giant leap forward in pharmacotherapy.

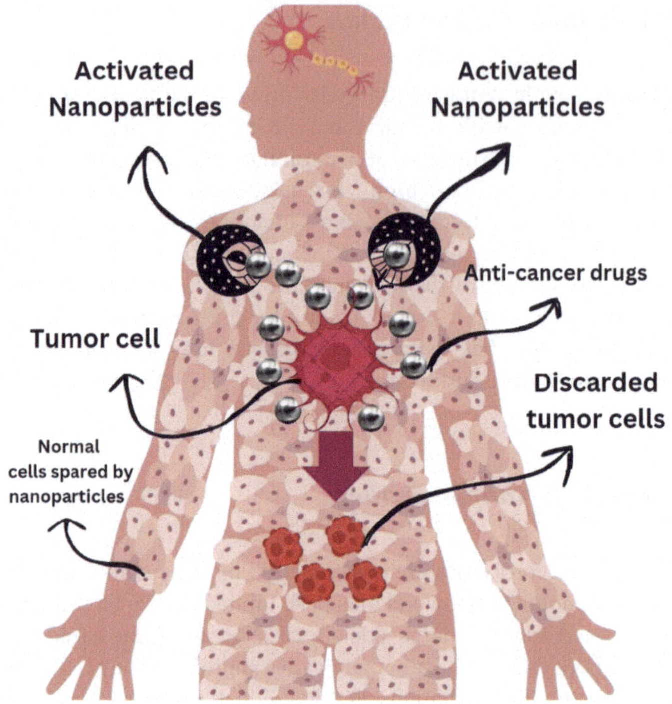

Fig. 6 pH-sensitive nanoparticles—releasing anti-cancer drugs

7.9 Nanobots for Diagnostics and Therapy

The emergence of biocompatible nanobots in the field of theranostics was another innovation in the aspect of diagnosis and treatment at the same time. Thus, multi-functional nanobots can diagnose diseases in real-time and at the same time deliver targeted therapies, making this type of treatment virtually painless and highly effective [144]. Nanobots can be programmed to move within the human body, search and target pathologic cells, and carry imaging and treatment functions as well. For instance, Magnetic nanobots can be targeted to particular locations in the body using the help of magnetic fields. Upon reaching the targeted location, such nanorobots can engage in tasks such as performing MRI scans covering the tumor tissues or applying localized treatments such as hyperthermia or drug disbursement [145].

Diagnostics is the most promising field of the application of nanobots, and targeted disease identification can be facilitated by such developments. Thus, if nanobots are given the role of early-phase biomarkers related to diseases or other dysfunctions of the body, the efficacy of the operation itself can ultimately be increased alongside corresponding patient outcomes. In the subject of cancer treatment for example, nanobots that themselves have fluorescent markers can be used to zero in on the cancer cells for a better imaging and treatment plan without having to do a lot of invasive procedures on the patient and without harming the healthy tissues or cells [146].

In therapeutic applications, nanobots can selectively transport drugs concerned for the diseased site and release them with high concentration at the site and thus they decrease the detrimental effects in the system. Also, they can be set up to exert physical motion on the target area for example to drill into tumor cells to implant drugs inside them or to make use of heat to kill the cancer cells. These capabilities are much advanced in the realm of minimally invasive surgery and targeted therapy and can provide new hope for patients suffering from difficult diseases.

7.9.1 Bioprinting with Smart Nanomaterials

Bioprinting technology is an advanced research area of tissue engineering and stem cell research in which layers of living cells are built up to create functional tissues using bioinks. The integration of smart nanomaterials into the bio-inks can take bioprinting to the next level of evolution wherein micro to macro tissues can be created that nearly resemble the natural tissues of the organs. Thanks to the creation of smart nanomaterials, bio-inks that would enhance tissue regeneration and the healing process can be produced. For example, antimicrobial nanomaterials can be used in bio-inks since wound infection is a major complication during healing [147]. Also, if nanomaterials exhibit the mechanical property of the tissue, they can offer structural support thus enabling bioprinted tissue to have improved functionality and stability. It is worthy of note that one of the most promising uses of bioprinting smart nanomaterials is in the tissue repair or replacement processes. Bioinks based on stem cells and growth factors, as well as nanomaterials designed to give structural and biochemical signals, make it possible to meticulously construct tissue constructs that would be effectively implanted back into the patient's body for regeneration of the damaged organs [148]. This approach has the potential to solve one of the major problems of the scarcity of donor organs and the whole field of organ transplantation. For example, organ-printed cardiac tissues integrated with conducting carbon nanotubes will enhance the development of functional gap junctions to excited cardiac tissues and ultimately help to recover defective cardiac tissues in patients with heart diseases [149]. Likewise, skin that is printed with nanomaterials for angiogenesis facilitates wound healing as well as tissue regeneration in burns [150]. Also, especially when employing smart nanomaterials, researchers can develop tissues that contain sensors for observing the health and efficiency of surrounding tissues in the process of bioprinting. These sensors could be used to acquire informative and potential complications data of the bioprinted tissues in terms of integration and functionality for further remedial action [151].

The new principles of nanobiocompatibility whereby nanomaterials are incorporated into the framework of healthcare, practices have introduced new and progressive approaches that offer a higher degree of effectiveness in the treatment, fewer side effects, along environmental conservation. Nanomedicine as well as the administration of drugs, specifically in the form of stimuli-responsive nanocarriers, goes a step further to provide targeted treatment as per the patient's biological context. Diagnostics and therapeutic nanobots show a notable improvement in the theragnostic in which the disease may be identified in real-time and treated at the same time, where necessary, through less invasive procedures [152]. Bioprinting of smart nanomaterials will be the next significant advancement in tissue engineering and regenerative medicine, which will result in the efficient development of tissues and organs that restore the usual structure and favor recovery processes [153].

Further sophistication in these technologies requires translational approaches into practice calling for a balanced and entirely beneficial merit for society. Thus, by prioritizing the means of ethical action that ensures long-term and sustainable development of the use of biocompatible nanomaterials, public well-being and humanity concerns could be sufficed. This is a nanoscale revolution in healthcare that is transitioning not just from a technological standpoint but from a global development perspective.

Swiss researchers have developed an environmentally friendly ink based on cellulose nanocrystals, a sustainable and renewable material derived from wood [154]. This innovative ink holds immense potential for 3D printing applications, particularly in the biomedical sector. By utilizing cellulose nanocrystals, the researchers have created a printing material that is both eco-friendly and versatile. The ink's properties allow for the precise and controlled deposition of material, making it suitable for creating intricate structures. This breakthrough has significant implications for producing biomedical implants and other healthcare products. The ability to 3D print with cellulose nanocrystals offers several advantages. Firstly, it promotes sustainability by reducing reliance on petroleum-based materials. Secondly, the biocompatibility of cellulose makes it an ideal choice for medical applications, minimizing the risk of adverse reactions. Finally, the precise control offered by 3D printing allows for the creation of customized and patient-specific implants. This development represents a significant step forward in the field of materials science and additive manufacturing [154]. By harnessing the power of nanotechnology and sustainable resources, researchers are paving the way for a greener and more sustainable future.

8 Conversion of Smart Nanomaterials to Achieve Sustainable Goals

8.1 The Social Components and Ethical Approach

The development and implementation of smart nanomaterials are moving at a very fast pace and they reflect the promise of great benefits In improving global sustainability. But for these technologies to be developed and used appropriately, one must assess the societal effects of these technologies. This chapter focuses on revealing the discussion of the environmental, health, and safety concerns for smart nanomaterials, the need for public engagement, and the overall ethical framework for the use and development of smart nanomaterials- precautionary, transparent, and more importantly, equitable approaches towards the understanding and usage of smart nanomaterials.

8.2 Environmental, Health, and Safety Considerations

Since smart nanomaterials are expected to be used in large-volume applications, the environmental impact of these applications during their entire life cycle, from manufacture to end of life, must be determined and managed. In many cases, the traditional risk assessments are inadequate for the identification of interactions between nanomaterials and biological and ecological systems. Life cycle assessments (LCAs) are much complete optimistic than those that focus on a single aspect of the nanomaterial as they include all the stages of its life cycle.

Environmental and health impacts of nanomaterials are another issue; this touches on toxicity and the ability of these materials to persist in the environment. It is noted that some nanomaterials have an exceptional state of stability and are rather resistant to degradation; nevertheless, their main disadvantage may be this durability. To offset these risks, researchers are paying more and more attention to the synthesis of biocompatible and biodegradable, smart nanomaterials. For example, the evolution in the production of biodegradable polymers and bio-based nanomaterials unveils new directions for the development of eco-friendly solutions.

Also, the synthesis and manufacturing of nanomaterials must face searching for their impact on the environment. The principles of applying green chemistry include the use of renewable feedstock and the generation of hazardous waste that is avoided in the synthesis of nanomaterials. For example, employing the extract from plants to synthesize nanoparticles not only decreases the use of toxic chemicals but also benefits from the feature that materials derived from plants can be biodegradable [155]. Environmental disposal and management at the end of the product's usefulness are also important. Recycling and reclamation solutions appropriate for nanomaterials may help decrease the extent of their harm to the environment. For example, nanomaterials in electronics and/or in medical applications such as implants need to be controllably recoverable and recyclable; otherwise, they will end up as endocrine disrupters perpetually in the environment [156].

8.3 Public Engagement and Open Dialogue

Thus, the emergence and use of smart nanotechnologies should not take place in a void. Public awareness should be encouraged to avoid misconceptions and keep people informed and fully involved in the progress to ensure that development is carried out properly. Functional interactions and engagement between science, policy, industry, and people can effectively reduce the information divide and explain nanotechnology. It is possible therefore to conclude that engagement strategies must not be conventional public meetings and consultations. By engaging in common decision-making, it can adopt technology-based forums, citizen science programs, and PTAs. For instance, participants in workshops and focus group research that involve people from different backgrounds can be of help in understanding the population's perception of nanomaterials.

Exchanges by others apart from the inventors about the advantages and dangers of nanomaterials are very important in informing the public. Public awareness, acceptance, health promotion initiatives, community activities, health-related educational activities, and schools, colleges/ universities awareness can make a difference. One way of doing this is to translate knowledge into different formats from complicated scientific ones to improved infographics, videos, and even exhibitions that would make nanotechnology more approachable and, in turn, make people understand its impact on sustainable development [157]. Additionally, recognizing that ELSI will be critical in getting the public's input and include the views of the society to avoid being countered in the future. Applied research in nanotechnology and a policy to share the results of the research with the public can eliminate biases that then lead to mistrust by the public towards nanotechnology projects.

8.4 The Ethical Landscape

Smart nanomaterials present various ethical issues that have called for elaborate frameworks for the conduct of research in this bracket. These frameworks must guarantee that nanotechnology will be positive to all people on this planet without being negative, misuse of this sense, and solve equity and accessibility challenges [158]. An example of a major ethical issue that may arise from the increased utilization of nanotechnology is the fact that the advantages enjoyed by the various parties involved must be shared fairly. It is possible that new types of nanomaterials, which are mentioned in the article, will deepen the existing divisions only if people cannot receive the same benefits from them. Achieving equal access will therefore need conscious efforts and measures put in place that will consider persons in the community who are socially, economically, and politically disadvantaged. For example, the use of low-cost nanotechnology solutions in the third world is relevant to reputable problems like the shortage of clean water and healthcare. The negative implications of nanotechnology include misuse specifically in surveillance, military use, and privacy also calls for ethical questioning. There needs to be sound and credible regulatory measures put in place to avert the creation and application of nanomaterials that can be risky to the lives of persons or the whole of society. Government solutions and cooperation can also be useful in establishing and implementing these norms, which make nanotechnology used for the improvement of the existence of human beings without engaging in pointless fights. Another ethical concern is related to future generations in terms of the existence and usage of nanotechnology. Inter-generation equity requires that the use of nanomaterials and the development of the resultant products should not jeopardize future generation's ability to meet fundamental human needs. This must happen with a precautionary principle wherein several aspects that could potentially be problematic are first investigated and controlled before widespread application.

Perhaps engagement of the public and ethical reasoning can be used in the development of these frameworks. Finding the ethical perspective brings diversity because all forms of values and concerns are sought. Ethical models should be in constant development based on scientific advancements and the changing society and basic principles of ethical models include being clear, answerable to the public, and engaging.

The incorporation of smart nanomaterials into various fields has presented a chance to create alterations that would sustainably benefit the globe. But to achieve this potential will entail adopting proper analysis of those technologies to the environmental, health, and social effects. Through performing life cycle assessments, involving the public in discussion, and creating solid ethical standards, we can guarantee that smart nanomaterials' innovation and use will be safe and useful for everyone.

Environmental, health, and safety aspects must remain central to the nanomaterial design where features such as biocompatibility and biodegradability should be pursued, as should lower the environmental impacts of production. Promoting and communicating before the occurrence of a public health threat builds confidence and supports, the management, and application of nanotechnology; ethical principles help manage nanotechnology proficiently and fairly. Altogether, they create the potential for hybrid smart nanomaterials to help a better future for current and future generations by building a sustainable and just society. It implies that for nanotechnology to achieve the benefits of change, the integration of the physical, biological, economic, social, and cultural properties of the technology, thorough careful analysis of the science must not be forgotten. The greatness of these materials must be weighed against the various possibilities of longevity and problems that may arise in the future. In this regard, life cycle assessments (LCAs) are needed. Consequently, an assessment of the environmental consequences of smart nanomaterials' life cycle can determine the risks and opportunities for development. LCAs assist in ascertaining that it is worth using these materials and that the social costs do not lead to the emergence of other unwanted side effects. It is also crucial to have appropriate legal demands regarding the smart nanomaterial's development and use. These frameworks should call for safety and sustainability from the concept development through research, production, marketing, and disposal. It becomes necessary for the regulatory agencies to consult with scientists, industries, and other people in society to formulate policies that adequately address both the health of the people and the environment.

Scientific and regulatory organizations should give guidelines or work hand in hand with other industries and the people to ensure that the health of the people and the environment is being preserved and maintained. Moreover, the ethical implication or consideration should be regarded as the core of development in nanoscale technology. It is the case that many approaches to the politics of technology and innovation work under the finiteness of risk governance, which is far from what is meant by responsible innovation, actively pursuing socially beneficial outcomes. This includes guaranteeing equal opportunity for smart nanomaterials' advantages and addressing the corresponding needs and concerns of communities for clear development [159].

Therefore, it can be stated that enhancing smart nanomaterials for the sake of sustainable objectives is a great opportunity to look forward to a sustainable future. These are some of the ideas that are considered radical given the traditional methodologies that have been used to formulate solutions to environmental issues hence the need to adopt these materials for the view they offer a new paradigm that is transformative of existing methods hence developing new solutions to match new problems. However, these innovations' success hinges on the state of responsible development, effective regulation of such innovative advancements, and ethics. The introduction of smart nanomaterials as tools for solving problems of various industries must be accompanied by a sense of responsibility to create a better future. Increasingly urgent is the need for proper responses to environmental problems, and drastic strategies have become evident. Smart nanomaterials enlighten humanity by presenting a vision of living in a sustainable society. By revealing the possible perspective, they hide and promoting the realization of the potential that underlines their production, there will be a certain chance to change the situation for the better and ensure the triumph of peace between humans and nature. The grass is still greener on the other side of this climb, and that climber is the next generation that should deserve a sustainable earth to live in.

9 Conclusion

Now that we are on the threshold of the new era on the wings of new smart nanomaterials, the way to achieving a sustainable future requires much more than scientific discoveries. There should be collective action in both interdisciplinary specialists, cross-country frameworks, and over the sectors of a society to ensure they are innovated and applied ethically. Summing up the key points to the discussion of this chapter, let us underline the role of collaboration, the issues that have to be faced shortly, along with the goals, that could be considered as a kind of legacy for future generations. International partnership is even more critical. Climate change, pollution, and resource depletion, global problems require global actions, and thus the strategies must be global to match the problems. International cooperation and joint projects, conferences, and exhibitions, inter alia, may help speed up the development and application of smart nanomaterials. For instance, open databases can contribute to knowledge and best practice sharing; this would allow researchers on a global level to work on one another's findings and knowledge or ideas. Government and businesses have the major responsibility of shaping the favorable environment, necessary for sound nanomaterials' advancement. The regulations laid down in one country may be ineffective or even harmful in another; therefore, the integration of the regulations will help to maintain safety and ethical standards and accelerate the development of new applications of nanotechnology in which the public will have confidence. Besides, it can establish research grants and tax exemptions for various industries to encourage the innovation process and solve society's problems. This is particularly true concerning public engagement in the society. This can be done by engaging the communities in advanced material discussions such as nanotechnology hence reducing undue alarms. The awareness raising through educational programs, open and effective information and communication, as well as the inclusion of the public in making decisions regarding the application of nanomaterials demonstrates that society would accept the application of nanomaterials if they understand the benefits that would accrue there from Pidgeon et al. [160]. Lack of safety and environmental consciousness is a big issue. Evaluations throughout the technology life cycle must be carried out to identify and prevent any health hazard to humans or harm to the environment by the use of nanomaterials. Creating materials that can merge with the living tissue and disintegrate at a certain point in time provides everlasting solutions but before adopting such technologies, they must undergo first, second, and third-degree tests to confirm their non-hazardous nature [161]. Concerning ethics and social responsibilities, the specifics should be assessed as possible inconveniences. Nanotechnology's gains should not deepen existing disparities for the benefits of the technology to be fully realized therefore, there is a need to get a solid policy for equity. Such practices could give rise to abuses hence requiring an ethical foundation in attempts to prevent abuse and to facilitate the development of clear guidelines concerning the distribution of gains. This entails global policies for technological development that incorporate the disadvantaged groups in society and the effects being brought about by the current innovations [162]. Nevertheless, it should be noted that the dream of a clean future full of smart nanomaterials is realistic and in the framework of these prospects. The proper development and utilization of such nanomaterials will go a long way in giving the next generation a clean, healthy, and resource-optimized society. Worry about tomorrow when solar cells created with the help of nanotechnology will collect energy from all the rays of sunshine and the world will be full of clean energy. Think about the water purification systems based on innovative nanofiltration membranes to filter out all the dangerous particles at the molecular level and provide people with clean water. Imagine an agricultural system in which the nanobots directly deliver the nutrients to the plant without wastage or, smart sensors on crop fields that provide real-time results on the soil health so that resource use can be maximized with efficiency in the farming process. In business, think of precise targeted, self-assembling drug delivery systems that adapt to patients' needs for optimal results and minimal reactions. For example, nanobots can detect diseases and treat them through special operations With improved accuracy or perhaps bio-tissues which are in a position of creating new tissues that replace the organ instead of having to rely on the donor tissues. These visions can be turned into reality if the research and invention are continued further. Today's investors realize that the exploration of nanotechnology reveals new opportunities and opens the door to breakthroughs that will help solve the problems of the day. While public and private organizations must patronize outstanding research as well as creation that pushes the boundaries, it is essential to do so.

Last, but not least, it is this journey to bequeath the world to the future generations in a better shape than it is now. The decisions that are being made currently will influence the destiny of tomorrow's generation. Thus, having exemplarily learned the possible benefits of smart nanomaterials, one can expect that with the help of further development of such technology, the world will be developed as a sphere of perfect synergy between the high quality of life and the absence of the threat of ecological catastrophe for generations to come.

Cellulose, the Earth's most abundant biopolymer, has emerged as a promising platform for developing sustainable and high-performance materials. By manipulating cellulose at the nanoscale, scientists have created materials with extraordinary properties, opening up new possibilities across various industries. From renewable energy and healthcare to packaging and construction, cellulosic nanomaterials offer a sustainable alternative to traditional, often petroleum-based materials. These materials, including cellulose nanocrystals (CNCs) and cellulose nanofibrils (CNFs), exhibit exceptional characteristics such as high strength, biodegradability, and renewability. Their versatility enables their integration into a wide range of applications, from advanced composites to water purification systems. Cellulosic nanomaterials contribute to a sustainable future by advancing SDGs 3, 6, 7, 9, 12, 13, and 15, including health, clean water, energy, industry, responsible consumption, climate action, and life on land. As research progresses, we can anticipate even more innovative and impactful uses for cellulosic nanomaterials, driving a sustainable future.

References

1. Channab, B.-E., El Idrissi, A., Essamlali, Y., Zahouily, M.: Nanocellulose: Structure, modification, biodegradation and applications in agriculture as slow/controlled release fertilizer, superabsorbent, and crop protection: a review. J. Environ. Manage. **352**, 119928 (2024). https://doi.org/10.1016/j.jenvman.2023.119928
2. Kargarzadeh, H., et al.: Recent developments in nanocellulose-based biodegradable polymers, thermoplastic polymers, and porous nanocomposites. Prog. Polym. Sci. **87**, 197–227 (2018). https://doi.org/10.1016/j.progpolymsci.2018.07.008
3. Hossen, S., et al.: Smart nanocarrier-based drug delivery systems for cancer therapy and toxicity studies: a review. J. Adv. Res. **15**, 1–18 (2019). https://doi.org/10.1016/j.jare.2018.06.005
4. Yu, Z.-L., Gao, Y.-C., Qin, B., Ma, Z.-Y., Yu, S.-H.: Revitalizing traditional phenolic resin toward a versatile platform for advanced materials. Acc. Mater. Res. **5**, 146–159 (2024). https://doi.org/10.1021/accountsmr.3c00194
5. Verma, V.K., Verma, S.: Applications and potential of advanced materials: an overview. Mater. Today Proc. (2024). https://doi.org/10.1016/j.matpr.2024.05.004
6. Kannan, K., Sadasivuni, K.K., Abdullah, A.M., Kumar, B.: Current trends in MXene-based nanomaterials for energy storage and conversion system: a mini review. Catalysts **10**, 495 (2020)
7. Abdel Maksoud, M.I.A., et al.: Advanced materials and technologies for supercapacitors used in energy conversion and storage: a review. Environ. Chem. Lett. **19**, 375–439 (2021). https://doi.org/10.1007/s10311-020-01075-w

8. Gómez-López, P., et al.: Nanomaterials and catalysis for green chemistry. Curr. Opin. Green Sustain. Chem. **24**, 48–55 (2020). https://doi.org/10.1016/j.cogsc.2020.03.001
9. Mitchell, S., Qin, R., Zheng, N., Pérez-Ramírez, J.: Nanoscale engineering of catalytic materials for sustainable technologies. Nat. Nanotechnol. **16**, 129–139 (2021). https://doi.org/10.1038/s41565-020-00799-8
10. Iravani, S., Varma, R.S.: Sustainable synthesis of cobalt and cobalt oxide nanoparticles and their catalytic and biomedical applications. Green Chem. **22**, 2643–2661 (2020). https://doi.org/10.1039/D0GC00885K
11. Sahu, T., et al.: Nanotechnology based drug delivery system: current strategies and emerging therapeutic potential for medical science. J. Drug Deliv. Sci. Technol. **63**, 102487 (2021). https://doi.org/10.1016/j.jddst.2021.102487
12. Zahin, N., et al.: Nanoparticles and its biomedical applications in health and diseases: special focus on drug delivery. Environ. Sci. Pollut. Res. **27**, 19151–19168 (2020). https://doi.org/10.1007/s11356-019-05211-0
13. Chandrakala, V., Aruna, V., Angajala, G.: Review on metal nanoparticles as nanocarriers: current challenges and perspectives in drug delivery systems. Emergent Mater. **5**, 1593–1615 (2022). https://doi.org/10.1007/s42247-021-00335-x
14. Naeem, M., et al.: Advances in colon-targeted nano-drug delivery systems: challenges and solutions. Arch. Pharmacal Res. **43**, 153–169 (2020). https://doi.org/10.1007/s12272-020-01219-0
15. Jain, K.K. (ed.): Drug Delivery Systems, pp. 1–54. Springer, New York (2020).
16. Rai, V., Acharya, S., Dey, N.: Implications of Nanobiosensors in Agriculture (2012). https://doi.org/10.4236/jbnb.2012.322039
17. Noah, N.M., Ndangili, P.M.: Green synthesis of nanomaterials from sustainable materials for biosensors and drug delivery. Sens. Int. **3**, 100166 (2022). https://doi.org/10.1016/j.sintl.2022.100166
18. Bharti, A., Jain, U., Chauhan, N.: From lab to field: nano-biosensors for real-time plant nutrient tracking. Plant Nano Biol., 100079. https://doi.org/10.1016/j.plana.2024.100079 (2024).
19. Padmanabhan, N.T., et al.: Graphene coupled TiO_2 photocatalysts for environmental applications: a review. Chemosphere **271**, 129506 (2021). https://doi.org/10.1016/j.chemosphere.2020.129506
20. Wang, G., Zhou, T., Wang, S. & Photocatalysts. Crystals 2023, 1109. Editorial: Semiconductor Photocatalysts. MDPI. https://doi.org/10.3390/cryst13071109 (2023).
21. Collins, A. & Tajvidi, M. Cellulose Nanofibrils Dewatered with Poly(Lactic Acid) for Improved Bio-Polymer Nanocomposite Processing. **14**, 1419 (2024).
22. Dufresne, A.: Nanocellulose: a new ageless bionanomaterial. Mater. Today **16**, 220–227 (2013). https://doi.org/10.1016/j.mattod.2013.06.004
23. Siró, I., Plackett, D.: Microfibrillated cellulose and new nanocomposite materials: a review. Cellulose **17**, 459–494 (2010). https://doi.org/10.1007/s10570-010-9405-y
24. Sun, Z., et al.: Complete lignocellulose conversion with integrated catalyst recycling yielding valuable aromatics and fuels. Nat. Catal. **1**, 82–92 (2018). https://doi.org/10.1038/s41929-017-0007-z
25. Das, A.K., Islam, M.N., Ashaduzzaman, M., Nazhad, M.M.: Nanocellulose: its applications, consequences and challenges in papermaking. J. Packag. Technol. Res. **4**, 253–260 (2020). https://doi.org/10.1007/s41783-020-00097-7
26. Balea, A., et al.: Industrial application of nanocelluloses in papermaking: a review of challenges, technical solutions, and market perspectives. Mol. (Basel, Switz.) **25** (2020). https://doi.org/10.3390/molecules25030526
27. Lin, N., Dufresne, A.: Nanocellulose in biomedicine: current status and future prospect. Eur. Polymer J. **59**, 302–325 (2014). https://doi.org/10.1016/j.eurpolymj.2014.07.025
28. Gopakumar, D.A., et al.: Nanoscale materials in water purification. In: Thomas, S., Pasquini, D., Leu, S.Y., Gopakumar, D.A. (eds.), pp. 59–85. Elsevier (2019)
29. Habibi, Y., Lucia, L.A., Rojas, O.J.: Cellulose nanocrystals: chemistry, self-assembly, and applications. Chem. Rev. **110**, 3479–3500 (2010). https://doi.org/10.1021/cr900339w

30. Abitbol, T., et al.: Nanocellulose, a tiny fiber with huge applications. Curr. Opin. Biotechnol. **39**, 76–88 (2016). https://doi.org/10.1016/j.copbio.2016.01.002
31. Qiao, C., Chen, G., Zhang, J., Yao, J.: Structure and rheological properties of cellulose nanocrystals suspension. Food Hydrocoll. **55**, 19–25 (2016). https://doi.org/10.1016/j.foodhyd.2015.11.005
32. George, J., Sabapathi, S.N.: Cellulose nanocrystals: synthesis, functional properties, and applications. Nanotechnol. Sci. Appl. **8**, 45–54 (2015). https://doi.org/10.2147/nsa.S64386
33. Rana, A.K., Thakur, V.K.: Impact of physico-chemical properties of nanocellulose on rheology of aqueous suspensions and its utility in multiple fields: a review. J. Vinyl Addit. Technol. **29**, 617–648 (2023). https://doi.org/10.1002/vnl.22006
34. Smriti, S.A., et al.: Recent developments of the nanocellulose extraction from water hyacinth: a review. Cellulose **30**, 8617–8641 (2023). https://doi.org/10.1007/s10570-023-05374-7
35. Czaja, W.K., Young, D.J., Kawecki, M., Brown, R.M., Jr.: The future prospects of microbial cellulose in biomedical applications. Biomacromolecules **8**, 1–12 (2007). https://doi.org/10.1021/bm060620d
36. Helenius, G., et al.: In vivo biocompatibility of bacterial cellulose. J. Biomed. Mater. Res. Part A **76**, 431–438 (2006). https://doi.org/10.1002/jbm.a.30570
37. Wasim, M., et al.: Development of bacterial cellulose nanocomposites: an overview of the synthesis of bacterial cellulose nanocomposites with metallic and metallic-oxide nanoparticles by different methods and techniques for biomedical applications. J. Ind. Text. **51**, 1886S–1915S (2022). https://doi.org/10.1177/1528083720977201
38. Yano, H., et al.: Optically transparent composites reinforced with networks of bacterial nanofibers. Adv. Mater. **17**, 153–155 (2005). https://doi.org/10.1002/adma.200400597
39. Nogi, M., Iwamoto, S., Nakagaito, A.N., Yano, H.: Optically transparent nanofiber paper. Adv. Mater. **21**, 1595–1598 (2009). https://doi.org/10.1002/adma.200803174
40. Svensson, A., et al.: Bacterial cellulose as a potential scaffold for tissue engineering of cartilage. Biomaterials **26**, 419–431 (2005). https://doi.org/10.1016/j.biomaterials.2004.02.049
41. Revin, V.V., et al.: Bacterial cellulose-based polymer nanocomposites: a review. Polymers **14**, 4670 (2022).
42. Portela, R., Leal, C.R., Almeida, P.L., Sobral, R.G.: Bacterial cellulose: a versatile biopolymer for wound dressing applications. Microb. Biotechnol. **12**, 586–610. https://doi.org/10.1111/1751-7915.13392 (2019)
43. Isogai, A., Saito, T., Fukuzumi, H.: TEMPO-oxidized cellulose nanofibers. Nanoscale **3**, 71–85 (2011). https://doi.org/10.1039/C0NR00583E
44. Tang, Z., Lin, X., Yu, M., Mondal, A.K., Wu, H.: Recent advances in TEMPO-oxidized cellulose nanofibers: oxidation mechanism, characterization, properties and applications. Int. J. Biol. Macromol. **259**, 129081 (2024). https://doi.org/10.1016/j.ijbiomac.2023.129081
45. Liu, S., Low, Z.-X., Xie, Z., Wang, H.: TEMPO-oxidized cellulose nanofibers: a renewable nanomaterial for environmental and energy applications. Adv. Mater. Technol. **6**, 2001180 (2021). https://doi.org/10.1002/admt.202001180
46. Pottathara, Y.B., Narwade, V.N., Bogle, K.A., Kokol, V.: TEMPO-oxidized cellulose nanofibrils–graphene oxide composite films with improved dye adsorption properties. Polym. Bull. **77**, 6175–6189 (2020). https://doi.org/10.1007/s00289-019-03077-3
47. Saito, T., Kimura, S., Nishiyama, Y., Isogai, A.: Cellulose nanofibers prepared by TEMPO-mediated oxidation of native cellulose. Biomacromolecules **8**, 2485–2491 (2007). https://doi.org/10.1021/bm0703970
48. Sehaqui, H., Liu, A., Zhou, Q., Berglund, L.A.: Fast preparation procedure for large, flat cellulose and cellulose/inorganic nanopaper structures. Biomacromolecules **11**, 2195–2198 (2010). https://doi.org/10.1021/bm100490s
49. Chen, C., et al.: A strategy to fabricate tough and strong bioinspired fiber with asymmetric structure and water-driven performance. Compos. Sci. Technol. **224**, 109464 (2022). https://doi.org/10.1016/j.compscitech.2022.109464

50. Chu, J.: "Nanofiber Yarn" Makes for Stretchy, Protective Artificial Tissue. https://news.mit.edu/2019/nanofiber-yarn-artificial-tissue-0423 (2019)
51. Matsumoto, K., et al.: Cellulose nanofiber-introduced continuous-ramie yarn-reinforced polylactic acid filament for 3D printing: novel fabrication process and mechanical properties. Compos. A Appl. Sci. Manuf. **176**, 107836 (2024). https://doi.org/10.1016/j.compositesa.2023.107836
52. Horta-Velázquez, A., Morales-Narváez, E.: Nanocellulose in wearable sensors. Green Anal. Chem. **1**, 100009 (2022). https://doi.org/10.1016/j.greeac.2022.100009
53. Suresh Khurd, A., Kandasubramanian, B.: A systematic review of cellulosic material for green electronics devices. Carbohydr. Polym. Technol. Appl. **4**, 100234 (2022). https://doi.org/10.1016/j.carpta.2022.100234
54. Zhu, C., Wu, J., Yan, J., Liu, X.: Advanced fiber materials for wearable electronics. Adv. Fiber Mater. **5**, 12–35 (2023). https://doi.org/10.1007/s42765-022-00212-0
55. Zhao, D., et al.: Cellulose-based flexible functional materials for emerging intelligent electronics. Adv. Mater. **33**, 2000619 (2021). https://doi.org/10.1002/adma.202000619
56. Meng, X., et al.: Rational design of cellulosic triboelectric materials for self-powered wearable electronics. Nano-Micro Lett. **15**, 124 (2023). https://doi.org/10.1007/s40820-023-01094-6
57. Xu, Y., et al.: Paper-based wearable electronics. iScience **24** (2021). https://doi.org/10.1016/j.isci.2021.102736
58. Tushar, S.I., et al.: Nanocellulose-based porous lightweight materials with flame retardant properties: a review. Carbohyd. Polym. **339**, 122237 (2024). https://doi.org/10.1016/j.carbpol.2024.122237
59. Li, A., Huber, T., Barker, D., Nazmi, A.R., Najaf Zadeh, H.: An overview of cellulose aerogels and foams for oil sorption: preparation, modification, and potential of 3D printing. Carbohyd. Polym. **343**, 122432 (2024). https://doi.org/10.1016/j.carbpol.2024.122432
60. Bidgoli, H., Mortazavi, Y., Khodadadi, A.A.: A functionalized nano-structured cellulosic sorbent aerogel for oil spill cleanup: synthesis and characterization. J. Hazard. Mater. **366**, 229–239 (2019). https://doi.org/10.1016/j.jhazmat.2018.11.084
61. Eichhorn, S.J., et al.: Current international research into cellulose as a functional nanomaterial for advanced applications. J. Mater. Sci. **57**, 5697–5767 (2022). https://doi.org/10.1007/s10853-022-06903-8
62. Nguyen, P.X.T., et al.: Recent developments in water treatment by cellulose aerogels from agricultural waste. IOP Conf. Ser. Earth Environ. Sci. **947**, 012011 (2021). https://doi.org/10.1088/1755-1315/947/1/012011
63. Huo, Y., et al.: Nanocellulose-based composite materials used in drug delivery systems. Polymers **14**, 2648 (2022)
64. Mazrouei-Sebdani, Z., Begum, H., Schoenwald, S., Horoshenkov, K.V., Malfait, W.J.: A review on silica aerogel-based materials for acoustic applications. J. Non-Cryst. Solids **562**, 120770 (2021). https://doi.org/10.1016/j.jnoncrysol.2021.120770
65. Mirtaghavi, A., Luo, J., Muthuraj, R.: Recent advances in porous 3D cellulose aerogels for tissue engineering applications: a review. J. Compos. Sci. **4**, 152 (2020)
66. Yin, J., et al.: Bioinspired and dual-functional nanocellulose aerogels for water disinfection and heavy metal removal. Nano Today **51**, 101918 (2023). https://doi.org/10.1016/j.nantod.2023.101918
67. Ahmad, A., et al.: Recent advances in nanocellulose aerogels for efficient heavy metal and dye removal. Gels **9**, 416 (2023)
68. Si, R., Pu, J., Luo, H., Wu, C., Duan, G.: Nanocellulose-based adsorbents for heavy metal ion. Polymers **14**, 5479 (2022)
69. Hamimed, S., et al.: Handbook of nanocelluloses: classification, properties, fabrication, and emerging applications. In: Barhoum, A. (ed.), pp. 1–42. Springer International Publishing (2020)
70. Baraka, F., Labidi, J.: The emergence of nanocellulose aerogels in CO_2 adsorption. Sci. Total Environ. **912**, 169093 (2024). https://doi.org/10.1016/j.scitotenv.2023.169093

71. Eldhose, M., Roy, R., George, C., Joseph, A.: Handbook of biopolymers. In: Thomas, S., Ajitha, A.R., Chirayil, C.J., Thomas, B. (eds.), pp. 1007–1032. Springer Nature Singapore (2023)
72. Pena-Pereira, F., Lavilla, I., de la Calle, I., Romero, V., Bendicho, C.: Detection of gases and organic vapors by cellulose-based sensors. Anal. Bioanal. Chem. **415**, 4039–4060 (2023). https://doi.org/10.1007/s00216-023-04649-z
73. Sehaqui, H., Salajková, M., Zhou, Q., Berglund, L.A.: Mechanical performance tailoring of tough ultra-high porosity foams prepared from cellulose I nanofiber suspensions. Soft Matter **6**, 1824–1832 (2010). https://doi.org/10.1039/B927505C
74. Zhang, L., et al.: Wear in-situ self-healing polymer composites incorporated with bifunctional microcapsules. Compos. Part B Eng. **232**, 109566 (2022). https://doi.org/10.1016/j.compositesb.2021.109566
75. Ong, X.-R., Chen, A. X., Li, N., Yang, Y.Y., Luo, H.-K.: Nanocellulose: recent advances toward biomedical applications. Small Sci. **3**, 2200076 (2023). https://doi.org/10.1002/smsc.202200076
76. Nicu, R., Ciolacu, F., Ciolacu, D.E.: Advanced functional materials based on nanocellulose for pharmaceutical/medical applications. Pharmaceutics **13** (2021). https://doi.org/10.3390/pharmaceutics13081125
77. Balalakshmi, C., Jeyachandran, S.: Handbook of nanocelluloses: classification, properties, fabrication, and emerging applications. In: Barhoum, A. (ed.), pp. 1–31. Springer International Publishing (2020).
78. Shi, Y., et al.: Functionalization of nanocellulose applied with biological molecules for biomedical application: a review. Carbohyd. Polym. **285**, 119208 (2022). https://doi.org/10.1016/j.carbpol.2022.119208
79. Djalal, T.: Nanocellulose as a promising sustainable material for biomedical applications. AIMS Mater. Sci. **5**, 201–205 (2018). https://doi.org/10.3934/matersci.2018.2.201
80. Ullah, M.W., Ul-Islam, M., Wahid, F., Yang, G.: Editorial: nanocellulose: a multipurpose advanced functional material, volume II. Front. Bioeng. Biotechnol. **10** (2022). https://doi.org/10.3389/fbioe.2022.931256
81. Leong, M.Y., Kong, Y.L., Harun, M.Y., Looi, C.Y., Wong, W.F.: Current advances of nanocellulose application in biomedical field. Carbohyd. Res. **532**, 108899 (2023). https://doi.org/10.1016/j.carres.2023.108899
82. Jose, J., Thomas, V., Vinod, V., Abraham, R., Abraham, S.: Nanocellulose based functional materials for supercapacitor applications. J. Sci. Adv. Mater. Devices **4**, 333–340 (2019). https://doi.org/10.1016/j.jsamd.2019.06.003
83. Amara, C., El Mahdi, A., Medimagh, R., Khwaldia, K.: Nanocellulose-based composites for packaging applications. Curr. Opin. Green Sustain. Chem. **31**, 100512 (2021). https://doi.org/10.1016/j.cogsc.2021.100512
84. Voisin, H., Bergström, L., Liu, P., Mathew, A.P.: Nanocellulose-based materials for water purification. Nanomater. (Basel, Switz.) **7** (2017). https://doi.org/10.3390/nano7030057
85. Wong, S., et al.: Fibrillar pharmacology of functionalized nanocellulose. Sci. Rep. **11**, 157 (2021). https://doi.org/10.1038/s41598-020-79592-5
86. Malekpour, K., Hazrati, A., Khosrojerdi, A., Roshangar, L., Ahmadi, M.: An overview to nanocellulose clinical application: biocompatibility and opportunities in disease treatment. Regen. Ther. **24**, 630–641 (2023). https://doi.org/10.1016/j.reth.2023.10.006
87. Badawy, A.H., El-Feky, M.S., El-Tair, A.M., Kohail, M.: Wood sawdust waste-derived nanocellulose as a versatile reinforcing agent for nano silica cement composites: a systematic study on its characterization and performance. Sci. Rep. **13**, 12811 (2023). https://doi.org/10.1038/s41598-023-39788-x
88. Akhlaghi, M.A., Bagherpour, R., Kalhori, H.: Application of bacterial nanocellulose fibers as reinforcement in cement composites. Constr. Build. Mater. **241**, 118061 (2020). https://doi.org/10.1016/j.conbuildmat.2020.118061
89. Aziz, M.A., Zubair, M., Saleem, M.: Development and testing of cellulose nanocrystal-based concrete. Case Stud. Constr. Mater. **15**, e00761 (2021). https://doi.org/10.1016/j.cscm.2021.e00761

90. Sahoo, N.G., Rana, S., Cho, J.W., Li, L., Chan, S.H.: Polymer nanocomposites based on functionalized carbon nanotubes. Prog. Polym. Sci. **35**, 837–867 (2010). https://doi.org/10.1016/j.progpolymsci.2010.03.002
91. Sajjad, U.: Nanotubes in aerospace: using CNTs to make satellites stronger. https://www.azonano.com/article.aspx?ArticleID=6489 (2023)
92. Chu, J.: A new approach to making airplane parts, minus the massive infrastructure; Carbon nanotube film produces aerospace-grade composites with no need for huge ovens or autoclaves. https://news.mit.edu/2020/carbon-nanotubes-making-airplane-aerospace-parts-1013 (2020)
93. Marchant, N.: 9 Things you didn't know could be made from wood. https://www.weforum.org/agenda/2021/04/surprising-everyday-items-wood/ (2021)
94. Ramchander, K.: Development of xylem-based water filters. M.Sc. Mechanical Engineering thesis, Massachusetts Institute of Technology (2016)
95. Chu, J.: MIT engineers make filters from tree branches to purify drinking water. https://news.mit.edu/2021/filters-sapwood-purify-water-0325#:~:text=MIT%20engineers%20have%20been%20investig (2021)
96. Othman, J.A.S., Ilyas, R.A., Nordin, A.H., Ngadi, N., Alkbir, M.F.M.: Recent advancements in bamboo nanocellulose-based bioadsorbents and their potential in wastewater applications: a review. Int. J. Biol. Macromol. **277**, 134451 (2024). https://doi.org/10.1016/j.ijbiomac.2024.134451
97. Liu, M., et al.: Highly permeable and regenerative microhoneycomb filters. ACS Appl. Mater. Interfaces **16**, 29177–29187 (2024). https://doi.org/10.1021/acsami.4c02697
98. WEF. This bag is partly made with bacteria. https://www.weforum.org/videos/bag-bacteria/ (2024)
99. Gibbons, M.: Wood-based computer chips could be the answer to the electronic waste crisis. https://qz.com/455962/wood-based-computer-chips-could-be-the-answer-to-the-electronic-waste-crisis (2015)
100. Strümpel, C., et al.: Modifying the solar spectrum to enhance silicon solar cell efficiency—an overview of available materials. Sol. Energy Mater. Sol. Cells **91**, 238–249 (2007). https://doi.org/10.1016/j.solmat.2006.09.003
101. Liang, H., et al.: Full-spectrum solar energy utilization and enhanced solar energy harvesting via photon anti-reflection and scattering performance using biomimetic nanophotonic structure. ES Energy & Environ. **8**, 29–41 (2020). https://doi.org/10.30919/esee8c456
102. Ganesan, A.A., Houtepen, A.J., Crisp, R.W.: Quantum dot solar cells: small beginnings have large impacts. Appl. Sci. **8**, 1867 (2018). https://doi.org/10.3390/app8101867
103. McConnell, I., Li, G., Brudvig, G.W.: Energy conversion in natural and artificial photosynthesis. Chem. & Biol. **17**, 434–447 (2010). https://doi.org/10.1016/j.chembiol.2010.05.005
104. Afre, R.A., Pugliese, D.J.M.: Perovskite solar cells: a review of the latest advances in materials, fabrication techniques, and stability enhancement strategies. Micromachines **15**, 192 (2024). https://doi.org/10.3390/mi15020192
105. Gaines, L., Richa, K., Spangenberger, J.J.M.E.: & Sustainability. Key issues for Li-ion battery recycling. **5**, E14 (2018). https://doi.org/10.1557/mre.2018.13
106. Yang, P., et al.: Ultrafast-charging supercapacitors based on corn-like titanium nitride nanostructures. Adv. Sci. **3** (2016). https://doi.org/10.1002/advs.201500299
107. Obiuto, N.C., Olajiga, O.K., Adebayo, R.A.: The role of nanomaterials in energy storage: a comparative review of USA and African development. World J. Adv. Res. Rev. **21**, 2073–2083 (2024). https://doi.org/10.30574/wjarr.2024.21.3.0925
108. Pakseresht, S., Celik, M., Guler, A., Al-Ogaili, A.W.M., Kallio, T.J.B.: Recent advances in all-solid-state lithium-oxygen batteries: challenges, strategies. Future **9**, 380 (2023). https://doi.org/10.3390/batteries9070380
109. Rathi, V., Brajpuriya, R., Gupta, R., Parmar, K., Kumar, A.: Graphene-Derived Composites: A New Frontier in Thermoelectric Energy Conversion (2024). https://doi.org/10.1039/D3YA00526G

110. Izquierdo-Monge, O., Peña-Carro, P., Hernández-Jiménez, A., Zorita-Lamadrid, A., Hernández-Callejo, L.: Methodology for energy management in a smart microgrid based on the efficiency of dispatchable renewable generation sources and distributed storage systems. Appl. Sci. **14**, 1946 (2024). https://doi.org/10.3390/app14051946
111. Tu, W.H., Zhao, Y., Chan, W.P., Lisak, G.: Reclaimed seawater discharge–desalination brine treatment and resource recovery system. Water Res. **251**, 121096 (2024). https://doi.org/10.1016/j.watres.2023.121096
112. Castelletto, S., Boretti, A.: Advantages, limitations, and future suggestions in studying graphene-based desalination membranes. RSC Adv. **11**, 7981–8002 (2021). https://doi.org/10.1039/D1RA00278C
113. Chen, S., et al.: Robust silica-graphene oxide nanocomposite membrane for separating NaCl salt from seawater under atmospheric pressure. Desalination **574**, 117208 (2024). https://doi.org/10.1016/j.desal.2023.117208
114. Li, J., et al.: Hexagonal Cu_2O-Pt alloyed nano-lilies mediated Fenton catalysis for recyclable degradation of organic dye. J. Water Process. Eng. **61**, 105364 (2024). https://doi.org/10.1016/j.jwpe.2024.105364
115. Van Nguyen, T.T., et al.: Magnetite nanoparticles decorated on cellulose aerogel for p-nitrophenol Fenton degradation: effects of the active phase loading, cross-linker agent and preparation method. Heliyon **9** (2023). https://doi.org/10.1016/j.heliyon.2023.e22319
116. Shukla, S., Sangwan, A., Pabreja, N., Gulati, S.: Nanosponges for Environmental Remediation, pp. 449–473. Springer (2023)
117. Huseien, G.F., et al.: Smart bio-agents-activated sustainable self-healing cementitious materials: an all-inclusive overview on progress, benefits and challenges. Sustainability **14**, 1980 (2022). https://doi.org/10.3390/su14041980
118. So, H.-M., et al.: Carbon nanotube based pressure sensor for flexible electronics. Mater. Res. Bull. **48**, 5036–5039 (2013). https://doi.org/10.1016/j.materresbull.2013.07.022
119. Awan, U., Sroufe, R.: Sustainability in the circular economy: insights and dynamics of designing circular business models. Appl. Sci. **12**, 1521 (2022). https://doi.org/10.3390/app12031521
120. Wang, C., Han, H., Wu, Y., Astruc, D.: Nanocatalyzed upcycling of the plastic wastes for a circular economy. Coord. Chem. Rev. **458**, 214422 (2022). https://doi.org/10.1016/j.ccr.2022.214422
121. Thiounn, T., Smith, R.C.: Advances and approaches for chemical recycling of plastic waste. J. Polym. Sci. **58**, 1347–1364 (2020). https://doi.org/10.1002/pol.20190261
122. Mishra, B., et al.: Valorization of agro-industrial biowaste to biomaterials: an innovative circular bioeconomy approach. Circ. Econ. 100050 (2023). https://doi.org/10.1016/j.cec.2023.100050
123. Venkateswaran, M.R., Khosravi, A., Zarepour, A., Iravani, S., Zarrabi, A.: Self-healing materials in biomedicine and the circular economy (2024). https://doi.org/10.1039/D4EN00235K
124. Arroyave, S., et al.: Evaluation and characterization of autonomous self-healing cementitious materials with low carbon footprint using hybrid organic/inorganic microcapsules (2023). https://doi.org/10.1016/j.matpr.2023.03.673
125. Ortega, S., et al.: Bottom-up engineering of thermoelectric nanomaterials and devices from solution-processed nanoparticle building blocks. Chem. Soc. Rev. **46**, 3510–3528 (2017). https://doi.org/10.1039/C6CS00567E
126. Anduix-Canto, C., et al.: Unraveling the mechanism of electrically induced adhesive debonding: a spectro-microscopic study. Adv. Mater. Interfaces **9**, 2101447 (2022). https://doi.org/10.1002/admi.202101447
127. Meng, K., et al.: Intelligent disassembly of electric-vehicle batteries: a forward-looking overview. Resour. Conserv. Recycl. **182**, 106207 (2022). https://doi.org/10.1016/j.resconrec.2022.106207
128. Fathi, Q.J., Sheidaee, E., Bazyar, P.: The Impact of Various Nanomaterials and Nano-Agrochemicals on Agricultural Systems (2023). https://doi.org/10.48309/jeires.2023.4.5

129. Mana, A., et al.: Sustainable AI-based production agriculture: exploring AI applications and implications in agricultural practices. Smart Agric. Technol. 100416 (2024). https://doi.org/10.1016/j.atech.2024.100416
130. Zhang, P., et al.: Nanotechnology and artificial intelligence to enable sustainable and precision agriculture. **7**, 864–876 (2021). https://doi.org/10.1038/s41477-021-00946-6
131. Hazafa, A. et al. in Insecticides-Impact and Benefits of Its Use for Humanity (IntechOpen, 2021).
132. Zhang, P., et al.: Strategies for Enhancing Plant Immunity and Resilience Using Nanomaterials for Sustainable Agriculture. (2024). https://doi.org/10.1021/acs.est.4c03522
133. Al-Asbahi, M. G., Al-Ofiry, B. A., Saad, F. A., Alnehia, A. & Al-Gunaid, M. Q. J. S. R. Silver nanoparticles biosynthesis using mixture of Lactobacillus sp. and Bacillus sp. growth and their antibacterial activity. **14**, 10224. https://doi.org/10.1038/s41598-024-59936-1 (2024).
134. Jabran, M. et al. Exploring the potential of nanomaterials (NMs) as diagnostic tools and disease resistance for crop pathogens. **11**, 75. https://doi.org/10.1186/s40538-024-00592-y (2024).
135. Dangi, K., Verma, A.K.J.M.T.P.: Efficient & eco-friendly smart nano-pesticides: Emerging prospects for agriculture. **45**, 3819–3824 (2021). https://doi.org/10.1016/j.matpr.2021.03.211
136. Thakur, M., Wang, B., Verma, M.L.: Development and applications of nanobiosensors for sustainable agricultural and food industries: Recent developments, challenges and perspectives. Environ. Technol. & Innov. **26**, 102371 (2022). https://doi.org/10.1016/j.eti.2022.102371
137. Yusof, N.A., Isha, A.: Nanomaterials for Agriculture and Forestry Applications, pp. 407–419. Elsevier (2020)
138. Uddin, M.J., Sherrell, J., Emami, A., Khaleghian, M. J. S. Application of artificial intelligence and sensor fusion for soil organic matter prediction. Sensors **24**, 2357 (2024). https://doi.org/10.3390/s24072357
139. de Oliveira, C., Mulinari, J., Reichert, F., Júnior, A.: Ciência, Tecnologia e Inovação: Do Campo à Mesa. In: Proceeding of International Agribusiness Congress pp. 25–27. Recife, Brazil
140. Ansah, J.P., Chiu, C.-T.: Projecting the chronic disease burden among the adult population in the United States using a multi-state population model. Front. Public Health **10**, 1082183 (2023). https://doi.org/10.3389/fpubh.2022.1082183
141. Shinn, J., Kwon, N., Lee, S.A., Lee, Y.: Smart pH-responsive nanomedicines for disease therapy. J. Pharm. Investig. **52**, 427–441 (2022). https://doi.org/10.1007/s40005-022-00573-z
142. Shi, Z. et al.
143. Das, K.P.: Nanoparticles and convergence of artificial intelligence for targeted drug delivery for cancer therapy: Current progress and challenges. Front. Med. Technol. **4**, 1067144 (2023). https://doi.org/10.3389/fmedt.2022.1067144
144. Londhey, P.M., Gholse, Y.N., Kasliwal, R.H.: Current novel updates on nanorobotics: application in disease diagnosis and treatment. https://doi.org/10.20959/wjpr20222-22757 (2021).
145. Belyanina, I., et al.: Targeted magnetic nanotheranostics of cancer. Molecules **22**, 975 (2017). https://doi.org/10.3390/molecules22060975
146. Augustine, R., et al.: Imaging cancer cells with nanostructures: prospects of nanotechnology driven non-invasive cancer diagnosis. Adv. Colloid Interface Sci. **294**, 102457 (2021). https://doi.org/10.1016/j.cis.2021.102457
147. Wang, S., Lee, J.M., Yeong, W.Y.: Smart hydrogels for 3D bioprinting. Int. J. Bioprinting **1**, 3–14. https://doi.org/10.18063/IJB.2015.01.005. (2015).
148. Gao, Q., Kim, B.-S., Gao, G.: Advanced strategies for 3D bioprinting of tissue and organ analogs using alginate hydrogel bioinks. Mar. Drugs **19**, 708 (2021). https://doi.org/10.3390/md19120708
149. Mani, M.P., et al.: A review on 3D printing in tissue engineering applications. J. Polym. Eng. **42**, 243–265 (2022). https://doi.org/10.1515/polyeng-2021-0059

150. Mota, C., Camarero-Espinosa, S., Baker, M.B., Wieringa, P., Moroni, L.: Bioprinting: from tissue and organ development to in vitro models. Chem. Rev. **120**, 10547–10607 (2020). https://doi.org/10.1021/acs.chemrev.9b00789
151. Keshavarz, M., et al.: Smart alginate inks for tissue engineering applications. Mater. Today Bio, 100829 (2023). https://doi.org/10.1016/j.mtbio.2023.100829
152. Domb, A.J., Sharifzadeh, G., Nahum, V., Hosseinkhani, H.: Safety evaluation of nanotechnology products. Pharmaceutics **13**, 1615 (2021). https://doi.org/10.3390/pharmaceutics13101615
153. Pandey, A., et al.: Black phosphorus as multifaceted advanced material nanoplatforms for potential biomedical applications. Nanomaterials **11**, 13 (2020). https://doi.org/10.3390/nano11010013
154. Siqueira, G., et al.: Cellulose nanocrystal inks for 3D printing of textured cellular architectures. Adv. Func. Mater. **27**, 1604619 (2017). https://doi.org/10.1002/adfm.201604619
155. Kadziński, M., et al.: Co-constructive development of a green chemistry-based model for the assessment of nanoparticles synthesis. Eur. J. Oper. Res. **264**, 472–490 (2018). https://doi.org/10.1016/j.ejor.2016.10.019
156. Khan, S.H.: Green nanotechnology for the environment and sustainable development. Green Mater. Wastewater Treat., 13–46 (2020). https://doi.org/10.1007/978-3-030-17724-9_2
157. Bowman, D.M., Hodge, G.: Nanotechnology: mapping the wild regulatory frontier. Futures **38**, 1060–1073. https://doi.org/10.1016/j.futures.2006.02.017
158. Oberdörster, G., Oberdörster, E., Oberdörster, J.: Nanotoxicology: an emerging discipline evolving from studies of ultrafine particles. Environ. Health Perspect. **113**, 823–839 (2005). https://doi.org/10.1289/ehp.7339
159. Frahm, N., Doezema, T., Pfotenhauer, S.: Fixing technology with society: the coproduction of democratic deficits and responsible innovation at the OECD and the European Commission. Sci. Technol. & Hum. Values **47**, 174–216 (2022). https://doi.org/10.1177/0162243921999100
160. Pidgeon, N., Harthorn, B., Satterfield, T.: Nanotechnology risk perceptions and communication: emerging technologies, emerging challenges. Risk Anal. Int. J. **31**, 1694–1700 (2011). https://doi.org/10.1111/j.1539-6924.2011.01738.x
161. Chhipa, H.C.: Nanofertilizers and nanopesticides for agriculture. Environ. Chem. Lett. **15**, 15–22 (2017). https://doi.org/10.1007/s10311-016-0600-4
162. Theus, A.S., et al.: Nanomaterials for bioprinting: functionalization of tissue-specific bioinks. Essays Biochem. **65**, 429–439 (2021). https://doi.org/10.1042/EBC20200095

Sustainable Cellulose Films as Packaging (Coating) Materials to Reduce Plastic Use

Mavia Rashid, Muhammad Arshad, and Ehsan Ul Haq

Abstract Initially, Petroleum based material such as plastics had been widely used as a dominating packaging material due to lightweight, durability, cost effectiveness and versatility. But, the increase in concern over environmental sustainability on a worldwide scale has led to a great deal of interest in creating sustainable packaging materials as substitutes for conventional petroleum-based ones. Cellulose has become a viable option for environmentally friendly packaging since it is made from renewable resources including wood, plants, and farm waste due to unique properties of cellulose, which make it a sustainable option for a range of packaging applications, including its renewability, biodegradability, and adaptability. It has been utilized in packaging in different forms such as nanocellulose fibers used as reinforcement in packaging, cellulose based films, cellulose fibers, packaging films containing cellulose crystals, etc. However, researchers from all over the world have aimed to overcome challenges faced by using cellulose as a packaging material in food and beverage industry such as sensitivity to moisture, cost considerations and inadequate barrier properties.

Keywords Petroleum based packaging · Sustainable packaging · Cellulose based packaging · Biodegradability · Food and beverage industry

1 Introduction

From the past few years, packaging of different products have been changed. The reason is consumer concerns about hazardous effect of plastic on living things and agrifood waste [1–4] Papers, metals, glass and plastics (e.g., polypropylene, polyethylene, and polyethylene terephthalate) are always used to pack Food products [5]. Plastics are responsible for their wide use in the industry related to food packaging due to durability, low production costs, good mechanical properties, and

M. Rashid · M. Arshad · E. U. Haq (✉)
Department of Metallurgical and Materials Engineering, Faculty of Chemical, Metallurgical and Polymer Engineering, University of Engineering and Technology, Lahore, Punjab, Pakistan
e-mail: amonehsan@uet.edu.pk

K. K. Kesari et al. (eds.), *Functionalized Cellulose Materials*, Engineering Materials,
https://doi.org/10.1007/978-3-031-76953-5_2

resistance to oil and chemicals, [6, 7]. Because of these qualities, plastics also persist a very long time in the environment and, if improperly recycled, accumulate as solid trash [8, 9]. When plastic is disposed of, its additives, plasticizers, and colorants cause major environmental problems [10]. However the primary cause of plastic pollution in the world is beverages and food packaging.

1.1 Packaging

Packaging is a science, art, and technology that are used to distribute, store, and preserve products as well as convey nutritional information and showcase them to consumers. The packaging material's standard serves as a bridge between the maker and the clients, ensuring the latter's confidence in the product [11]. Packaging materials such as Papers, metals, glass and plastics (e.g., polypropylene, polyethylene, and polyethylene terephthalate) are always used to pack Food products but they have some cons as well described in Table 1.

1.1.1 Need for Alternatives in Packaging

Concerns about prolonged synthetic plastic waste's negative environmental effects are becoming more widespread worldwide. Because there are insufficient landfills and disposal options, their decomposition results in the production of hazardous air pollution and costly recycling techniques, which fuels worldwide initiative to create biodegradable packaging materials that may serve all food-related purposes without endangering the environment [26].

1.1.2 Characteristics of Ideal Packaging Material

A perfect packing material must have the following properties:

- **Prevent Product Loss**: The product's design should offer protection and ease of handling throughout marketing, distribution, and transportation. The packages' dimensions form, and weight in particular should be taken into consideration.
- **Protection against Chemical and Physical Damage**: Such as that caused by water and water vapor, oxidation, light, and insects, must be provided for the product.
- **Maintaining the Product's Cleanliness**: It needs to act as a barrier against dirt and other impurities while maintaining the product's cleanliness.
- **Information regarding to Product Usage**: Give the information required to ensure proper usage of the product and its attractiveness to consumers [11, 26, 27]
- **Affecting Customer's Choice**

Table 1 Advantages and disadvantages of different materials used for packaging

Packaging material	Example	Advantages	Disadvantages	References
Plastic (Petroleum based): polyethylene, polystyrene, polypropylene, polyester	Food, textile, pharmaceutical industries	Sufficient barrier properties, a lightweight quality, cheap, ease in process ability, good liquid resistance, good strength	Causes air pollution due to release of greenhouse gas, unrecyclable which leads to oceans and land contamination	[12–15]
Paper	Furniture, drugs, construction and home related materials, manufacturing, electrical equipment, electronics and the food and beverage sectors	Safe, low cost, sustainable easy to handle, provides protection to products	Less liquid resistance, low strength, less chemical stability	[16–19]
Glass	Pharmaceutical companies, chemical, Food industries, etc.	Food hygiene and preservation, preservation of sensory qualities, quality control, and defense against chemical attacks	High weight, brittleness, and susceptibility to changes in temperature, low resistance to extremely high temperature	[20, 21]
Metal (Lead, chromium, tin, and aluminum	Food cans, lids, caps, drink cans	Inexpensive, flexible, lightweight, recyclable, and highly heat resistant	High levels of toxicity, carcinogenicity are linked to Chromium, lead, etc.	[22–25]

(a) **Identification**: The package's appearance in terms of making the brand easily recognized.
(b) **Pay Attention**: When consumers do not have any knowledge related to brand, packages with graphics on them use colors and shapes to draw their attention.
(c) **Communication**: How the package looks can also affect how the main product is rated. The package endorsing the idea that, even in the absence of creation, brand identity and image may be improved through packages [28].

1.2 Flexible Packaging

Flexible packaging (FP) as an expanding sector of the manufacturing industry, offers efficient methods for handling paper, plastic, glass and aluminum foil, to fulfil the demands of different products' packaging [10]. Flexible films typically possess the following qualities like cheaper price, superior barrier against gas and moisture, thermally resistant, Strength damp and dry, lightweight and convenient for the producer, distributor, and customer. It also adds less weight and takes up minimal space when being stored and distributed since they conform closely to the shape of the meal [29]. Generally speaking, synthetic thermoplastic polymers like PVC, Polyolefin, and Polystyrene comprise all flexible films. But they also have negative impact on environment. Every year, countless tons of waste made of plastic, which includes packaging, carrying bags, cartons and rubbish sacks, are dug in landfills across the globe. The manufacturing of hazardous chemicals, the depletion of resources, and the creation of solid waste as a result of inadequate recycling and the lack of availability of sustainable packaging materials are among major environmental issues linked to FP materials [30]. The dispersal of garbage made of FP materials causes problems such soil erosion, animal consumption, and obstruction of drainage systems. Plants, animals, and people are all impacted by the air, water, and soil pollution, which is brought on by the presence of macro, micro, and nano plastics. Fundamentally, freshwater toxicity, eutrophication, ozone depletion, acidification, climate change, and marine pollution are all caused by FP [31]. Expanding the study and creation of bio based packaging led to a long-term solution to all the problems [32].

1.3 Bio-polymer Based Films

The need for packaging materials composed of biodegradable, renewable biological resources has grown as a result of all of considerations that are needed to attain sustainability. Packaging film is a common tool used by food packaging companies to protect food from environmental contaminants. Natural macromolecules derived from living things are called biopolymers. They can be substituted for materials produced from petroleum in the utilization and generation of packaging films. Numerous examples of these films made of proteins, lipids, polysaccharides, and their mixtures may be found in the literature [33]. A number of significant obstacles to the widespread consumption of biopolymer films include their inadequate gas barrier qualities and mechanical properties and restricted capacity to create films as compared to conventional synthetic materials. Another drawback of the processes for mining, purification, transformation, and recovery is the high cost of production. Commercializing unique biopolymers with enhanced qualities and new functions for packaging films, coatings, and textile applications has garnered a lot of attention. There are various ways to make biopolymers: either directly from natural sources that consist of proteins and polysaccharides as well as the polymerization of

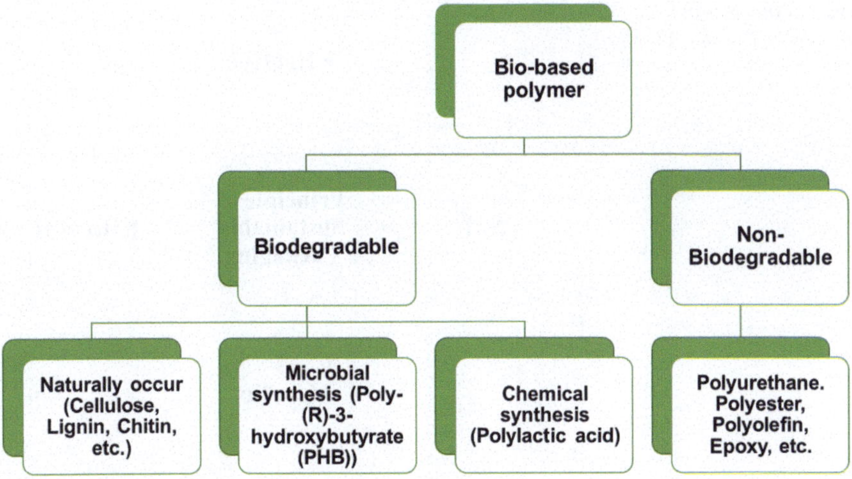

Fig. 1 Classification of bio-based polymers

monomers obtained from biomass. Among the most prevalent natural materials in the world, cellulose, has gained popularity as an environmental friend (biodegradable and renewable) material that is used for packaging [34]. Figure 1 displays the classification of biopolymers in the basis of biodegradability.

2 Sustainable Substitute for Packaging

2.1 Sustainable Packaging

Sustainability can be broadly classified into three dimensions: the environment, the economy, and human well-being. These three areas—fair burden-sharing and social fairness, for example—can be seen as ways to enhance human well-being and preserving the ecosystem's resiliency [35, 36]. When less virgin resources are used and post-consumed materials can be recyclable or reusable, packaging materials are considered sustainable. The criteria that determine material sustainability range from the manufacture to the processing of end-of-life, from financial to ecological, encompassing expenses and consequences, the usefulness of visual aspects, and from local to global implications [37].

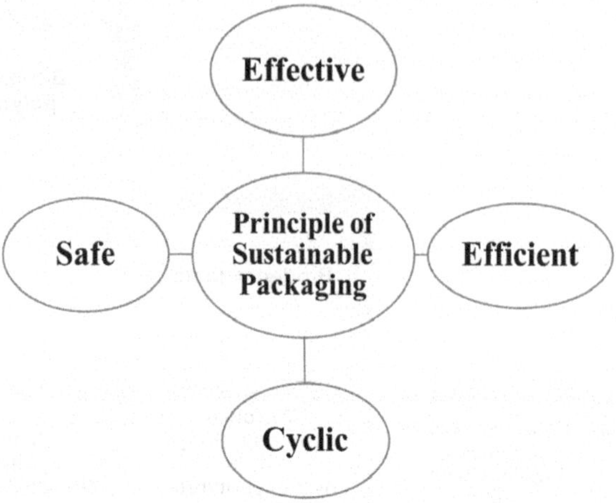

Fig. 2 Principles of sustainable packaging [38]

2.1.1 Principle

Four key principles form the foundation of the Sustainable Packaging Alliance (SPA) presentation shown in Fig. 2. It states that a balanced package should be safe (non-toxic and harmless because they won't harm people) and effective (both functionally and financially), efficient (in terms of using energy and financial resources), cyclical (allowing for repurposing via both industrial and natural methods), and both individuals and ecosystems [38].

2.2 Cellulose as a Solution to Plastic Pollution

In the packaging business, cellulose-based materials are used for flexible and rigid packaging, primary and secondary packages, and wrapping materials and containers.

2.2.1 Cellulose and Cellulose Derivatives

The most significant biopolymer is cellulose, which is made of d-glucose repeat units joined by β-1,4 glyosidic linkages (Fig. 3). It is made up of crystalline structure and closely packed chains of cellulose polymers that are resistant to dissolution in watery solutions. The only solution to make cellulose more water soluble is to treat it with an alkali to enlarge its structure, which then reacts with chloroacetic acid, propylene oxide, and methyl chloride to form methylcellulose, carboxymethyl cellulose, hydroxypropylmethylcellulose, and hydroxypropylcellulose [39–42]. The most popular source of cellulose and its derivatives is plants. Other sources of Cellulose are sugarcan, bamboo, cotton, bacteria, algae, etc. [43].

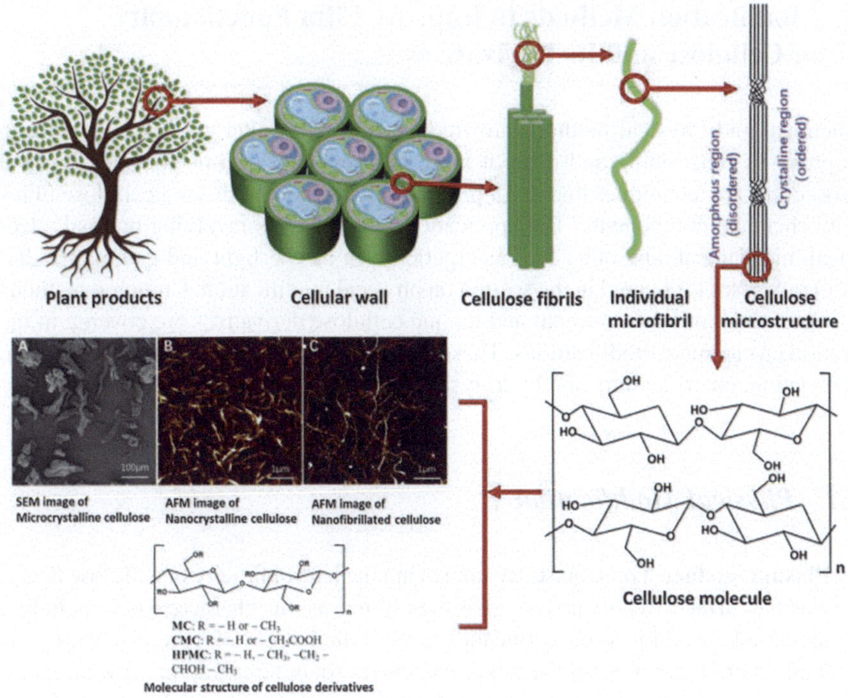

Fig. 3 Cellulose sources, microstructure, chemical structure, SEM, AFM image of microcrystalline, nano crystalline and nano fibrillated cellulose [47]

2.2.2 Characteristics of Cellulose Based Films

Cellulose is a good film-making materials; it is

- Translucent
- Reasonably flexible
- Resistant to fats and oils (Hydrophobes)
- Have no flavour or smell
- Water insoluble
- Biodegradable [44–46].

3 Modification Methods to Improve Film Functionality of Cellulose and Its Derivatives

Chemical and Physical methods are frequently used to alter cellulose in order to improve its functioning and make it more industry- and film-responsive. Physical procedures are techniques that don't primarily depend on modifying cellulose fibres with chemical components. The application of techniques involving thermal, electrical, mechanical, and other high-energetic (such as UV light and gamma irradiation) processes is covered in the section on physical modification. Common methods for obtaining both conventional and unique cellulose derivatives are covered in the section on chemical modifications. These methods include silylation, carbamylation, acetylation, etherification, and hydrolysis.

3.1 *Physical Modification*

- **Plasma Surface Treatment**: By improving the hydrophilicity of cellulose fibers, plasma surface treatment can give their films useful characteristics including increased air and oxygen permeability. With the help of electric discharge, this method produces ionized gas under vacuum at room temperature. Low temperature, low pressure, atmospheric glow discharge, and atmospheric pressure plasma jet techniques can all be used to produce ionized gas. The combination of heavy particles, electrons, and free radicals in the ionized gas facilitates the alteration of cellulose fibers and fiber aggregates in films. The changes can improve microcrystalline cellulose, eliminate low molecular weight contaminants from nanocellulose fibers, and cause defibrillation [48]. By producing polar functional groups, plasma treatment can enhance the hydrophilicity of the fibers and improve cellulose coating layer adherence. Plasma treatment was also demonstrated to improve the crosslinking and entanglements of cellulose nanofibrils, increase the amount of impermeable crystals, which limits the molecular migration of oxygen to boost the oxygen barrier properties of microfibrillar cellulose films [49].
- **UV radiation Therapy**: The use of electromagnetic rays with wavelengths ranging between 100–400 nm is known as UV radiation therapies. Using an ultraviolet laser as a source of light, cellulose is treated with sufficient energy in this process to change the molecular connections in the top layer. In summary, UV light causes cellulose fiber surfaces to produce more carbonyl groups and becomes more polar on the surface. Furthermore, encouraging crosslinking between fibril units can boost the cellulose films' and sheets' strength and gas barrier capacity. As an illustration, cellulose sheets exposed to UV light demonstrated higher tensile strength because the radiation permits intermolecular cross-linking between nearby cellulose molecules [50].

- **Thermal Treatment**: Thermal treatment can be applied either before or after another treatment to improve the carbon–oxygen ratio, eliminate unwanted materials, and increase the crystallinity of fibres of cellulose. These changes have the potential to raise the acidity and enhance the gas barrier qualities of the surfaces of fibres that are utilized to make sheets. As an illustration, It has been demonstrated that heating tempo-oxidized nanofiber sheets made of cellulose to a temperature of up to 145 °C greatly reduces their oxygen permeability and water retention when compared to the parental sheets made that do not undergo any post-heat treatment [51]. Furthermore, heat treatment has the ability to cause the pores between cellulose nanofibers to close. This can be explained by the reorientation of Para crystalline and amorphous regions into crystalline regions [52].

3.2 Chemical Modification

- **Esterification**: In the esterification or acylation process, organic acid and alcohol undergo chemical reaction to form water and ester. Aliphatic and aromatic chemicals can cause esterification reactions on the outermost layer of cellulose fibres or on cellulose polymer chains. There are two types of cellulose esters: inorganic (cellulose xanthate, cellulose nitrate, cellulose sulfate, and cellulose phosphate) and organic (cellulose acetate, cellulose acetate butyrate, and cellulose acetate propionate) [53].
- **Hydrolysis**: Hydrolysis is the process of partially depolymerizing cellulose using diluted mineral acid. The highly crystalline, micron-sized portions of microfibrils of cellulose are called microcrystalline cellulose (MCC). Hydrochloric acid can be used to hydrolyze partially depolymerized α-cellulose, producing MCC. Because of its low density, physicochemical stability, elasticity, wide surface area, and biodegradability, MCC is favored as a biocomposite in a variety of application fields, including pharmacology and medicine [54].
- **Etherification**: The etherification of cellulose requires homogeneous or heterogeneous interactions with etherifying agents, such as halogenated alkanes, epoxides, silyl chlorides, alkyl, bromides, or vinyl compounds. Cellulose ethers feature hydroxyl groups containing alkyl halide structures and are categorized as ionic alkyl ethers, silyl ethers, and nonionic alkyl ethers. They are produced on a massive scale and are among the most prevalent derivatives of cellulose. Commercial variants of carboxymethyl cellulose, hydroxyethyl cellulose, methylcellulose, ethyl cellulose, and hydroxypropyl cellulose are utilized in a variety of industries. Water-soluble, tasteless, non-flammable, and non-toxic characteristics are represented by cellulose ethers. These characteristics allow them to be applied in textile and packaging applications, as well as a thickening, water binder, handling assistance in food and pharmaceutical formulations [55].
- **Carbamylation**: The hydroxyl group of cellulose reacts with isocyanate to cause carbamylation of the material. Thermal stability and reactivity of nanofibers can

be enhanced through carbamylation. The polarity of cellulose fibres is altered by carbamylation [56].

- **Silylation**: The process of silylation involves the insertion of multifunctional silane compounds into cellulose fibres or micro- and nanocrystals, such as methyltrimethoxysilane, aminopropyltriethoxysilane, alkoxysilane, and methacrylopropyltrimethoxysilane. Depending on the particular groups (OH, CH, or ∞COOH) in the glucose unit, silyation offers distinct properties. The cellulose fibres' strength, crosslinking, and crystallinity can all be efficiently increased by the silylation procedure. Silylation can enhance the functional qualities of films, including their water vapour barrier and thermal resistance, much like other cellulose derivatives [57].

4 Types of Cellulose Based Packaging Films

Cellulose has several benefits when it comes to packaging, such as being affordable, biodegradable, lightweight, and sustainable (Fig. 4) [58]. As a multipurpose biopolymer, cellulose is used to create films for packaging, membranes for filters, hydrogels for absorbents, aerogels for insulation, and fibres for textiles and reinforcements [43]. Nanocellulose has been deployed as films in the form of CNFs and CNCs. They serve as reinforcements in packaging, enhancing the films' gas barrier qualities. Unfortunately, nanocellulose has weak water vapour barrier qualities because it absorbs moisture. From agricultural waste, cellulose fibres can be extracted to create biodegradable composites with suitable mechanical qualities for packaging [58].

The unique qualities of nanocellulose-based coatings, such as their biodegradability, transparency, barrier, antibacterial, and mechanical characteristics [59]

4.1 Composite Films Containing Nano Crystals of Cellulose

The coating reinforced with CNC performed similarly to the coating reinforced with Montmorillonite nanoclay, despite the latter offering better fruit weight persistence. More recent research has shown that CNCs are frequently utilized as reinforcement in biopolymeric matrices to create composite based packaging films. Despite the hydrophilic nature of CNC and most biopolymers, research has demonstrated that the resulting nanocomposite enhanced WVBP in a variety of biopolymer matrices [60, 61]. In addition to strengthening the film's cohesion, the formation of hydrogen bonds between the different biopolymers and CNC prevents O–H groups from leaking water. Furthermore, a high degree of CNC crystallinity improves the thermo-mechanical characteristics of the nanocomposite based films [62]. To improve WVBP and OBP, distribution of NC is uniform within the biopolymer based matrix must be guaranteed. Using surfactants, which prevent CNC agglomeration and function as a compatibilizer between the hydrophobic matrix and NC is one approach that may

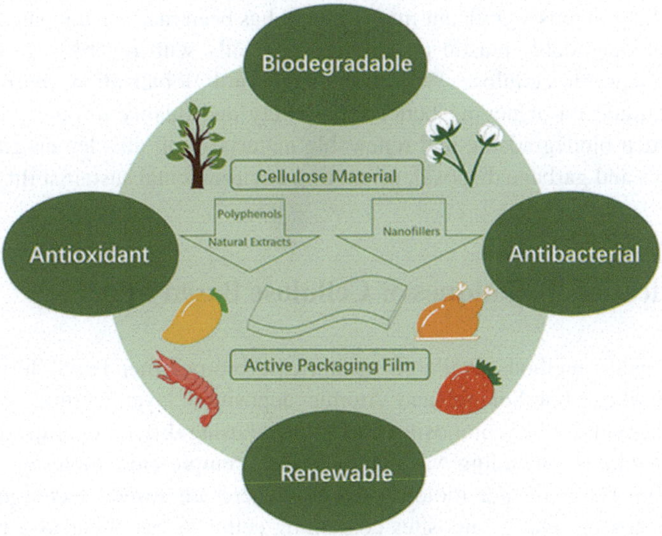

Fig. 4 Cellulose as a sustainable solution in packaging industry [58]

be taken. There is proof that composites with CNC as filler have improved moisture barrier qualities and decreased water vapor permeability (WVP) [63]. Strong CNF films can be produced by vacuum filtration or spray coating [64]. Although they can be mixed in directly with the food, antioxidants are best integrated into the film used for packaging.

4.2 *Cellulose Nanofibrils Based Films*

The creation of cellulose nanofibrils (CNF), also called as nanocellulose, usually involves high pressure fluidization or mechanical grinding. CNF is made up of long (several μm) and very thin (~20 nm) fibrils. At low concentrations (less than 2%), it forms a translucent gel-like substance that can be used to create dense, homogeneous, and ecologically acceptable films [65]. Many raw materials can be used to make CNF. Recently, pulps from hardwood, softwood, sugar beet, hemp, and bananas have been used to create CNF films. Crosslinking further decreased permeability and adsorption, while high humidity conditions also significantly improved the oxygen barrier's qualities. Recently, superior CNF films with outstanding technical qualities have been produced using a roll-to-roll pilot line in a semi-industrial setting. The created technique is simple to scale up and advances CNF film manufacturing toward commercialization [66–68]. Cellulose nanofibers have been extensively used in diverse industries, such as medical [69], packaging [70], paper and films [71], electronics based industry [72], and outer membranes [73], due to their promising features. The previous significant improvement has been made in the past few years to create biodegradable plastic packaging, especially with regard to biomaterials based on renewable cellulose. With the trend toward globalization, these progress are concentrated on achieving better food safety and quality through packaging. Utilizing such biodegradable and renewable materials will also lessen greenhouse gas balances and garbage disposal, promoting environmental sustainability.

5 Fabrication Techniques of Cellulose Based Films

There are many methods available to fabricate bio-polymer based films (cellulose based films) but Sol gel and Atomic deposition layer method are mostly used. While natural fiber composites can be made using different composite manufacturing methods, including vacuum bagging, compression molding, injection molding, and resin transfer molding (RTM). There are several techniques available for processing nanocomposites containing cellulose nanofiber as a reinforcement, but the frequently cited one-pot synthesis methods in recent literature are cast-drying, vacuum assisted filtration, electro-spinning, layer-by-layer (LbL) assembly, and freeze drying wet-spinning, and different patterning techniques [74–77].

5.1 Sol–Gel Technique

Sol–gel coatings are hybrid materials made from a sol, or colloidal solution, which acts as a precursor. The materials can be ceramic or biologically modified. Polycondensation between the hydroxyl groups generated by alkoxysilane hydrolysis and the residual alkoxy groups frequently results in the formation of a polymeric network, or gel. By spraying the sol–gel coatings onto the base paper, a coating of two to five grams per square meter is achieved. The wettability and consumption of base paper were shown to be more effectively reduced by coatings with a longer alkyl chain [78]. Other barrier qualities were also enhanced by sol–gel hybrid coatings applied on PLA laminated paperboard using a roll-to-roll pilot line [65]. Furthermore, PE-coated paper has been activated using the atmospheric plasma pretreatment before the coatings are applied. Hydrocarbons consisting of long chains or amino group functionalized sol-gels were sprayed, dried at 110°C, and then cured. While hydrophilic coatings consisting of amino groups decreased oxygen transfer rates, particularly at low relative humidity, hydrophobic coatings with long hydrocarbon chains proved to be excellent barriers against grease penetration. Performances of both coatings were notably enhanced by plasma activation [79]. On the other hand, the water vapour barrier qualities only saw slight improvements [80].

5.2 *ALD*

The self-limiting gas–solid interactions that underpin the atomic layer deposition (ALD) technique are layer-by-layer deposition method which is a surface-controlled method. It works well for producing biopolymers including PLA, cellulose nanofibrils, galactoglucomannan, and pectin as well as barrier coatings that are inorganic like SiO_2, Al_2O_3, and ZnO on a variety of substrates [10, 81–85]. Using this thin-film layer deposition method, a variety of materials can be coated with high-quality finishes [86, 87]. Even at comparatively low temperatures, layer-by- layer development ensures accurate control over the thickness of the film.

6 Applications of Cellulose Based Films

6.1 *Food Industry*

Food packaging includes everything that surrounds or holds food, including boxes, bags, straws, plates, cups, and wraps. Packaging is crucial in the food business to maintain the nutritional value of food and beverages. It is primarily required to preserve food safety, shield food from microbial contaminations, stop food waste or loss, extend food shelf life, and tells customers about the products. One of the most commonly found natural materials in the world, cellulose, has gained popularity as an environmentally benign material that may be used to packaging (Fig. 5). Because cellulose acetate and cellulose triacetate have lower gas and moisture barrier qualities than other derivatives, they are also more frequently utilized as rigid wrapping films [70, 71].

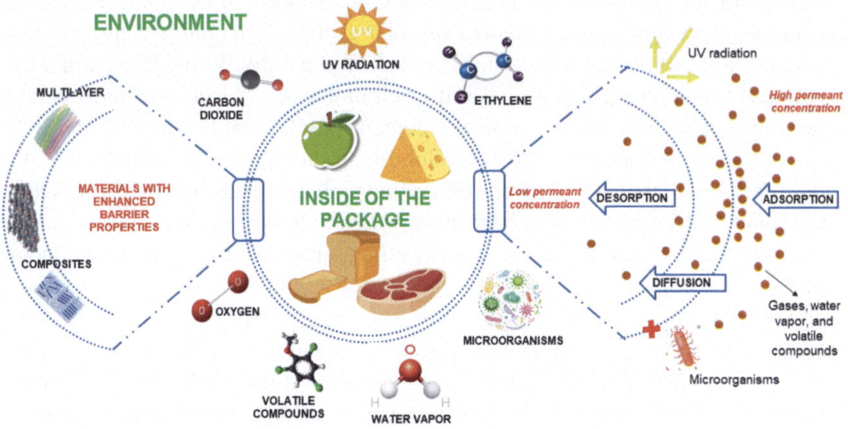

Fig. 5 Cellulose based films application in food industry [78]

6.1.1 Characteristics of Cellulose Based Films Used in Food and Beverage Industry

- **Oxygen Barrier characteristics**

  Polymeric polymers, both biodegradable and non-biodegradable, are frequently utilized in packaging films. However, they are not very effective at preserving an oxygen barrier; they are not used for very perishable products or products that are particularly prone to oxidation. Because of their hydrogen bonding, biodegradable polymers offer excellent oxygen barrier qualities in some situations, but in humid environments, their high water absorption capacity causes a noticeable decrease in such capabilities. This significantly restricts their use as materials for food packaging. Excellent oxygen barrier qualities of Nanocelluloses (NCs) have been observed to be comparable to other polymers in a dry condition [88]. NCs have the ability to form hydrogen bonds both with themselves and other biopolymers. NCs can create hydrogen bonds with other biopolymers as well as with themselves to generate long, rod-like particles that interact to form a dense network that ultimately results in a specific laminar structure. The narrow pore size of this makes it difficult for different gas molecules to flow through [89]. Because cellulose nanoparticles have a strong oxygen barrier, they can be used to make composite films with other polymers and as coatings for films with a low oxygen barrier. Cellulose nanofilms (CNF), bacterial cellulose (BC) and cellulose nanocrystals (CNC) films are based on pure cellulose nanoparticles. A range of processes, such as suspension drying, shear coating, vacuum filtering, hot pressing, and spin coating, are used to manufacture these dense, smooth-surfaced, solid films [90].
- **Properties associated with the Water Vapour Barrier**

  The transfer of water vapor between food and its surroundings has a major impact on food shelf life. Water loss must be avoided with fresh foods like meat and vegetables, but environmental moisture exposure must be avoided with dry goods like bread and puffy snacks [91]. Despite not being a water-soluble polymer, cellulose does have a large number of polar groups on its top that interact with water molecules to create hydrogen bonds. Consequently, during storage and use, cellulosic materials gradually absorb or eliminate water from the surroundings until they reach equilibrium moisture content (EMC) that is consistent with the surrounding environment [92]. The solubility of the film is also influenced by the hydrogen bonds that exist between the cellulose chains. These bonds are broken by the water molecules that have been adsorbed, which loosens the cellulose chains' intermolecular connections, promote swelling, and raise free volume producing plasticizing impact. This naturally alters the form of the film material created and degrades the food product's quality. The gas diffusion rate rises as the cohesiveness of the film is hampered by water molecules at higher humidity levels [93]. Factors affecting the water vapor transmission rate (WVTR) of polymeric composite films are temperature, pressure, crystallinity (the amorphous areas of the polymer are usually the areas where diffusion and adsorption primarily take place.), hydrophilicity, film thickness as well as density, structure, and pore size.

Foods must not be exposed to moisture while being transported, stored, or used with films to increase their service life.

- **Mechanical properties**

  In practical applications, the mechanical qualities of a material directly dictate its range of service life. As a result, in packing, the mechanical characteristics of sheets are crucial. In terms of packing materials, the mechanical characteristics of a material are often assessed using the following parameters: Young's modulus, ultimate tensile strength (UTS), stress (σ), and strain (ε). Laminated films are mostly made from cellulose-based materials. Cellulose modifies the mechanical qualities of the sheet and aids in the mixing of the polymers [94].

- **Antioxidant and UV Barrier Characteristics**

  Antioxidants' physicochemical properties can change how food tastes to the senses, even though some antioxidants can bind to food directly. Consequently, in order to regulate the release of these antioxidants, a packaging method is required. Food lipid oxidation can be postponed in part by antioxidant substances. Some manmade and natural compounds demonstrate significant antioxidant characteristics, such as essential oils, polypyrrole, coffee leaves and berry polyphenols. Plant tannins possessing phenolic hydroxyl based structures act as naturally occurring antioxidants by scavenging reactive oxygen species and thwarting cellular oxidation. So films based on cellulose and anti-oxidants are prepared to avoid food macromolecules' oxidation [94–97].

  Because UV photons are more energetic than visible light, organic components do not have extremely strong covalent bonds. Hence, in certain polymeric materials exposed to UV rays coming from sunlight, UV photons have the ability to disrupt the covalent bonds, causing physical and/or chemical changes that ultimately lead to their degradation. These changes in chemical composition can cause antioxidants to break down, proteins, lipids, and vitamins to oxidize, as well as cause texture and color changes and off-tastes to emerge [98]. As a result, the quality of packaged goods is greatly impacted by the materials' capacity to block light. This highlights how important it is for food packaging to have antioxidant and radical-scavenging properties. UV absorbers are required to improve the performance of CNF based films like protection form UV rays, which generally perform poorly in this regard. So pretreatment of the films has been done to achieve good UV barrier.

- **Heat related properties**

  In some uses of food packaging, particularly where reheatable food is involved, the packaging needs to hold up well to high temperatures and extended heating periods. The method of cellulose extraction (extracting methods and acid–base methods) also affects thermal stability of film. High thermal stability was demonstrated by the regenerated films made from cellulose extracted using the extraction process. Films containing cellulose provides good thermal stability to the packaging [99, 100].

- **Anti-Bacterial properties**

  Food quality and freshness are impacted by environmental microbe exposure [101]. Therefore, food degradation brought on by microorganisms can be avoided with the consumption of food packaging materials with antimicrobial qualities. The biodegradable film made of cellulose and thiazolidine has strong antibacterial action and a very high UV blocking percentage (99.8%). Anti-bacterial characteristics of Cellulose based films are further modified by cross linking cellulose with amino acid [102].

6.2 Medical Packaging

The market for pharmaceutical packaging has grown steadily over the past few years due to factors such as aging populations, chronic illnesses, changing personal lifestyles, and rising wages in developing nations. Packaging used in medicine meets the demands for product quality, identification, protection, information, security and convenience, with prompt and dependable solutions [65]. Applications for cellulosic fiber networks as raw materials include pharmaceutical and medical packaging. The outer containers for capsules and tablets that are packed in board cartons, such as cartons, boxes, blisters, envelopes, and strips, are prepared using these sheets [79]. Because of their ability to increase display area, improve stock item display, and offer physical protection—especially for things like metal collapsible tubes—cartons are mostly utilized in pharmaceutical packaging. Moreover, the secretion of medications from capsule and tablet formulations has been altered by the use of cellulose. Additionally, it functions as emulsifying and suspension agents, rheology and thickening control agents for water retention, layer formation, and improvement in adhesive strength, aids in tablet binding [80].

7 Measurements for the Betterment of Sustainability

The potential depletion of virgin materials might come from excessive usage. Another issue resulting from a shortage of virgin materials is the lack of raw resources for the creation of packaging materials. Environmental problems around the world are exacerbated by the exploitation of raw resources for packaging (from their extraction to their refined result). Reusing waste materials to make comparable or alternative products is necessary to ensure sustainable packaging. Many studies have been conducted on biodegradable polymers, including lipids, polysaccharides, proteins, and vegetal sources (such as starch, collagen, whey protein, cellulose, maize zein, chitosan, waxes, etc.). Packaging materials made of these kinds of materials ought to be promoted. The government should also create and implement laws governing the use of 100% recyclable and biodegradable materials. In the same way, packaging material manufacturers ought to strive toward a 100% recyclable product. The percentage of packaging materials that can be recycled is the ideal indicator of how sustainable those materials are. As a result, producers of packaging materials ought to sendeavor to employ sustainable, biodegradable materials with high recycling rates.

8 Conclusion

Renewable, biodegradable, and ecologically friendly packaging films that employ natural polymers as their basic ingredients have drawn increased attention in an effort to lessen the pollution that synthetic packaging films made from petroleum-based resources bring to the environment. The future of cellulose looks quite promising because of its unique microstructure, unique characteristics, and most abundant natural source in the domain of food and beverage based packaging materials. In short, reducing and minimizing the impact of packaging waste in surrounding is made possible by the development and use of sustainable cellulose films as packaging materials.

References

1. White, A., Lockyer, S.: Removing plastic packaging from fresh produce–what's the impact? Nutr. Bull. **45**, 35–50 (2020)
2. Singh, P., Wani, A.A., Langowski, H.C. (eds.): Food Packaging Materials: Testing Quality Assurance. CRC Press (2016)
3. Khalid, M.Y., Arif, Z.U.: Novel biopolymer-based sustainable composites for food packaging applications: a narrative review. Food Packag. Shelf Life **33**, 100892 (2022)
4. Han JH: Chapter 1-A review of food packaging technologies and innovations. In Innovations in Food Packaging. Edited by Han JH. Academic Press; 2014:3–12.
5. Sundqvist-Andberg, H., Åkerman, M.: Sustainability governance and contested plastic food packaging–an integrative review. J. Clean. Prod. **306**, 127111 (2021)
6. Emblem, A., Emblem, H.: 13-Plastics properties for packaging materials. In Packaging Technology (pp. 287–309). Woodhead Publishing (2012)
7. Teck Kim, Y., Min, B., Won Kim, K.: Chapter 2–general characteristics of packaging materials for food system. In Han, J.H. (ed.) Innovations in Food Packaging (pp. 13–35). Academic Press (2014)
8. Andrady, A.L., Neal, M.A.: Applications and societal benefits of plastics. Philos. Trans. R. Soc. B Biol. Sci. **364**(1526), 1977–1984 (2009)
9. Ashok, A., Mathew, M., Rejeesh, C.R.: Innovative value chain development of modified starch for a sustainable environment: a review. Int. J. Polym. Sci. Eng. **2**(1), 20–32 (2016)
10. Thompson, R.C., Swan, S.H., Moore, C.J., et al.: Our plastic age. Philos. Trans. R. Soc. B Biol. Sci. **364**, 1973–1976 (2009)
11. Ojha, A., et al.: Food packaging–materials and sustainability-a review. Agric. Rev. 2015 **36**(3), 241–245
12. Pal, M., Devrani, M., Hadush, A.: Recent developments in food packaging technologies. Beverage Food World **46**, 21–25 (2019)

13. MacArthur, E., Waughray, D., Stuchtey, M.: Rethinking plastics, starting with packaging. Cologny World Econ. Forum, Switz., 1–206 (2016)
14. Rutkowska, M., Heimowska, A., Krasowska, K., Janik, H.: Biodegradability of polyethylene starch blends in sea water. Polish J. Environ. Stud. **11**, 267–272 (2002)
15. Chinaglia, S., Tosin, M., Degli-Innocenti, F.: Biodegradation rate of biodegradable plastics at molecular level. Polym. Degrad. Stab. **147**, 237–244 (2018)
16. Cela, E., Kaneko, S.: Determining the effectiveness of the Danish packaging tax policy: the case of paper and paperboard packaging imports. Resour. Conserv. Recycl. **55**, 836–841 (2011)
17. Chen, W., Wang, X., Tao, Q., Wang, J., Zheng, Z., Wang, X.: Lotus-like paper/paperboard packaging prepared with nano-modified overprint varnish. Appl. Sur. Sci. **266**, 319–325 (2013)
18. Muller, J., González-Martínez, C., Chiralt, A.: Combination of poly (lactic) acid and starch for biodegradable food packaging. Materials **10**, 952 (2017)
19. Walsh, H., Kerry, J.P.: Packaging of ready-to-serve and retail-ready meat, poultry and seafood products. In: Kerry, J.P. (ed.) Advances in Meat, Poultry and Seafood Packaging, pp. 406–436. Woodhead Publishing, Sawston, UK (2012)
20. Kobayashi, M.L.: Glass packaging properties and attributes. In: Reference Module in Food Science. Elsevier, Amsterdam (2016)
21. Girling, P.: Food Packaging Technology (pp. 152–173). Blackwell Publishing, London, UK; CRC Press, Boca Raton, FL, USA (2003)
22. Whitaker, R.: Metal packaging and chemical migration into food. In: Barnes, K.A., Sinclair, C.R., Watson, D.H. (eds.) Chemical Migration and Food Contact Materials, pp. 251–270. Woodhead Publishing, Sawston, UK (2007)
23. Rodushkin, I., Magnusson, A.: Aluminium migration to orange juice in laminated paperboard packages. J. Food Compos. Anal. **18**, 365–374 (2005)
24. Annette, F.: Future Food and Drinks Packaging. Simon Fraser University, Burnaby, BC, Canada (2008)
25. Alamri, M.S., Qasem, A.A.A., Mohamed, A.A., Hussain, S., Ibraheem, M.A., Shamlan, G., Alqah, H.A., Qasha, A.S.: Food packaging's materials: a food safety perspective. Saudi J. Biol. Sci. **28**, 4490–4499 (2021)
26. Mitrus, M., Wojtowicz, A., Moscicki, L.: Biodegradable polymers and their practical utility. In: Janssen L.P.B.M., Moscicki, L. (eds.) Thermoplastic Starch: A Green Material for Various Industries. Wiley-VCH Verlag gmbh & Co. Kgaa, Weinheim, Germany (2009). https://doi.org/10.1002/9783527628216.ch1
27. Packaging Material for Food, Practical Action Technical Brief
28. Underwood, R.L., Ozanne, J.L.: Is your package an effective communicator? A normative framework for increasing the communicative competence of packaging. J. Mark. Commun. **4**(4), 207–220 (1998)
29. Kliopova-Galickaja, I., Kliaugaite, D.: VOC emission reduction and energy efficiency in the flexible packaging printing processes: analysis and implementation. Clean Technol. Environ. Policy **20**, 1805–1818 (2018)
30. Daniela, G., Ovidiu, T.: Biopolymers used in food packaging a review. Acta Universitatis CibiniensisSeries E, Food Technology (2012)
31. Ahamed, A., Veksha, A., Giannis, A., Lisak, G.: Flexible packaging plastic waste—environmental implications, management solutions, and the way forward. Curr. Opin. Chem. Eng. **32**, 100684 (2021)
32. Ahamed, A., Veksha, A., Yin, K., Weerachanchai, P., Giannis, A., Lisak, G.: Environmental impact assessment of converting flexible packaging plastic waste to pyrolysis oil and multi-walled carbon nanotubes. J. Hazard. Mater. **390**, 121449 (2020)
33. Dang, S., Chu, L.: Evaluation framework and verification for sustainable container management as reusable packaging. J. Bus. Res. **69**(5), 1949–1955 (2016). https://doi.org/10.1016/j.jbusres.2015.10.086
34. Song, T., Qian, S., Lan, T., et al.: Recent advances in bio-based smart active packaging materials. Foods **2022**, 11 (2022). https://doi.org/10.3390/foods11152228

35. Unilever. Waste and Packaging. https://www.unilever.com/sustainable-living/thesustainable-living-plan/reducing-environmentalimpact/waste-and-packaging. Last accessed 2016
36. Otto, S., Strenger, M., Maier-Nöth, A., Schmid, M.: Food packaging and sustainability–consumer perception vs. correlated scientific facts: a review. J. Clean. Prod. **298**, 126733 (2021)
37. Martin, J., Henrichs, T., Francis, C., Hoogeveen, Y., Kazmierczyk, P., Pignatelli, R., Speck, S.: Environmental Indicator Report 2012: Ecosystem Resilience and Resource Efficiency in a Green Economy in Europe. European Environment Agency, Copenhagen, Denmark (2012)
38. Reichert, C.L., Bugnicourt, E., Coltelli, M.B., Cinelli, P., Lazzeri, A., Canesi, I., Braca, F., Martínez, B.M., Alonso, R., Agosti-nis, L., et al.: Bio-based packaging: Materials, modifications, industrial applications and sustainability. Polymers **12**, 1558 (2020)
39. Xie, K., Tu, H., Dou, Z.L., Liu, D.Y., Wu, K., et al.: The effect of cellulose molecular weight on internal structure and properties of regenerated cellulose fibers as spun from the alkali/urea aqueous system. Polymer **215**, e123379 (2021). https://doi.org/10.1016/j.polymer.2021.123379
40. Singha, A.S., Thakur, V.K.: Saccaharum cilliare fiber reinforced polymer composites. E-J. Chem. **5**, 782–791 (2008). https://doi.org/10.1155/2008/941627
41. Thakur, V.K., Singha, A.S., Thakur, M.K.: Biopolymers based green composites: mechanical, thermal and physico-chemical characterization. J. Polym. Environ. **20**, 412–421 (2012). https://doi.org/10.1007/s10924-011-0389-y
42. Thakur, V.K., Singha, A.S.: Mechanical and water absorption properties of natural fibers/polymer biocomposites. Polym.-Plast. Technol. Eng. **49**, 694–700 (2010). https://doi.org/10.1080/03602551003682067
43. Singha, A.S., Thakur, V.K.: Fabrication and characterization of H. sabdariffa fiber-reinforced green polymer composites. Polym.-Plast. Technol. Eng. **48**, 482–487 (2009). https://doi.org/10.1080/03602550902725498
44. Singha, A.S., Thakur, V.K.: Synthesis, characterization and study of pine needles reinforced polymer matrix based composites. J. Reinf. Plast. Compos. **29**, 700–709 (2010). https://doi.org/10.1177/0731684408100354. (INTERNATIONAL JOURNAL OF POLYMER ANALYSIS AND CHARACTERIZATION 11)
45. Sadeghi-Varkani, A., Emam-Djomeh, Z., Askari, G.: Physicochemical and microstructural properties of a novel edible film synthesized from Balangu seed mucilage. Int. J. Biol. Macromol. **108**, 1110–1119 (2018). https://doi.org/10.1016/j.ijbiomac.2017.11.029
46. Dehghani, S., Hosseini, S.V., Regenstein, J.M.: Edible films and coatings in seafood preservation: a review. Food Chem. **240**, 505–513 (2018). https://doi.org/10.1016/j.foodchem.2017.07.034
47. Singha, A.S., Thakur, V.K.: Synthesis and characterization of short grewia optiva fiber-based polymer composites. Polym. Compos. **31**, 459–470 (2010). https://doi.org/10.1002/pc.20825
48. He, X., Lu, W., Sun, C., Khalesi, H., Mata, A., Andaleeb, R., Fang, Y.: Cellulose and cellulose derivatives: different colloidal states and food-related applications. Carbohydr. Polym. **255**, 117334 (2021). ISSN 0144-8617, https://doi.org/10.1016/j.carbpol.2020.117334
49. Panaitescu, D.M., Vizireanu, S., Nicolae, C.A., et al.: Treatment of nanocellulose by submerged liquid plasma for surface functionalization. Nanomaterials **2018**, 8 (2018). https://doi.org/10.3390/nano8070467
50. Syverud, K., Stenius, P.: Strength and barrier properties of MFC films. Cellulose **16**, 75–85 (2009)
51. Zaman, H.U., Khan, M.A., Khan, R.A.: Improvement of mechanical properties of jute fibers-polyethylene/polypropylene composites: effect of green dye and UV radiation. Polymer—Plastic Technol. Eng. **48**, 1130–1138 (2009)
52. Xia, J., Zhang, Z., Liu, W., et al.: Highly transparent 100% cellulose nanofibril films with extremely high oxygen barriers in high relative humidity. Cellulose **25**, 4057–4066 (2018)
53. Chen, Y.M., Wan, J.Q., Huang, M.Z., et al.: Influence of drying temperature and duration on fiber properties of unbleached wheat straw pulp. Carbohyd. Polym. **85**, 759–764 (2011)

54. Attari, N., Hausler, R.: Reinforcing effects of fibrous and crystalline nanocelluloses on cellulose acetate membranes. Carbohydr. Polym. Technol. Appl. **5**, 100281 (2023)
55. Rojas-Lema, S., Terol, J., Fages, E., et al.: Microencapsulation of copper(ii) sulfate in ionically cross-linked chitosan by spray drying for the development of irreversible moisture indicators in paper packaging. Polym. (Basel) **2020**, 12 (2020). https://doi.org/10.3390/POLYM12092039
56. Ostrovskii, D., Kjøniksen, A.L., Nyström, B., et al.: Association and thermal gelation in aqueous mixtures of ethyl(hydroxyethyl)cellulose and ionic surfactant: FTIR and Raman study. Macromolecules **32**, 1534–1540 (1999)
57. Missoum, K., Sadocco, P., Causio, J., et al.: Antibacterial activity and biodegradability assessment of chemically grafted nanofibrillated cellulose. Mater. Sci. Eng. C **45**, 477–483 (2015)
58. Jankauskaite, V., Balčiunaitiene, A., Alexandrova, R., et al.: Effect of cellulose microfiber silylation procedures on the properties and antibacterial activity of polydimethylsiloxane. Coatings **2020**, 10 (2020). https://doi.org/10.3390/COATINGS10060567
59. Shah, U., Naqash, F., Gani, A., Masoodi, F.A.: Art and science behind modified starch edible films and coatings: a review. Compr. Rev. Food Sci. Food Saf. **15**, 568–580 (2016). https://doi.org/10.1111/1541-4337.12197
60. Kerch, G.: Chitosan films and coatings prevent losses of fresh fruit nutritional quality: a review. Trends Food Sci. Technol. **46**, 159–166 (2015). https://doi.org/10.1016/j.tifs.2015.10.010
61. Nazrin, A., Sapuan, S.M., Zuhri, M.Y.M., Ilyas, R.A., Syafiq, R., Sherwani, S.F.K.: Nanocellulose reinforced thermoplastic starch (TPS), polylactic acid (PLA), and polybutylene succinate (PBS) for food packaging applications. Front. Chem. **8**(4), 213 (2020). https://doi.org/10.3389/fchem.2020.00213
62. Dhar, P., Bhardwaj, U., Kumar, A., Katiyar, V.: Poly(3-hydroxybutyrate)/cellulose nanocrystal films for food packaging applications: barrier and migration studies. Polym. Eng. Sci. **55**(10), 2388–2395 (2015). https://doi.org/10.1002/pen.24127
63. George, J., Siddaramaiah.: High performance edible nanocomposite films containing bacterial cellulose nanocrystals. Carbohydr. Polym. **87**(3), 2031–2037 (2012). https://doi.org/10.1016/j.carbpol.2011.10.019
64. Huq, T., Salmieri, S., Khan, A., Khan, R.A., Le Tien, C., Riedl, B., et al.: Nanocrystalline cellulose (NCC) reinforced alginate based biodegradable nanocomposite film. Carbohyd. Polym. **90**(4), 1757–1763 (2012). https://doi.org/10.1016/j.carbpol.2012.07.065
65. Paunonen, S.: Strength and barrier enhancements of cellophane and cellulose derivative films: a review. BioResources **8**(2), 3098–3121 (2013)
66. Fang, Z., Li, B., Liu, Y., Fang, Z., Li, B., Liu, Y., et al.: Critical role of degree of polymerization of cellulose in super-strong nanocellulose films. Matter **2**(4), 1000–1014 (2020). https://doi.org/10.1016/j.matt.2020.01.016
67. Zhang, K., Barhoum, A., Xiaoqing, C., Haoyi, L., Samyn, P.: Cellulose nanofibers: fabrication and surface functionalization techniques. In: Barhoum, A., Bechelany, M., Makhlouf, A. (eds.) Handbook of Nanofibers. Springer, Cham (2019). https://doi.org/10.1007/978-3-319-42789-8_58-1
68. Frank, B.P., Durkin, D.P., Caudill, E.R., Zhu, L., White, D.H., Curry, M.L., et al.: Impact of silanization on the structure, dispersion properties, and biodegradability of nanocellulose as a nanocomposite filler. ACS Appl. Nano Mat. **1**(12), 7025–7038 (2018). https://doi.org/10.1021/acsanm.8b01819
69. Phanthong, P., Reubroycharoen, P., Hao, X., Xu, G., Abudula, G., Guan, G.: Nanocellulose: extraction and application. Carbon Resour. Convers. **1**, 32–44 (2018)
70. Shen, Y., Seidi, F., Ahmad, M., Liu, Y., Saeb, M.R., Akbari, A., Xiao, H.: Recent advances in functional cellulose-based films with antimicrobial and antioxidant properties for food packaging. J. Agric. Food Chem. **71**(44), 16469–16487 (2023). https://doi.org/10.1021/acs.jafc.3c06004
71. Cherian, B.M., et al.: Isolation of nanocellulose from pineapple leaf fibres by steam explosion. Carbohydr. Polym. **81**(3), 720–725 (2010)

72. Lavoine, N., et al.: Elaboration of a new antibacterial bio-nano-material for food-packaging by synergistic action of cyclodextrin and microfibrillated cellulose. Innov. Food Sci. Emerging Technol. **26**, 330–340 (2014)
73. Charani, P.R., et al.: Production of microfibrillated cellulose from unbleached kraft pulp of Kenaf and Scotch Pine and its effect on the properties of hardwood kraft: microfibrillated cellulose paper. Cellulose **20**(5), 2559–2567 (2013)
74. Nogi, M., Yano, H.: Transparent nanocomposites based on cellulose produced by bacteria offer potential innovation in the electronics device industry. Adv. Mater. **20**(10), 1849–1852 (2008)
75. Sehaqui, H., Salajkova, M., Zhou, Q., Berglund, L.A.: Mechanical performance tailoring of tough ultra-high porosity foams prepared from cellulose nanofiber suspensions. Soft Matter **6**, 1824–1832 (2010)
76. Kulachenko, A., Denoyelle, T., Galland, S., Lindström, S.B.: Elastic properties of cellulose nanopaper. Cellulose **19**, 793–807 (2012)
77. de Mesquita, J.P., Donnici, C.L., Pereira, F.V.: Biobased nanocomposites fromlayer-by-layer assembly of cellulose nanowhiskers with chitosan. Biomacromol **11**, 473–480 (2010)
78. Versino, F., Ortega, F., Monroy, Y., Rivero, S., López, O.V., García, M.A.: Sustainable and bio-based food packaging: a review on past and current design innovations. Foods **12**, 1057 (2023). https://doi.org/10.3390/foods12051057
79. Zadbuke, N., et al.: Recent trends and future of pharmaceutical packaging technology. J. Pharm. Bioallied Sci. **5**(2), 98 (2013)
80. Kamel, S., et al.: Pharmaceutical significance of cellulose: a review. Express Polym. Lett. **2**(11), 758–778 (2008)
81. Dhandapani, E., Suganthi, S., Vignesh, S., Dhanalakshmi, M., Kalyana Sundar, J.K.: Fabrication and physicochemical assessment of L-histidine cross-linked PVA/CMC bio-composite membranes for antibacterial and food-packaging applications. Mater. Technol. **37**(2), 124–134 (2020)
82. Wang, S., Mahlberg, R., Nikkola, J., Mannila, J., Jämsä, S., Ritschkoff, A., Peltonen, J.: Surface characteristics and wetting properties of sol-gel coated base paper. Surf. Interface Anal. **44**, 539–547 (2012). https://doi.org/10.1002/sia.3841
83. Nikkola, J., Vartiainen, J., Vähä-Nissi, M., Nättinen, K.: Novel packaging materials using functional thin films and coatings. In: European Coatings Conference-Packaging and Can Coatings, Düsseldorf, 11–12 June 2013, Vincentz Network GmbH & Co. (2013)
84. Nikkola, J.: Raising the barriers. Eur. Coat. J., 41–47 (2011)
85. Nättinen, K., Nikkola, J., Minkkinen, H., Heikkilä, P., Lavonen, J., Tuominen, M.: Reel-to-reel inline atmospheric plasma deposition of hydrophobic coatings. J. Coat. Technol. Res. **8**, 237–245 (2011). https://doi.org/10.1007/s11998-010-9292-z
86. Flieger, M., Kantorová, M., Prell, A., et al.: Biodegradable plastics from renewable sources. Folia Microbiol. **48**(1), 27–44p (2003)
87. Wang, Y., Aurelio, D., Li, W., Tseng, P., Zheng, Z., Li, M., Kaplan, D.L., Liscidini, M., Omenetto, F.G.: Adv. Mater. **29** (2017). https://doi.org/10.1002/adma.201702769
88. Thielemans, W., Warbey, C.R., Walsh, D.A.: Permselective nanostructured membranes based on cellulose nanowhiskers. Green Chem. **11**(4), 531–537 (2009)
89. Wang, J.W., Gardner, D.J., Stark, N.M., Bousfield, D.W.: Moisture and oxygen barrier properties of cellulose nanomaterial-based film. ACS Sustain. Chem. & Eng. **6**(1), 49–70 (2018)
90. Wu, Y.J., Liang, Y.Y., Mei, C.T., Cai, L.P., Nadda, A.: Advanced nanocellulose-based gas barrier materials: Present status and prospects. Chemosphere **286**, e131891 (2022)
91. Jafarzadeh, S., Mohammadi Nafchi, A., Salehabadi, A., Oladzad-Abbasabadi, N., Jafari, S.M.: Application of bio-nanocomposite films and edible coatings for extending the shelf life of fresh fruits and vegetables. Adv. Coll. Interface. Sci. **291**, e102405 (2021)
92. Huang, G.X., Dong, H.L., Zhang, Z.P., Bian, H.Y., Yong, Q.: Procuring the nano-scale lignin in prehydrolyzate as ingredient to prepare cellulose nanofibril composite film with multiple functions. Cellulose **27**, 9355–9370 (2020)

93. Tong, R.P., Chen, G.X., Tian, J.F., He, M.H.: Highly transparent, weakly hydrophilic and biodegradable cellulose film for flexible electroluminescent devices. Carbohyd. Polym. **227**, e115366 (2020)
94. Xie, Y.J., Pan, Y.F., Cai, P.X.: Hydroxyl crosslinking reinforced bagasse cellulose/polyvinyl alcohol composite films as biodegradable packaging. Ind. Crops Prod. **176**, e114381 (2022)
95. Zhang, W.J., Zhang, Y.Q., Cao, J.K., Jiang, W.B.: Improving the performance of edible food packaging films by using nanocellulose as an additive. Int. J. Biol. Macromol. **166**, 288–296 (2021)
96. Huang, X.D., Ji, Y.Z., Guo, L.K., Xu, Q.H., Jin, L.Q.: Incorporating tannin onto regenerated cellulose film towards sustainable active packaging. Ind. Crops Prod. **180**, e114710 (2022)
97. Kriechbaum, K., Bergstrom, L.: Antioxidant and UV-blocking leather-inspired nanocellulose-based films with high wet strength. Biomacromol **21**(5), 1720–1728 (2020)
98. Wu, M., Yang, J., Chen, S.L., Lu, P., Wang, R.F.: TOCNC-G-PEI nanoparticle encapsulated oregano essential oil for enhancing the antimicrobial activity of cellulose nanofibril packaging films. Carbohyd. Polym. **274**, e118654 (2021)
99. Guo, R.S., Hu, K., He, P., Ni, Y.H., Liu, K.: Photochromic nanocellulose composite films with excellent anti-UV capacity. Appl. Phys. A **126**(10), e812 (2020)
100. Yadav, M.: Study on thermal and mechanical properties of cellulose/iron oxide bionanocomposites film. Compos. Commun. **10**, 1–5 (2018)
101. Suwanprateep, S., Kumsapaya, C., Sayan, P.: Structure and thermal properties of rice starch-based film blended with mesocarp cellulose fiber. Mater. Today Proc. **17**, 2039–2047 (2019)
102. Ali, M.A.S.S., Jimat, D.N., Nawawi, W.M.F.W., Sulaiman, S.: Antibacterial, mechanical and thermal properties of PVA/starch composite film reinforced with cellulose nanofiber of sugarcane bagasse. Arab. J. Sci. Eng. **47**(5), 5747–5754 (2021)

Collagen Fusion with Natural Biopolymers in Food Packaging: An In-Depth Review from Conceptualization to Consumer Application

Malavika Jayaprakash, Syed Mukhameel Jani Basha, Shamita Jaiganesh, Selva Sasi Kumar Senthil Kumar, Lavanya Jayaraman, and Gaurav Mudgal

Abstract As demand for sustainable and eco-friendly packaging grows, industries are studying biopolymers for food packaging. From the stage of conceptualization to the stage of consumer applications in the real world, this review paper focuses on novel combinations of collagen and a variety of natural biopolymers. It offers an in-depth analysis of the process. The review begins with collagen's biopolymer properties and then explores many natural biopolymers suitable for collagen fusion. By examining synergistic effects, the paper shows how each biopolymer improves blend performance. In this study, the influence of processing parameters on the formation of films as well as the mechanical, barrier, and thermal properties of collagen-natural biopolymer films is investigated systematically. The in-depth study includes food industry case studies and consumer perception and sensory attributes of collagen-natural biopolymer packaging. To understand this field's challenges and opportunities, regulatory and compliance issues are examined. To promote sustainable packaging, collagen-natural biopolymer films are evaluated for environmental impact and biodegradability. Cellulose is a promising natural biopolymer for food packaging applications due to its biodegradability, renewability, and excellent barrier properties. When combined with collagen, cellulose can enhance the mechanical strength and stability of the packaging material. Collagen-natural biopolymer films are compared to conventional packaging materials for performance, cost, and environmental impact. This review concludes with current research, challenges, and

M. Jayaprakash · S. M. J. Basha · S. Jaiganesh · S. S. K. S. Kumar · L. Jayaraman (✉)
Department of Biotechnology, School of Bioengineering, SRM Institute of Science and Technology, Kattankulathur, Chennai 603203, Tamil Nadu, India
e-mail: lavanyaj@srmist.edu.in

G. Mudgal (✉)
Center for Waste Management and Renewable Energy, Saveetha Dental College and Hospitals, Saveetha Institute of Medical and Technical Sciences, Saveetha University, Chennai 600077, India
e-mail: gauravmdgl@gmail.com

University Institute of Biotechnology, Chandigarh University, Mohali 140413, Punjab, India

K. K. Kesari et al. (eds.), *Functionalized Cellulose Materials*, Engineering Materials,
https://doi.org/10.1007/978-3-031-76953-5_3

future directions in collagen fusion with natural biopolymers for food packaging. This review aims to inform researchers, industry professionals, and policymakers about sustainable food packaging.

Keywords Film formation · Biodegradability · Eco friendly packaging · Consumer perception · Cost analysis · Blend performance

Highlights:

- Packaging made of renewable materials instead of petroleum is more environmentally friendly.
- Using collagen generated from fish offal to improve sustainability and package integrity.
- Combining collagen with different biopolymers to expand the uses of biomaterials.

1 Introduction

The utilization of petroleum-based polymers in food packaging has ignited significant apprehension due to its potential as an environmental pollutant. Consequently, a discernible shift is underway, steering away from these polymers in favor of renewable and biodegradable packaging materials, particularly those derived from polysaccharides and proteins [23]. In the realm of material development, natural polymers emerge as a compelling opportunity, driven by their inherent structural and physical attributes. Beyond their tangible properties, these polymers present a unique blend of safety, wide availability, biocompatibility, and biodegradability. The intricate structure of natural polymers not only underscores their versatility but also positions them as promising candidates for sustainable materials. Their innate safety profile lends credence to their applicability in various industries, assuring users of materials that prioritize health and well-being. In this context, many valuable opportunities arise from the often neglected by-products that result from both the slaughter of livestock and the processing of fish. The industrial processing of livestock produces various residual materials, such as waste viscera, fat, skin, bones, and feet. In the context of industrial fishing, the discarded remnants consist of muscle, skin, fins, bones, viscera, and scales. Significantly, these secondary products are particularly noteworthy due to their plentiful and varied collection of invaluable substances. Proteins like collagen obtained from these sources can function as essential constituents as they show good barrier and mechanical properties for packaging materials.

Cellulose is a low-cost material with strong mechanical properties and a high demand in the food and textile industry. It is extracted from agricultural waste, including fruits, vegetables, cotton stalks, and forest residues, and is gaining attention in the food packaging industry due to its unique characteristics, biodegradability, and outstanding physical and mechanical properties. Cellulose and its derivatives, when blended with antibacterial polymers, form composite films. Cellulose also has high thermal resistance, UV protection, antioxidant and antibacterial properties, making it a valuable material in the food packaging industry [20].

Collagen possesses a remarkable capability to create robust, insoluble fibers with exceptional tensile strength. Purified collagen finds widespread application in pharmaceutical and cosmetic formulations, owing to its unique properties. Similarly, gelatin, derived from collagen through hydrolysis, is extensively utilized in the food industry. While traditional sources of collagen primarily involve pig skin and cowhide, alternative sources such as fish offal, including bones, skin, scales, and fins, have gained attention. Fish collagen differs from mammalian collagen in several aspects; notably, it exhibits lower gelling and melting temperatures but boasts comparatively higher viscosities when compared to its bovine counterparts.

Collagen, an essential individual from the biopolymer family, assumes a critical part in giving principal underlying and mechanical help inside the creature's body. This assorted family contains 29 hereditarily particular collagen types, each contributing interestingly to different tissues and works [3]. Among the significant sorts, Type I collagen holds conspicuousness, normally tracked down in skin, ligament, and bone tissues. Eminently, Type I collagen remains as the most predominant and broadly concentrated on individuals from the collagen family, tracking down wide applications in biomedical fields. The underlying creation of Type I collagen is described by chains, each composed of 1000 amino acids, framing a α-helix. Moreover, three α-helices meet up to make a triple helical construction. These designs unpredictably collect into collagen fibrils and filaments, framing the premise of collagen's assorted functionalities in supporting tissues and organs. Customarily, collagen extraction has prevalently revolved around land creature skins, with cows and pigs being normal sources. These conventional techniques have long filled in as essential repositories for acquiring collagen.

In logical exploration, an unmistakable center has been coordinated towards rodent tail ligament collagen. This specific kind is liked because of its dissolvability in acidic pH conditions and a denaturation temperature of roughly 40 °C. These attributes make rodent tail ligament collagen especially appropriate for explicit logical applications, underlining its importance in exploratory examinations and examinations [24], without a doubt, collagen stands apart as an enamoring biopolymer with enormous expected in the planning of polymer mixes. The logical writing widely investigates mixes including collagen, with engineered polymers as well as with different other regular polymers. Especially fascinating are the mixes of collagen and chitosan, two broadly used regular polymers in the advancement of biomaterials [37]. These biopolymer mixes broaden their utility past examination research facilities, tracking down applications as bundling materials. The domain of the logical creative mind is unfathomable, taking into consideration the production of mixes including biopolymers that may not normally coincide under ordinary circumstances. In research center settings, collagen can be consistently mixed with chitosan, elastin, keratin, silk fibroin, hyaluronic corrosive, and many more [40].

2 Intrinsic Properties of Collagen as a Biopolymer

Collagen is the predominant structural protein found in animals, making up around 30% of the total protein content. It is present in various tissues and organs such as skin, tendons, bones, cartilage, and ligaments [12]. Among the approximately 28 identified types, type I collagen stands out as the most abundant variant [13]. Collagen is composed of three α-chains that come together to form a triple helix domain. This domain can arrange itself into various shapes and structures in living organisms. Although collagen is well-known for its strong structure, the process of extracting it can cause the breaking of bonds within and between molecules, leading to the formation of gelatin, which is a partially broken-down version of collagen [34].

Collagen not only serves as a structural support but also possesses notable functional properties due to its protein side chains. These properties include the ability to form gels, emulsify substances, create foam, and form films. Collagen and its derivative, gelatin, are essential components in a wide range of food systems and even in edible packaging materials due to their versatile functionalities [6]. Furthermore, collagen fibers, specifically sourced from bovines, have been increasingly used in the meat industry for creating artificial casings, demonstrating the flexible and versatile nature of collagen in various industrial settings.

The molecular structure of collagen is defined by its Madras Triple Helix Geometry, which consists of coils that are approximately 300 nanometers in length, breadth, and thickness. These coils form a triple helix with a width of 1.5 nanometers. The distinctive structure of collagen, which has an average density of approximately 85.5 coils per inch, highlights its remarkable mechanical durability and ability to recover from stress [33]. By amalgamating collagen's intrinsic properties with its functional versatility, this protein emerges not only as a structural cornerstone in biological systems but also as a versatile resource with myriad industrial applications.

Collagen's primary variety, exemplified by different sorts like sort I and type II, considers mind boggling subatomic courses of action custom fitted to explicit tissues, displaying its adaptability. This primary flexibility is urgent in clinical and tissue designing applications, where collagen's inborn biocompatibility guarantees its acknowledgment by the human body, assuming a vital part in different restorative settings. In addition, collagen's biodegradability adds a harmless to the ecosystem aspect, as it goes through regular debasement over the long haul [29].

The momentous rigidity shown by collagen filaments is irreplaceable for maintaining the primary respectability of fundamental tissues like skin, ligaments, and bones. This strength is a demonstration of the viability of collagen in keeping up with the mechanical security of different organic designs. The triple helix structure comprising the collagen spine fundamentally adds to its soundness and flexibility, building up its reasonableness for a scope of uses.

Collagen's hydrophilic nature upgrades its utility by making it capable at drawing in water particles. This hydrophilicity demonstrates worthwhile applications where dampness retention or the detailing of hydrogels is alluring. Moreover, collagen's capacity to help cell grip and expansion positions it as a central participant in tissue designing and regenerative medication, encouraging the improvement of biocompatible frameworks for cell development. While collagen is by and large portrayed by low immunogenicity, cautious contemplations are vital while using collagen from various sources to guarantee similarity and relieve possible immunological reactions. A nuanced comprehension of these inherent properties engages specialists to tailor collagen-based materials for a wide range of uses, going from clinical inserts to creative and reasonable bundling arrangements.

3 Collagen Based Food Packaging Films

Under certain circumstances, a significant quantity of collagen molecules demonstrates the capacity to autonomously arrange themselves into various forms while maintaining their triple-helical structures. Nevertheless, the extraction process disrupts the covalent and hydrogen bonds that stabilize the collagen triple helix. This disruption leads to the creation of gelatin, which is a mixture of polypeptides derived from the partial degradation of collagen. Collagen and gelatin have functional properties, including the ability to form films, create foam, emulsify, and gel. This makes them suitable for a variety of food products and edible packaging materials.

Edible packaging materials, such as those made from collagen, have become essential in the worldwide food industry. These materials, primarily made from hide or pig skins, have exceptional abilities as delivery systems. They are strengthened by improved barrier and mechanical properties, which enhance the protection of food products. In addition to preservation, their functionality encompasses the manipulation of sensory experiences, allowing for the controlled release of active ingredients to enhance flavors and provide nutritional advantages. It is worth mentioning that sausages commonly make use of collagen casings, which function as edible films. Collagen casings have been found to exhibit a notable decrease in biogenic amines during fermentation, leading to an improvement in the quality attributes of sausages when compared to traditional pork casings. In addition, collagen casings can be equipped with additional features, such as antimicrobial properties.

In addition, the combined advantages of mixing different molecules in film production allow for the utilization of the unique properties of each polymer used. The blending process enables the tailoring of polymer properties to fulfill specific demands while maximizing the overall performance of the material. Collagen, an important agricultural commodity, is currently being thoroughly investigated for its potential in the development of films. Collagen's plentiful supply, exceptional compatibility with living organisms, and ability to break down naturally make it a cost-effective option with great potential for diverse uses.

Collagen fiber is known for its exceptional tensile strength, remarkable toughness, and ease of processing. These outstanding mechanical properties are a result of its unique helical rod structure, abundant peptide bonds, and the presence of amino and hydroxyl groups. The inherent characteristics of collagen highlight its potential for commercial utilization in packaging applications.

4 Collagen and Its Derivatives

Fish-derived biopolymers are intrinsically biodegradable, which means they can be degraded over time by biological processes into simpler chemicals. This property is critical for decreasing environmental pollution and dealing with the growing concerns about non-biodegradable packaging materials. Collagen and gelatin derived from fish waste are biocompatible, which means they can be used with living tissues and organisms [18]. This feature makes them suitable for use in direct contact with food products, as they represent a low risk of adverse reactions or contamination.

Fish-derived biopolymers provide barrier qualities against gasses, moisture, and other environmental variables, thereby protecting packaged food goods from external contamination and extending shelf life. These materials can be customized to meet specific barrier needs depending on the type of food and storage conditions. Collagen films are not intended to replace non-edible packaging; rather, they are used to give strong protection against insects, oxidation, microorganisms, and other elements that may compromise the product's quality and shelf life. Collagen derivatives include gelatin, collagen peptides, and hydrolyzed collagen. These derivatives are formed through a variety of procedures, including hydrolysis and denaturation, which provide molecules with changed functionality [36].

5 Gelatin

Gelatin is primarily formed by a process known as hydrolysis, which includes breaking down the collagen found in animal skin, bones, and connective tissues. This procedure can be accomplished using a variety of ways, including acid or alkaline hydrolysis, enzymatic hydrolysis, and thermal hydrolysis. Acid or alkaline hydrolysis is the process of processing raw collagen with acid or alkali solutions for a set period of time at controlled temperatures [36]. Enzymatic hydrolysis uses proteolytic enzymes to cleave peptide links inside collagen molecules. Thermal hydrolysis is the process of heating collagen under controlled conditions in order to degrade its protein structure. The hydrolysis process degrades collagen fibers into smaller peptides and amino acids, giving gelatin its characteristic gel-forming qualities.

6 Collagen Peptides

Collagen peptides, also known as collagen hydrolysates or collagen hydrolysate, are formed by further hydrolysis of gelatin or collagen. This additional hydrolysis step degrades the gelatin or collagen into smaller peptide fragments and free amino acids, yielding a product with increased solubility and bioavailability. The hydrolysis process can be tuned to provide precise molecular weight distributions and functional qualities.

Enzymatic hydrolysis is a typical method for producing collagen peptides, in which proteolytic enzymes are utilized to cleave the peptide bonds in collagen molecules [36]. Collagen peptides are commonly employed as dietary supplements and functional ingredients in a wide range of food and beverage items, such as protein powders, nutritional bars, sports drinks, and beauty supplements.

7 Hydrolyzed Collagen

Hydrolyzed collagen is related to collagen peptides and is manufactured using similar hydrolysis techniques. However, hydrolyzed collagen may be subjected to different processing conditions or have various molecular properties when compared to collagen peptides. Hydrolyzed collagen, like collagen peptides, is made by breaking down collagen proteins into smaller peptide fragments and amino acids to increase solubility and bioavailability [36].

Although research into the advantages of hydrolyzed collagen is ongoing, some studies indicate that it may improve skin health, joint pain, and muscle mass. However, further research is required to properly comprehend its potential benefits and mechanisms of action. While hydrolyzed collagen supplements are usually regarded safe for most people, those with certain dietary limitations or allergies should proceed with caution and check with a healthcare expert before incorporating them into their routine. Furthermore, like with any supplement, it is critical to select a high-quality product from a reliable company.

8 Casting Techniques

Several techniques are used to prepare food packaging films made with biopolymers, taking into account their distinct properties. The methods discussed are solution casting, extrusion, electrospinning and compression molding, which are used to shape biopolymers into films, pellets, and desired shapes, respectively. Each method provides unique benefits in terms of effectiveness, scalability, and the characteristics of the film, guaranteeing adaptability and durability in food packaging options.

9 Solvent Casting Method

The solvent casting process is a popular method for creating food packaging films from biopolymers. It entails dissolving specific biopolymers, such as collagen or chitosan, in an appropriate solvent to produce a uniform solution. To improve film qualities, additives such as plasticizers or antibacterial agents might be used. The solution is then cast onto a flat surface and allowed to dry, leaving a continuous layer.

The solvent evaporates during drying, leaving behind a solid coating of the required thickness and characteristics. The resulting films can be further conditioned and are appropriate for use in food packaging applications due to their simplicity, adaptability, and capacity to incorporate diverse biopolymers and additives.

10 Extrusion Method

The blown-extrusion technology transforms the ingredients into amorphous and homogenous films suited for food packaging [21]. Extrusion is a primary polymer processing technique that is commonly used commercially. It is essential for changing the structure of materials to improve their physicochemical properties. This process is divided into three main zone [35],

- Feeding Zone: Raw materials are introduced into the extruder and compressed. This initial stage lays the groundwork by preparing the materials for subsequent processing.
- Kneading Zone: The materials are then transferred to the kneading zone, where they are thoroughly mixed and blended to achieve uniformity and consistency. This zone is critical for maintaining a uniform composition throughout the material.
- Heating Zone: When the materials reach the heating zone at the exit, controlled heating is used to soften the material and facilitate shape. The controlled heating procedure is critical to getting the correct physicochemical characteristics in the finished product.

By passing through these separate zones, the extrusion process allows for the creation of polymer films with improved qualities suited for a variety of applications, including the development of edible coatings for food products. This technology enables the transformation of raw materials into functional and high-quality films with improved properties, making it an important approach in the polymer processing sector [35].

11 Compression Method

Compression molding is a manufacturing process that shapes materials like polymers or composites into solid objects within a mold under high pressure and temperatures. The process starts with preparing the material, which may include combining polymer pellets or composite preforms to create the necessary characteristics. The prepared material is carefully placed in a mold cavity to ensure good alignment and distribution [7].

Subsequently, high pressure is applied to the mold to compress the material uniformly and compactly inside the cavity. At the same time, high temperatures are employed to soften the material and make it flexible for shaping. This combination of pressure and heat allows the material to take the shape of the mold and adapt to its precise intricacies.

Once the material has been formed to the desired dimensions, the mold is chilled to help the item solidify and maintain structural integrity. This cooling step is critical in determining the final shape and qualities of the item. Once the material has set, the molded object is expelled from the mold, signaling the end of the operation.

In summary, compression molding is a precise and regulated manufacturing process that includes material preparation, shape under pressure and heat, chilling for solidification, and ejecting the molded object. This procedure allows for the creation of things with precise shapes and features matched to their intended application [7].Compression molding is ideal for creating thicker films with specialized forms and textures.

12 Electrospinning Method

Electrospinning is a versatile and novel process for creating nanofibers with diameters ranging from a few nanometers to several micrometers. This approach involves applying a high-voltage electric field to a polymer solution or melt, causing the material to form a jet, which is subsequently stretched and formed into ultrafine fibers as it travels toward a grounded collector [42]. A polymer solution is formed by dissolving a polymer in an appropriate solvent. The qualities of the nanofibers produced are influenced by the polymer and solvent used. The polymer solution is loaded into a syringe or reservoir that is attached to a metallic needle or spinneret. A high voltage is supplied between the needle and a ground collector. When an electric field is produced, the surface tension of the polymer solution overcomes electrostatic forces, resulting in a cone-shaped droplet known as the Taylor cone at the needle's tip. As the electric field intensity increases, a charged jet is blasted from the Taylor cone and extends toward the collector. As the polymer jet travels toward the collector, the solvent evaporates, and solidifies into nanofibers (Table 1).

Table 1 Comparing the merits and demerits of various film casting techniques

Method	Advantages	Disadvantages	References
Solvent casting	This method enables the integration of additives, plasticizers, or active chemicals into the film matrix to enable antibacterial characteristics or increased mechanical strength	It requires a drying interval for solvent evaporation and film formation, resulting in lengthier processing times than other film production methods	[26, 30]
Extrusion	Short processing time and low energy consumption Improved mechanical and optical properties of films	Raw materials must be temperature-tolerant and low in moisture	[25, 35]
Compression	It is a cost-effective manufacturing process This process can produce parts with intricate details and complex shapes	Compression molding has longer cycle times than other molding techniques, which might affect overall production efficiency	[7, 8]
Electrospinning	Electrospun nanofibers exhibit a high surface area-to-volume ratio. Nanofiber porosity can be controlled using electrospinning parameters to achieve customizable permeability in active food packaging	Electrospinning has a lower productivity than standard film creation technologies. Electrospinning requires precise control of parameters, which might be difficult to optimize	[22, 42]

13 Collagen Blended Films

Collagen blend films are biodegradable packaging materials that combine collagen with other ingredients to improve their qualities and functioning. By combining collagen with other materials such as polysaccharides, proteins, or lipids, the resulting films can have higher mechanical strength, barrier qualities, and utility for a variety of packaging purposes. These blend films can be customized to meet specific flexibility, transparency, water vapor resistance, and antibacterial needs. The inclusion of other components such as chitosan, starch, alginate, or essential oils might improve the performance of collagen blend films in terms of increasing packaged product shelf life, inhibiting microbiological development, and preserving food quality.

13.1 Collagen and Sodium Alginate

Collagen-based sausage casings offer excellent taste and transparency. The main disadvantage of collagen-based food packaging is its low mechanical strength. To address this issue, a new food packaging film was created with collagen as the major component and sodium alginate as a reinforcement agent. Glutaraldehyde was employed as a crosslinking agent to enhance the strength of collagen films.

Collagen was extracted from pig skin. Collagen-based food packaging films were made from collagen, starch, sodium carboxymethyl cellulose, and sodium alginate. Glutaraldehyde was used as a cross-linking agent, and glycerol as a humectant..The study examined how the amount of sodium alginate affected surface characteristics, tensile strength, and break elongation. The optimal quantity of sodium alginate was determined. Adding sodium alginate significantly increased the tensile strength and elongation at break of collagen composite films.. Adding sodium alginate improves intermolecular interactions and results in a more uniform structure. The combination of these elements significantly improved the mechanical characteristics of the collagen composite film. The novel collagen film produced in this study seems uniform, translucent, and smooth, with no bubbles. Higher tensile strength (TS) and lower elongation at break (*E*) compared to individual collagen and sodium alginate films. The mechanical resistance and plasticity of the films help to ensure their longevity and handling during packaging and storage. The barrier properties of this film seemed to have water vapor permeability issue and also oxygen permeability issue).By adding SiO2 as a nanoparticle can improve to overcome these barrier properties.SiO2 nanoparticles can help to improve the barrier characteristics of the films, making them more effective in inhibiting the passage of gasses and liquids [23]. Collagen/sodium alginate blend films are good moisture and oxygen barriers, which can contribute to food degradation [39]. These coatings contribute to the long-term freshness and quality of food goods by reducing moisture penetration and oxygen exposure. The barrier qualities of these coatings help to extend the shelf life of perishable goods by limiting microbial growth, oxidation, and moisture-related deterioration. This lengthening of shelf life improves the overall quality and safety of packaged foods.

13.2 Collagen and Chitosan

The production of collagen-chitosan blend films has various advantages over traditional packaging materials, including improved mechanical qualities, barrier properties, and biodegradability. These films have the potential to help reduce environmental pollution while also promoting sustainable food practices [5]. Collagen/chitosan mixes can be valuable for food bundling readiness. For this reason, fish handling waste acquired in food readiness, for instance, fish skin and bones, is the best collagen source [28].

Collagen was extracted from fish waste by an acid solubilization technique. This will most likely include treating the fish waste with an acidic solution to dissolve the collagen protein, followed by purification methods to separate the collagen from other waste components [5]. The obtained collagen was combined with chitosan, a biopolymer generated from seafood waste, to create a composite material appropriate for food packaging. The blending procedure was expected to entail combining collagen and chitosan solutions, followed by casting or extrusion to generate films.

The collagen-chitosan mix films were effectively produced, indicating good compatibility between the two biopolymers. The films have favorable mechanical, barrier, and biodegradable qualities, making them potential candidates for food packaging applications. Acid-soluble collagen derived from fish waste can be successfully combined with chitosan to produce biodegradable food packaging films.

These films have potential advantages such as sustainability, biocompatibility, and functional qualities, which will help to build eco-friendly packaging solutions for the food sector [5]. By adding chitosan collagen, the blend of films' thermal stability improved, as evidenced by an increase in the temperature at which they disintegrated [14]. This improvement in thermal stability was attributed to the interaction via hydrogen bonding and electrostatic interactions. The blend films demonstrated enhanced light barrier characteristics in the ultraviolet (UV) region. The water solubility of the blend films reduced, indicating that they were less likely to dissolve in water.

Chitosan, derived from crustacean shells, is a linear polysaccharide with antimicrobial properties, offering safety and quality preservation in food packaging. Its versatility allows modification into various forms like gels, films, and nanoparticles. As the second most abundant polysaccharide, chitosan's bio-functional attributes make it preferred for sustainable packaging. Combining it with other materials enhances functional properties, contributing to improved food quality and extended shelf life [17].

The concentrate on antibacterial properties and miscibility of chitosan/collagen mixes included mixing collagen with chitosan at various deacetylation degrees in arrangements. To survey miscibility, natural consistency, and differential filtering calorimetry investigations were led on collagen, chitosan, and collagen/chitosan examples. Key discoveries included subatomic level miscibility among collagen and chitosan particles across all pieces and deacetylation degrees. Fourier change infrared examinations uncovered that the level of held triple helix structures in collagen particles inside collagen/chitosan examples diminished as chitosan contents expanded. Critically, it's prominent that the collagen utilized in this study was obtained from pigskin. This data adds to how we might interpret the atomic associations, miscibility, and underlying changes in chitosan/collagen mixes, with possible ramifications for applications in materials with antibacterial properties [40].

Overall, the addition of chitosan improved the blend films' mechanical strength, thermal stability, and light barrier qualities, and reduced water solubility, making them better suited for food packing.

13.3 Collagen and Pectin

Pectin is a complex polysaccharide with a branched structure, consisting of galacturonan fragments, rhamnose, galactose, arabinose, xylose, and other neutral sugars. The matrix includes cellulose and hemicellulose, contributing to cell structure. Pectin has an undefined molecular weight, similar to other polysaccharides, and its structure involves a polysaccharide backbone of a-(1–4)-linked D-galacturonic acid. The preparation of different substructural components is influenced by the source and extraction method. However, it's important to note that commercial extraction methods can lead to significant degradation and alterations in the neutral properties of the sugar side chains.

Pectin, a versatile substance, is sourced from a wide array of fruits and by-products. Commercially, it is commonly extracted from fruits like citrus and apples. However, various other sources, including mango peel, sugar beet pulp, sunflower head, soybean hull, Akebia trifoliata peel, passion fruit peel, peach pomace, banana peel, and more, contribute to the abundance of this essential polysaccharide. The diverse origins of pectin highlight its availability and applicability across different industries [19].

Collagen and pectin films are edible food packaging materials. They are manufactured by combining collagen and pectin. Collagen-pectin films incorporated with Melissa officinalis, a herbaceous perennial in the mint family (Lamiaceae) is incorporated to improve the barrier and antioxidant properties [4]. The films are made by using film casting method. The antioxidant properties of pectin/collagen hydrogel-based films supplemented with Melissa officinalis essential oil were assessed utilizing the ABTS and DPPH inhibition percentage assays. These experiments evaluated the films' ability to suppress the ABTS and DPPH radicals, which are routinely employed to evaluate the antioxidant potential of materials. The experiments quantify the films' ability to scavenge free radicals, demonstrating their potential to prevent food oxidation and deterioration. The results demonstrated that adding Melissa officinalis essential oil to the films increased their antioxidant qualities, making them appropriate for use in active food packaging materials. The essential oil enhances the flexibility and thickness of the films, while also increasing their antioxidant activity. The barrier characteristics of pectin/collagen hydrogel-based films containing Melissa officinalis essential oil were found to be improved. The films exhibited superior barrier qualities against water vapors, showing that the addition of Melissa officinalis essential oil to the composite hydrogel films improved barrier performance.

13.4 Collagen with Whey and Methylcellulose

Collagen, whey protein, and methylcellulose are among the most important agro-based raw ingredients being researched for film production, with the potential for commercial use in package development. For the production of biodegradable food packaging materials, methylcellulose, whey protein, and collagen are combined. The specific qualities of these ingredients are examined, as well as their potential for use in combinations for packaging purposes. The films are created by homogenizing film-forming solutions containing glycerol in a thermostatic bath with a mechanical stirrer. The films' barrier properties are assessed based on their thickness, mechanical properties, and other features such as water solubility, tensile strength, elongation at break, and resistance to water vapor. These qualities are crucial in determining the efficiency of films in packaging applications because they influence the films' capacity to protect food products from external factors such as moisture and oxygen.

When methyl cellulose is added to the film, it improves its temperature resistance, resulting in more cohesive structures. The addition of methyl cellulose improves the film's mechanical qualities, such as tensile strength, as well as its elongation at break. Furthermore, methylcellulose increases the film's transparency, making it suited for generating translucent materials that allow visual inspection of the packaged product.

Under specific conditions, certain globular proteins can undergo unfolding and cross-linking, forming new polymers that serve as excellent ingredients for food packaging films and coatings. These cross-linked protein films are recognized for their heightened stability and extended durability, surpassing counterparts based on polysaccharides. Whey protein, with its beneficial functionalities, has attracted attention as a material for food packaging due to superior optical and barrier properties compared to existing biopolymers. The unique characteristics of whey protein position it as a promising candidate for the development of high-performance and durable food packaging materials [32]. In the context of whey protein composition, it typically contains various globular proteins, such as approximately 19% a-lactalbumin (a-La), 57% b-lactoglobulin (b-Lg), 7% bovine serum albumin, 13% various immunoglobulins, and 4% polypeptide protease-peptone. Among these proteins, b-Lg plays a crucial role in gelation and compaction in whey protein, making up a significant portion. Films and coatings made from whey protein are well-known substitutes for synthetic polymers in the packaging of food products [31].

Adding whey to the film improves its oxygen, scent, and lipid barrier qualities. Whey protein has been shown to contribute to the film's color properties as well as increase its mechanical strength. Blending whey protein films with other polymers results in increased tensile strength and a bigger overall color difference (ΔE^*). The inclusion of whey protein also alters the film's solubility and swelling index, which affects its affinity for water and integrity in humid products. Overall, integrating whey protein into the film can improve numerous properties, making it an important component in the development of sustainable packaging materials. Whey protein films are biodegradable, which can assist reduce pollution and waste accumulation, contributing to sustainable packaging methods [26]. Whey proteins have functional qualities that allow them to interact with other components in the film formulation, resulting in cohesive and stable film structures. Whey protein's functional characteristics contribute to the overall performance of the packaging film, ensuring optimal food protection and preservation.

13.5 Collagen with Casein and Keratin

The cross-linked collagen fiber-based film including casein, keratin, and soy protein isolate is a specialty packaging material intended to improve the structure-stability and packaging properties of collagen fiber [41]. This film is generated by crosslinking collagen fibers with transglutaminase (TGase), which improves its stability and strength. By integrating proteins such as casein, keratin, and soy protein isolate in particular quantities into the collagen fiber matrix, the film gains special features that make it suited for food packaging applications. The film has superior barrier qualities, shielding food goods from moisture, gasses, and exterior pollutants. It improves mechanical strength, ensuring the safe handling and transportation of packed foods. The film has improved thermal stability, which allows it to endure temperature fluctuations during storage and shipping. The composition of the film can be changed by modifying the type and concentration of proteins supplied, allowing for customization depending on unique food packaging needs. Collagen-based films are biodegradable and environmentally benign, providing a sustainable alternative to typical plastic packaging materials.

Adding casein to a crosslinked collagen fiber-based film can improve thermal stability, mechanical strength, barrier qualities, and customization choices, making the film more versatile and suited for a wide range of food packaging applications [41].

Incorporating keratin into the crosslinked collagen fiber-based film can improve elongation-at-break, structural variety, customizing options, functional benefits, and potential bioactive qualities [41].

13.6 Collagen with Polyhydroxyalkanoates(Pha)

This polymer family exhibits noteworthy properties for packaging when compared to low-density polyethylene. Polyhydroxybutyrate (PHB), known for its low water vapor permeability, finds common applications in mass shrinkage packaging and flexible intermediate containers for food packaging[2]. Additionally, copolymers of hydroxybutyrate and hydroxyvalerate, specifically incorporating poly(b-hydroxybutyrate-co-b-hydroxyvalerate) (PHBg), are commercially available. These copolymers possess thermoplastic properties similar to polypropylene and exhibit good mechanical properties (Table 2).

The introduction of 3Hg or 4Hb comonomers in the PHA chains significantly influences their mechanical behavior. The addition of comonomers leads to increased toughness and decreased stiffness and tensile strength. Presently, PHA serve as effective alternatives to some petrochemical-based polymers, offering comparable physical and chemical characteristics [2].

In maintainable food bundling, an imaginative methodology includes mixing collagen, known for its biodegradability and film-framing properties, with polyhydroxyalkanoates (PHAs), perceived for biodegradability and biocompatibility. While PHAs offer admirable highlights, including biodegradability and biocompatibility, their application is upset by creation expenses and restricted mechanical properties.

To address these difficulties, a fibrogenic arrangement integrates gelatin and cheddar whey, basic parts close by PHAs. Gelatin, known for incredible handling

Table 2 Characteristics and barrier properties of the blended films

S. No.	Film	Characteristics	References
1	Collagen—chitosan	Strong tensile strength Good water solubility Good thermal stability	[14, 15]
2	Collagen-sodium alginate	Flexibility Thermal stability Good compatibility Rapid biodegradability	[23, 39]
3	Collagen—pectin	Good antioxidant activity Strong tensile strength Good transparency	[4]
4	Collagen- whey	Oxygen property Odor barrier Temperature resistance	[26]
5	Collagen—keratin	Rapid biodegradability Biocompatibility Good moisture property	[11, 41]
6	Collagen- casein	Thermal stability Improved mechanical strength	[41]
7	Collagen—PHA	Strong tensile strength Rapid biodegradability	[1]

qualities, and cheddar whey, a result of the dairy business, synergistically improve optical, mechanical, and primary film properties.

Glycerol is utilized as a plasticizer. Multi-facet films, especially with higher gelatin focus (5% w/v), show better ascribes than standard PHA films. The arrangement fundamentally further develops water fume penetrability, stretching, and haziness. A film with 5% gelatin, cheddar whey, and PHAs shows remarkable qualities: water fume penetrability of 0.45 g mm/m^2/d/kPa, an extension of 2.18%, and obscurity of 14.5%.

Morphological examination uncovers a homogeneous surface without breaks, featuring primary honesty. Warm examination highlights polymeric part miscibility. While this study doesn't specify collagen, joining biopolymers to improve reasonable food bundling films gives a promising structure. Investigating collagen in comparative definitions could make hearty, eco-accommodating bundling materials [1].

14 Applications of Collagen-Based Films

Collagen-based films are used as edible coatings on food goods to increase shelf life, keep them fresh, and prevent microbiological growth. Collagen films are biodegradable and environmentally friendly, making them ideal for sustainable food packaging. Barrier Properties: Collagen films can act as an oxygen and moisture barrier, preserving food quality and freshness. Antimicrobial Properties: Collagen-based films can be treated with antimicrobial agents to improve food safety and shelf life. Active Packaging: Collagen films can be engineered to release bioactive chemicals like antioxidants or antimicrobials, which improve food quality and safety. Collagen-based casings serve an important role in meat production because they provide flexibility, protection, consistency, and edibility. Their application in the meat business improves processing efficiency, product quality, and the entire consumer experience [18].

The cross-linked collagen fiber-based films containing casein, keratin, and soy protein isolate can be tailored to individual food packaging needs by varying the kind and concentration of proteins used. This customization enables personalized solutions that answer the specific needs of various food products, ensuring that the packaging provides the best barrier properties, mechanical strength, thermal stability, and other desirable features. Manufacturers may design packaging materials that are well-suited for a variety of food items by fine-tuning the film composition, thereby increasing shelf-life, safety, and presentation [41].

Collagen-based films and coatings can be designed to include antibacterial and antioxidant ingredients. Antimicrobial components help to suppress the growth of bacteria that might cause food spoilage or contamination, improving the safety and quality of food items. On the other hand, the antioxidant qualities of these films and coatings aid in the reduction of oxidative damage to food, which can result in a loss of quality and flavor. By combining these active qualities with collagen-based materials, the overall biological features of the food are improved, resulting in a longer shelf life for the items [9].

15 Innovative Processing Techniques

In the domain of collagen-based bundling materials, creative handling strategies assume a critical part in upgrading the properties and execution of the end results. Novel handling techniques plan to enhance the construction, usefulness, and adaptability of collagen-based films.

15.1 3D Printing

3D Imprinting in Collagen-Based Bundling: 3D printing, or added substance fabricating, includes the layer-by-layer testimony of collagen-based materials to make complex three-layered structures. This procedure offers accuracy and customization in planning bundling with explicit shapes and functionalities [10, 16].

Benefits: Handcrafted Bundling Plans: 3D printing considers the making of customized and complex plans, giving unrivaled adaptability in bundling feel and usefulness. Decreased Material Waste: The exact testimony of materials layer by layer in 3D printing limits material waste, lining up with maintainability objectives in bundling improvement. Potential for Making Complex Calculations: The inborn layering cycle of 3D printing empowers the creation of intricate calculations that might be testing or difficult to accomplish through other assembling techniques, introducing open doors for inventive and proficient bundling arrangements. This coordination of 3D printing with collagen-based materials implies a union of cutting-edge fabrication procedures and biomaterials, offering a pathway for economical, adjustable, and complicatedly planned bundling arrangements.

15.2 High-Pressure Processing

High-pressure Processing (HPP) is exposing materials to raised pressures, regularly in the scope of 100–1000 megapascals (MPa). With regards to collagen-based materials, this strategy is utilized to actuate modifications in the construction and properties of collagen films. The utilization of high tension can prompt changes in the sub-atomic course of action and association inside the collagen grid [43].

The course of HPP can affect the thickness and strength of collagen films by impacting elements like atomic pressing, intermolecular corporations, and the general morphology of the material. The high tension applied during handling can cause primary changes, possibly bringing about a smaller and firmly bound collagen organization. This upgraded thickness can add to worked on mechanical strength and sturdiness of collagen films.

Specialists and producers investigate high-pressure handling as a way to upgrade the exhibition of collagen-based materials, fitting their properties to meet explicit application prerequisites. The strategy offers a pathway to tweak the qualities of collagen films for different purposes, including food bundling, biomedical applications, and different fields where collagen's novel properties are utilized.

16 Future Prospects

On-going studies into collagen and collagen peptides is likely to stimulate the development of novel applications and formulations for collagen films in the food sector. Researchers are looking into novel ways to use collagen and its derivatives to provide functional and sustainable food packaging solutions [6]. Scientists hope to improve the functional qualities and performance of collagen-based films in a variety of culinary applications by exploring novel processing techniques and experimenting with alternative constituent combinations.

Processing technique advancements include improvements to the procedures used to extract, clean, and process collagen derived from animal sources [6]. These developments may result in the creation of collagen films with higher mechanical strength, flexibility, and barrier characteristics. Researchers can adjust the properties of collagen films to specific requirements for different food products by modifying the production settings.

Furthermore, experimenting with different constituent combinations can improve the usefulness and versatility of collagen films. By integrating additives, bioactive chemicals, or other polymers into collagen formulations, researchers can develop composite materials with better qualities such as greater tensile strength, moisture resistance, and shelf life. These novel constituent combinations may broaden the possible applications of collagen films in food packaging and preservation [6].

By incorporating nanomaterials into collagen/sodium alginate blend films, manufacturers can create advanced nanocomposite films with superior barrier properties, mechanical strength, and antimicrobial effects, providing innovative solutions for food packaging applications that require high-performance and sustainable packaging materials.

Traditional plastics made from fossil fuels contribute to environmental degradation and resource depletion. Using collagen-based materials in the creation of biodegradable plastics and sustainable packaging can reduce the environmental impact of plastic waste, resulting in a more sustainable and eco-friendly approach to packaging [27].

The shift to collagen-based biodegradable plastics and sustainable packaging is consistent with rising customer demand for environmentally friendly and sustainable products. Businesses that provide biodegradable, renewable, and environmentally friendly packaging solutions can appeal to environmentally concerned consumers and contribute to a more sustainable future [27].

Overall, continued research into collagen and collagen peptides paves the path for the creation of sophisticated collagen-based films with improved functional capabilities.

Researchers are prepared to uncover new potential for collagen films in the food sector by using cutting-edge processing techniques and experimenting with novel constituent combinations, providing sustainable and effective solutions for food packaging, preservation, and quality improvement [6] (Fig. 1).

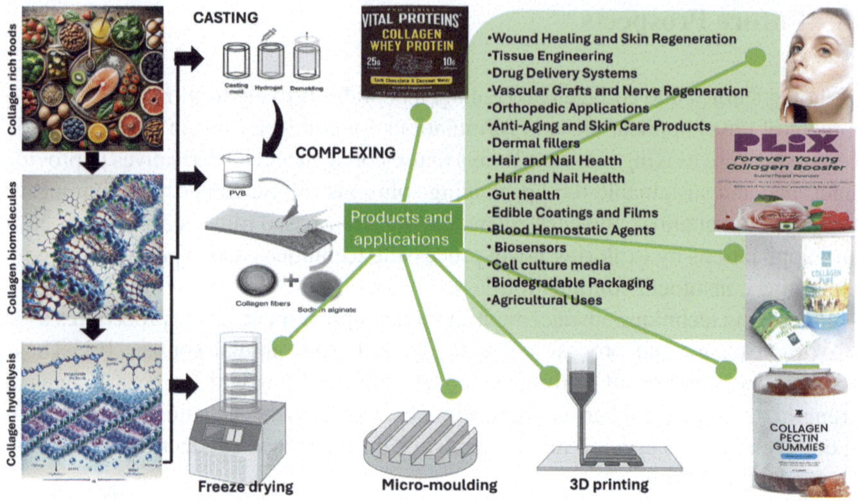

Fig. 1 Compelling view of various collagen-biopolymer fusion products and their applications

17 Conclusion

Collagen blend films represent a significant improvement in the realm of bio-based packaging solutions, with considerable potential for the future of sustainable packaging practices. These films are made by combining collagen, a natural protein, with other ingredients to produce a flexible material with a variety of benefits. Collagen blend films address environmental problems while also improving package performance in a number of ways. Collagen blend films excel at improving the quality and durability of packed goods. These films' superior qualities help to increase shelf life, protect items from external causes, and maintain overall product quality. This is especially important in the food industry, where freshness, safety, and preservation are top priorities. The incorporation of cellulose into collagen-based food packaging materials offers a promising solution for sustainable and biodegradable packaging. By adjusting the cellulose content, the mechanical strength, barrier properties, and biodegradability of the material can be tailored to meet specific application requirements. With the potential to replace traditional plastics and reduce environmental impact, cellulose-collagen composite materials demonstrate significant potential for future development in the food packaging industry. Further research and optimization of cellulose content and formulation will be crucial to unlock the full potential of these biodegradable materials and contribute to a more circular and sustainable food packaging ecosystem.

As the demand for sustainable packaging solutions grows, collagen blend films stand out as a creative and environmentally responsible option. They represent a harmonic combination of natural components and cutting-edge technology, providing a comprehensive solution to modern packaging challenges. Collagen blend films pave the way for a more sustainable and efficient packaging landscape by fulfilling the changing needs of packaging methods in a variety of industries, including food and others. In conclusion, collagen blend film is consistent with the increasing consumer preference for environmentally friendly items that promote a circular economy.

References

1. de Andrade, C.S., Fonseca, G.G., Mei, H.I., Fakhouri, F.M.: Development and characterization of multilayer films based on polyhydroxyalkanoates and hydrocolloids. J. Appl. Polym. Sci. **134**(6) (2017). https://doi.org/10.1002/APP.44458
2. Barletta, M., Puopolo, M., Tagliaferri, V., Vesco, S.: Graphene-modified poly(lactic acid) for packaging: material formulation, processing and performance. J. Appl. Polym. Sci. **133**(2) (2016). https://doi.org/10.1002/APP.42252
3. Bella, J.: A new method for describing the helical conformation of collagen: dependence of the triple helical twist on amino acid sequence. J. Struct. Biol. **170**(2), 377–391 (2010). https://doi.org/10.1016/J.JSB.2010.02.003
4. Bhatia, S., Al-Harrasi, A., Alhadhrami, A.S., Shah, Y.A., Kotta, S., Iqbal, J., Anwer, Md.K., Nair, A.K., Koca, E., Aydemir, L.Y.: Physical, chemical, barrier, and antioxidant properties of pectin/collagen hydrogel-based films enriched with Melissa officinalis **9**(7), 511 (2023). https://doi.org/10.3390/GELS9070511
5. Bhuimbar, M.V., Bhagwat, P.K., Dandge, P.B.: Extraction and characterization of acid soluble collagen from fish waste: development of collagen-chitosan blend as food packaging film. J. Environ. Chem. Eng. **7**(2), 102983 (2019). https://doi.org/10.1016/J.JECE.2019.102983
6. Cao, C., Xiao, Z., Ge, C., Wu, Y.: Animal by-products collagen and derived peptide, as important components of innovative sustainable food systems—a comprehensive review. Crit. Rev. Food Sci. Nutr. **62**(31), 8703–8727 (2022). https://doi.org/10.1080/10408398.2021.1931807
7. Ceballos, R.L., von Bilderling, C., Guz, L., Bernal, C., Famá, L.: Effect of greenly synthetized silver nanoparticles on the properties of active starch films obtained by extrusion and compression molding. Carbohyd. Polym. **261**, 117871 (2021). https://doi.org/10.1016/J.CARBPOL.2021.117871
8. Chuaynukul, K., Nagarajan, M., Prodpran, T., Benjakul, S., Songtipya, P., Songtipya, L.: Comparative characterization of bovine and fish gelatin films fabricated by compression molding and solution casting methods. J. Polym. Environ. **26**(3), 1239–1252 (2018). https://doi.org/10.1007/S10924-017-1030-5/FIGURES/7
9. Coppola, D., Oliviero, M., Vitale, G.A., Lauritano, C., D'Ambra, I., Iannace, S., de Pascale, D.: Marine collagen from alternative and sustainable sources: extraction, processing and applications **18**(4), 214 (2020). https://doi.org/10.3390/MD18040214
10. David, G.: Collagen-based 3D structures—versatile, efficient materials for biomedical applications In: Biopolymer-Based Formulations: Biomedical and Food Applications, pp. 881–906 (2020). https://doi.org/10.1016/B978-0-12-816897-4.00035-7
11. Dou, Y., Zhang, L., Zhang, B., He, M., Shi, W., Yang, S., Cui, Y., Yin, G.: Preparation and characterization of edible dialdehyde carboxymethyl cellulose crosslinked feather keratin films for food packaging. Polymers **12**, 158 (2020). https://doi.org/10.3390/POLYM12010158
12. Gisbert, V.G., Benaglia, S., Uhlig, M.R., Proksch, R., Garcia, R.: High-speed nanomechanical mapping of the early stages of collagen growth by bimodal force microscopy. ACS Nano **15**(1), 1850–1857 (2021). https://doi.org/10.1021/ACSNANO.0C10159/SUPPL_FILE/NN0C10159_SI_005.AVI
13. Hennet, T.: Collagen glycosylation. Curr. Opin. Struct. Biol. **56**, 131–138 (2019). https://doi.org/10.1016/J.SBI.2019.01.015

14. Hou, C., Gao, L., Wang, Z., Rao, W., Du, M., Zhang, D.: Mechanical properties, thermal stability, and solubility of sheep bone collagen-chitosan films. J. Food Process Eng **43**(1), e13086 (2020). https://doi.org/10.1111/JFPE.13086
15. Jiang, Y., Lan, W., Sameen, D.E., Ahmed, S., Qin, W., Zhang, Q., Chen, H., Dai, J., He, L., Liu, Y.: Preparation and characterization of grass carp collagen-chitosan-lemon essential oil composite films for application as food packaging. Int. J. Biol. Macromol. **160**, 340–351 (2020). https://doi.org/10.1016/J.IJBIOMAC.2020.05.202
16. Kim, M., Choe, Y.E., Kim, G.H.: Injectable hierarchical micro/nanofibrous collagen-based scaffolds. Chem. Eng. J. **365**, 220–230 (2019). https://doi.org/10.1016/J.CEJ.2019.02.044
17. Lagarón, J.M., López-Rubio, A., Fabra, M.J.: Active edible films: current state and future trends. J. Appl. Polym. Sci. **133**(2). https://doi.org/10.1002/APP.42631
18. Lionetto, F., Corcione, C.E.: Recent applications of biopolymers derived from fish industry waste in food packaging. Polymers **13**(14) (2021). https://doi.org/10.3390/POLYM13142337
19. Liu, X., Song, R., Zhang, W., Qi, C., Zhang, S., Li, J.: Development of eco-friendly soy protein isolate films with high mechanical properties through HNTs, PVA, and PTGE synergism effect **7**(1), 1–9 (2017). https://doi.org/10.1038/srep44289
20. Liu, Y., Ahmed, S., Sameen, D.E., Wang, Y., Lu, R., Dai, J., Li, S., Qin, W.: A review of cellulose and its derivatives in biopolymer-based for food packaging application. Trends Food Sci. Technol. **112**, 532–546 (2021). https://doi.org/10.1016/j.tifs.2021.04.016
21. Lopes, A.C., Barcia, M.K., Veiga, T.B., Yamashita, F., Grossmann, M.V.E., Olivato, J.B.: Eco-friendly materials produced by blown-film extrusion as potential active food packaging. Polym. Adv. Technol. **32**(2), 779–788 (2021). https://doi.org/10.1002/PAT.5130
22. Lyszczarz, E., Brniak, W., Szafraniec-Szczęsny, J., Majka, T.M., Majda, D., Zych, M., Pielichowski, K., Jachowicz, R.: The impact of the preparation method on the properties of orodispersible films with aripiprazole: electrospinning vs. casting and 3D printing methods **13**(8), 1122 (2021). https://doi.org/10.3390/PHARMACEUTICS13081122
23. Marangoni Júnior, L., Rodrigues, P.R., da Silva, R.G., Vieira, R.P., Alves, R.M.V.: Improving the mechanical properties and thermal stability of sodium alginate/hydrolyzed collagen films through the incorporation of SiO2. Curr. Res. Food Sci. **5**, 96–101 (2022). https://doi.org/10.1016/J.CRFS.2021.12.012
24. Miles, C.A., Avery, N.C., Rodin, V.V., Bailey, A.J.: The increase in denaturation temperature following cross-linking of collagen is caused by dehydration of the fibres. J. Mol. Biol. **346**(2), 551–556 (2005). https://doi.org/10.1016/J.JMB.2004.12.001
25. Phothisarattana, D., Wongphan, P., Promhuad, K., Promsorn, J., Harnkarnsujarit, N.: Blown film extrusion of PBAT/TPS/ZnO nanocomposites for shelf-life extension of meat packaging. Colloids Surf. B **214**, 112472 (2022). https://doi.org/10.1016/J.COLSURFB.2022.112472
26. Pirsa, S., Sharifi, K.A.: A review of the applications of bioproteins in the preparation of biodegradable films and polymers. J. Chem. Lett. **1**, 47–58
27. Puscaselu, R., Gutt, G., Amariei, S.: Rethinking the future of food packaging: biobased edible films for powdered food and drinks **24**(17), 3136 (2019). https://doi.org/10.3390/MOLECULES24173136
28. Raghavankutty, M., Jose, G.M., Sulaiman, M., Kurup, G.M.: Evaluating the biocompatibility of marine-derived chitosan–collagen polymeric blends for biomedical applications (2017). https://doi.org/10.1177/0883911517747892
29. Revell, C.K., Jensen, O.E., Shearer, T., Lu, Y., Holmes, D.F., Kadler, K.E.: Collagen fibril assembly: new approaches to unanswered questions. Matrix Biology Plus **12**, 100079 (2021). https://doi.org/10.1016/J.MBPLUS.2021.100079
30. Roy, S., Rhim, J.W.: Preparation of bioactive functional poly(lactic acid)/curcumin composite film for food packaging application. Int. J. Biol. Macromol. **162**, 1780–1789 (2020). https://doi.org/10.1016/J.IJBIOMAC.2020.08.094
31. Schmid, M., Müller, K.: Whey protein-based packaging films and coatings. In: Whey Proteins: From Milk to Medicine, pp. 407–437 (2019). https://doi.org/10.1016/B978-0-12-812124-5.00012-6

32. Schmitt, C., Turgeon, S.L.: Protein/polysaccharide complexes and coacervates in food systems. Adv. Coll. Interface. Sci. **167**(1–2), 63–70 (2011). https://doi.org/10.1016/J.CIS.2010.10.001
33. Shenoy, M., Abdul, N.S., Qamar, Z., Al Bahri, B.M., Al Ghalayini, K.Z.K., Kakti, A.: Collagen structure, synthesis, and its applications: a systematic review. Cureus **14**(5) (2022). https://doi.org/10.7759/CUREUS.24856
34. Sorushanova, A., Delgado, L.M., Wu, Z., Shologu, N., Kshirsagar, A., Raghunath, R., Mullen, A.M.: The collagen suprafamily: from biosynthesis to advanced biomaterial development. Adv. Mater. **31**(1), 1801651 (2019). https://doi.org/10.1002/ADMA.201801651
35. Suhag, R., Kumar, N., Petkoska, A.T., Upadhyay, A.: Film formation and deposition methods of edible coating on food products: a review. Food Res. Int. **136**, 109582 (2020). https://doi.org/10.1016/J.FOODRES.2020.109582
36. Tang, C., Zhou, K., Zhu, Y., Zhang, W., Xie, Y., Wang, Z., Zhou, H., Yang, T., Zhang, Q., Xu, B.: Collagen and its derivatives: from structure and properties to their applications in food industry. Food Hydrocolloids **131**, 107748 (2022). https://doi.org/10.1016/J.FOODHYD.2022.107748
37. Taravel, M.N., Domard, A.: Collagen and its interaction with chitosan: II. Influence of the physicochemical characteristics of collagen. Biomaterials **16**(11), 865–871 (1995). https://doi.org/10.1016/0142-9612(95)94149-F
38. View of fabrication of biopolymer-based piezoelectric thin film from chitosan using solvent casting method. https://semarakilmu.com.my/journals/index.php/applied_sciences_eng_tech/article/view/1157/638. Accessed 16 Feb 2024
39. Wang, Z., Hu, S., Wang, H.: Scale-up preparation and characterization of collagen/sodium alginate blend films. J. Food Qual. **1**, 4954259 (2017). https://doi.org/10.1155/2017/4954259
40. Wei, W., Zhou, Y.H., Chang, H.J., Yeh, J.T.: Antibacterial and miscibility properties of chitosan/collagen blends. J. Macromol. Sci. **54**(2), 143–158 (2015). https://doi.org/10.1080/00222348.2014.987097
41. Wu, X., Luo, Y., Liu, Q., Jiang, S., Mu, G.: Improved structure-stability and packaging characters of crosslinked collagen fiber-based film with casein, keratin and SPI. J. Sci. Food Agric. **99**(11), 4942–4951 (2019). https://doi.org/10.1002/JSFA.9726
42. Zhang, C., Li, Y., Wang, P., Zhang, H.: Electrospinning of nanofibers: potentials and perspectives for active food packaging. Compr. Rev. Food Sci. Food Saf. **19**(2), 479–502 (2020). https://doi.org/10.1111/1541-4337.12536
43. Zhu, Y., Yan, Y., Yu, Z., Wu, T., Bennett, L.E.: Effects of high pressure processing on microbial, textural and sensory properties of low-salt emulsified beef sausage. Food Control **133**(2), 108596 (2022). https://doi.org/10.1016/J.FOODCONT.2021.108596

Cellulose-Based Packaging: An Emerging Sustainable Packaging Option for Future Use

Pavada Madhusudan Rao, Anuvansh Omveer Singh, K. Sriniketha Narayanan, Rayearth Shankar, Gaurav Mudgal, and Panjanathan Radha

Abstract Cellulose is becoming an exciting option for tomorrow's packaging applications because it is found in large quantities in the walls of plant cells and is abundant in natural resources as well. Different from the other packages that damage ecosystems, cellulose stands out because it comes from nature and decomposes easily. Additionally, cellulose greatly broadens popular use through its mechanical advantages such as rigidity, durability, and resistance to permeation. The land use and the carbon footprint necessary for traditional packaging can be decreased by producing cellulose-based materials from agricultural waste or by-products. Cellulose packaging products are biodegradable and compostable enabling them to merge smoothly with current recycling practices hence reducing disposal of renewable waste. The packaging industry is adapting methods for packaging materials so that they can be sustainable according to consumer preferences. The consumer demand is now more often for environmentally friendly packed items and that is why cellulose remains among the most preferable options because it is adaptable and sustainable.

Keywords Cellulose · Renewable · Biodegradable · Recycling · Packaging

P. M. Rao · A. O. Singh · K. Sriniketha Narayanan · R. Shankar · P. Radha (✉)
Bioprocess Principles Laboratory, Department of Biotechnology, College of Engineering and Technology, SRM Institute of Science and Technology, Kattankulathur, Chengalpattu District, Chennai 603203, Tamil Nadu, India
e-mail: radhap@srmist.edu.in

G. Mudgal (✉)
Center for Waste Management and Renewable Energy, Saveetha Dental College and Hospitals, Saveetha Institute of Medical and Technical Sciences, Saveetha University, Chennai 600077, India
e-mail: gauravmdgl@gmail.com

University Institute of Biotechnology, Chandigarh University, Mohali 140413, Punjab, India

R&D, Vanproz Agrovet LLP, Bagroda Industrial Area, Bhopal 462026, Madhya Pradesh, India

K. K. Kesari et al. (eds.), *Functionalized Cellulose Materials*, Engineering Materials,
https://doi.org/10.1007/978-3-031-76953-5_4

1 Introduction

Plastic materials are all around us, starting from the toys we play with to the packaging materials we use. The primary materials that make up the composites are organic polymers, which consist mainly of chains of atoms of carbon [1]. Despite playing such a vital part in modern life, there is a marked dissatisfaction with the prevalence of plastics—the negative consequences they have on the environment. Plastic pollution has become a seemingly continuous global issue that is very destructive to the environment. Every year several million tons of plastic are sent to landfills and the environment at large cannot get rid of them, so they lay there for many years without decomposing. These then break even more into microplastics, which penetrate the food cycle to infect humans and the world with untold misery. To the point of threat, could be anything at a rate, where some plastic products come with harmful chemicals, such as Bisphenol A, which is one of the endocrine disruptors causative of developmental problems, cancer, and various other health complications as well. While people who take action to encourage plastic recycling deserve praise there is still a sorrowful fact that most plastic containers cannot be recycled. In today's context of a culture that has been for quite a while "throw-away" dominated by plastic packaging, the irreparable practices need new revolutionary resolutions. The result of these changes will be a higher number of environmentally conscious customers, as sustainable alternative use will be promoted to prevent a clean future [2]. Sugars are moonlighting bioresources over the planet for their multiple aspects of use and application, beyond a carbon source for energy for all life forms, as a sweetener, as signaling molecules, attachment factors, as cushioning molecules for proteins and so on [1, 2, 3–9]. Cellulose is an example that has come to the fore, and it is considered a key entity in the process of material syntheses. The films made of cellulose constitute a novel approach to packaging, which will then replace traditional packaging materials with eco-friendly alternatives. Cellulose films, which can be mainly derived from plant sources, such as wood pulp or even cotton lint, become the main contributing factor for their antibiotic properties and biocompatibility with body tissues, which are balanced with environmental preservation and sustainability objectives [3]. In the quest for green packaging, cellulose has emerged to guarantee a high jump toward making the packaging green. Development and research studies are further fine-tuning cellulose packaging to improve its efficiency and overall performance, driving innovations that open up avenues for developing strong, agile, sustainable packaging solutions in line with the changing requirements of the industry. Eventually, technology would mature, cellulose production would scale seamlessly, and the positioning of cellulose films as the most cost-efficient alternative to traditional plastics would herald a new age of sustainability-oriented practices in packaging.

Concerning the potential held by cellulose films in transforming the packaging industry, it is just huge and tantalizingly propitious toward a far more sustainable future. Advanced, innovative films from renewable sources with biodegradable properties can bring huge potential for diminishing the ecological footprint and achieving a circular economy. Performance, durability, and cost-effectiveness will, with continued research and development, be continually improved upon, so cellulose films will no doubt be established forthwith as new benchmarks of responsible packaging in industries as diverse as food and drink to pharmaceuticals. It may be that in shifting to cellulose films [4], we can entail a more radiant and environmentally sensitive realm where convenience and sustainability bestride together in perfect harmony to make just truly a long-lasting legacy for responsible packaging practices.

2 Exploring the Versatility of Cellulose

Cellulose, the most abundant organic compound found on earth, holds significant importance for plant cells due to its essential role in providing structure and support to their cellular framework. This versatile compound plays a crucial part in maintaining the structural integrity of plants, acting as a key component in their cell walls. Additionally, cellulose's incredible strength and durability make it an optimal choice for various applications, particularly in the realm of sustainable packaging [5]. Cellulose-based materials are paving the way for eco-friendly alternatives in packaging solutions, promoting environmental sustainability and reducing the ecological footprint of industries. The renewable and biocompatible characteristics of cellulose further contribute to its potential in advancing innovative technologies and sustainable practices.

2.1 *Morphology of Cellulose*

Cellulose, a vital polysaccharide essential for the structural integrity of plant cell walls, is comprised of intricate chains of β-D-glucose molecules joined together through specific glycosidic bonds [6]. These bonds, specifically of the β $(1 \rightarrow 4)$ type, play a critical role in forming the polysaccharide's distinct linear structure. Within this structural framework, each β-D-glucose unit functions as a pivotal link, facilitated by the β $(1 \rightarrow 4)$ glycosidic bonds that connect the first carbon atom (C1) of one glucose unit to the fourth carbon atom (C4) of the subsequent unit. This arrangement compels the glucose units to adopt a chair conformation, emphasizing the linear nature of cellulose. Moreover, the presence of hydrogen bonds between cellulose units functions as essential connectors, acting like sturdy bridges that hold the chains together. These bonds contribute significantly to the creation of a rigid and exceedingly strong crystalline structure in cellulose [7]. The remarkable durability and inflexibility of cellulose can be attributed to the combined effects of its linear arrangement and the extensive hydrogen bonding evident within its structure. Of particular significance are the hydrogen bonds formed between the hydroxyl groups

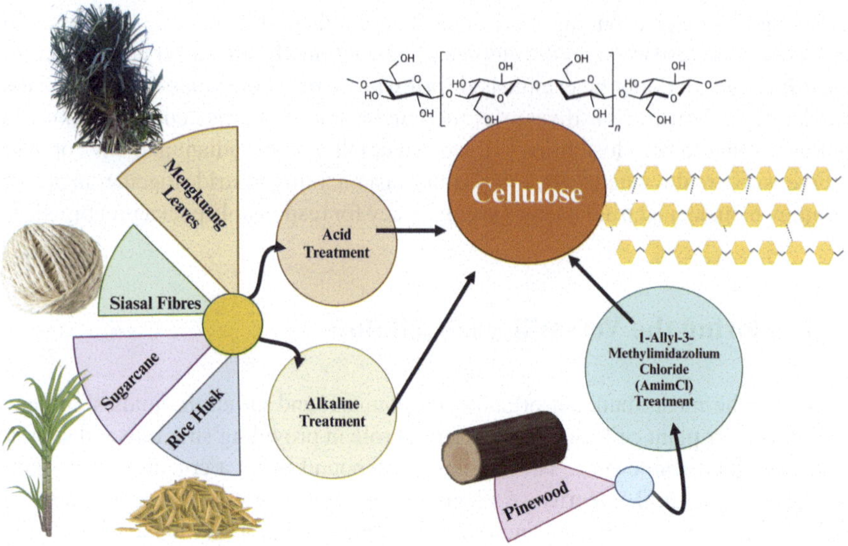

Fig. 1 Morphology of cellulose

on adjacent cellulose chains, further reinforcing the cohesion of the crystalline structure. Despite the impressive strength and rigidity conferred by these bonding interactions, cellulose also exhibits regions of disorder known as amorphous regions. The position of these amorphous regions with the highly ordered crystalline segments contributes to the unique mechanical properties observed in cellulose, making it a versatile and indispensable component in various biological processes (Fig. 1).

2.2 Understanding Various Forms of Cellulose

Cellulose is not a single entity it has multiple forms varying in different properties depicted in Fig. 2 which are also described here as follows [8]:

- **Cellulose I**: It is the most common form of cellulose which has a highly crystalline structure, contributing to strength and rigidity. It makes up the primary structural component of plant cell walls. It exists in two subcategories:

  Cellulose Iα: This form predominates in bacteria and algae. It has a Triclinic unit cell which means that the cell doesn't have equal angles between axes and contains only one cellulose chain per unit cell. These are metastable, that is they can convert to the more stable Iβ form, under heat treatment.

  Cellulose Iβ: This form is found in higher plants like wood. It has a Monoclinic unit cell, which means it has equal angles between two axes and contains two parallel cellulose chains per unit cell. These are thermodynamically stable forms of cellulose I, which is the preferred structure under normal conditions.

- **Cellulose II**: It is less common in nature but is found in cotton and regenerated cellulose products like rayon. This form is produced when cellulose I undergoes specific chemical treatments. The conversion from cellulose I to II is irreversible. This has a slightly different arrangement of the cellulose chains compared to Cellulose I, resulting in a less rigid structure. Still relatively strong and insoluble in water. Used in textiles, food casings, and other applications.
- **Cellulose III**: It is produced by treating cellulose with strong chemicals. It has the most disordered structure of the three and is the least stable. There are chemically modified forms of cellulose for specific applications. Examples include:
  - **Methylcellulose**: It is water-soluble, forming clear viscous solutions or gels, and absorbs moisture from the environment. It is used as a thickener, emulsifier, and stabilizer in various products in food and pharmaceutical products.
  - **Carboxymethyl cellulose (CMC)**: It can be either white or slightly yellow, odorless, and tasteless powder. It is anionic due to carboxymethyl groups, attracting positively charged molecules. It is hygroscopic, shows good film-forming properties, and acts as a lubricant.
  - **Cellulose acetate**: It is insoluble in water, but soluble in some organic solvents. It can be softened and molded under heat. It is water-resistant and has good barrier properties [9] and various commercial sources provide them as membranes susually used in dialysis and/or as tools to purify or separate biomolecules like proteins using the versatile molecular weight cut-off (MWCO) of the pores [10]. It is non-toxic but not as readily biodegradable as other cellulose derivatives. It is used to produce fibers for textiles like rayon as well.
  - **Cellulose Nanofibrils (CNFs)**: These are extremely thin fibers extracted from cellulose. The high surface area promotes interaction with other materials [10]. It is lightweight and biocompatible. These are thread-like structures derived from cellulose [11], possessing remarkable strength-to-weight ratios exceeding some high-performance plastics.
  - **Bacterial Cellulose**: This highly pure form is produced by some bacteria through fermentation. It has high crystallinity and mechanical strength [12]. It boasts water-holding capacity[1] and tensile strength compared to plant-derived cellulose [13], making it suitable for biomedical applications and wound dressings.

2.3 Properties of Cellulose

The unique structure of cellulose possesses unique properties which are as follows:

- **Strength**: The linear structure and hydrogen bonding network make cellulose incredibly strong which makes them resist tensile stress [14]. This property makes it perfect for providing structural support in plant cell walls and making them stand

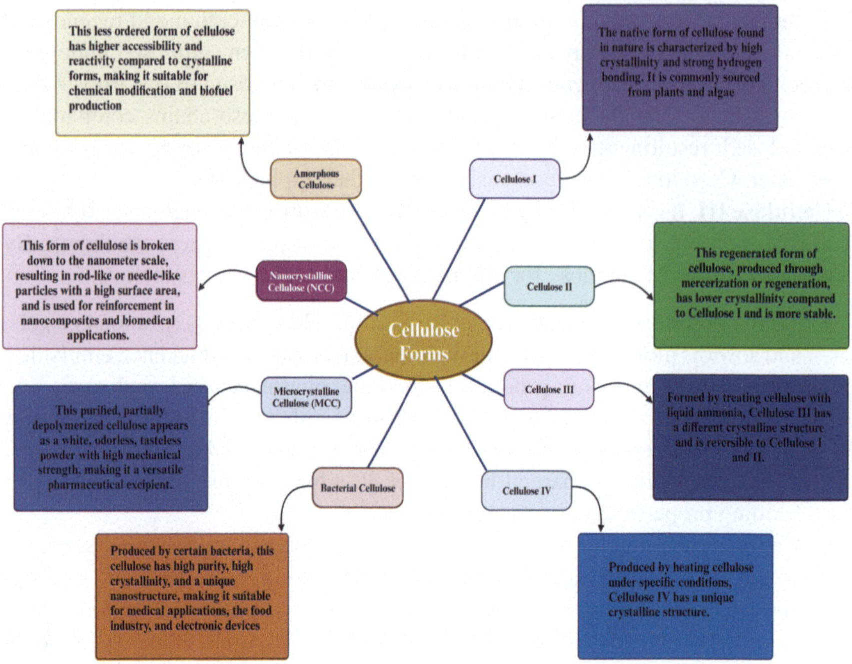

Fig. 2 Forms of cellulose

upright in wind and rain. It is also used in many man-made products like textiles, paper, and building materials.

- **Appearance**: Cellulose is white, odorless, and tasteless. It is a natural polymer that consists of long chains of glucose units linked together by beta-1,4 glycosidic bonds. These chains are arranged in a crystalline structure that gives cellulose its strength and rigidity.
- **Hygroscopicity**: It exhibits hygroscopic behavior, that is it can absorb water vapor from the surrounding environment. This is because the hydroxyl groups (OH groups) on the cellulose chains can form hydrogen bonds with water molecules [15]. The amount of moisture that cellulose absorbs depends on the relative humidity of the environment. At high humidity, cellulose can absorb up to 20% of its weight in water. This is important for maintaining plant cell wall integrity.
- **Insolubility in Water**: The network of hydrogen bonds in cellulose creates a strong affinity for water molecules. However, these bonds prevent individual cellulose molecules from dissolving in water. This characteristic is crucial for plant cell walls, as it helps them maintain their structure and rigidity even in the presence of water.
- **Thermal Stability**: Cellulose has a relatively high melting point, but it begins to decompose at lower temperatures. At even higher temperatures, cellulose will break down into smaller molecules this is called pyrolysis. The thermal stability of cellulose can be improved by treatment with certain chemicals.

- **Chemical Reactivity**: It is resistant to many chemicals, and can be modified through various chemical treatments. Cellulose can form hydrogen bonds with a variety of other molecules. This makes it a good material for use in composite materials, where it can act as a binder to hold other materials together.
- **Non-toxic**: Cellulose is a non-toxic material. It is not harmful to human or animal tissues, making it safe for various applications, like food packaging and textiles.
- **Biocompatibility**: Cellulose exhibits good biocompatibility, it is generally well-tolerated by living organisms [16]. This makes it suitable for biomedical applications such as wound dressings and artificial implants.
- **Biodegradability**: Cellulose is a natural polymer, and under the right conditions, it can be broken down by microorganisms present in the environment. This is an important property, as it helps to reduce the amount of waste that goes into landfills [16]. The biodegradability of cellulose also makes it a renewable resource.

3 Cellulose Extraction Methods

3.1 Physical Methods for Cellulose Extraction

The extraction of cellulose is essential for producing various products and often involves the separation of non-cellulosic components like lignin, hemicellulose, and other impurities. While traditional chemical methods have been widely used for this purpose, there is a growing interest in exploring physical extraction techniques due to their advantages. These alternative methods are gaining traction as they eliminate the need for harsh chemicals, making them more environmentally friendly and sustainable. Physical extraction methods typically involve a series of mechanical applications to break down the raw material efficiently. Techniques such as grinding, high-shear homogenization, and ultrasonic treatment are commonly employed in this context (Fig. 3). Overall, the combination of mechanical applications in physical extraction methods offers a promising alternative to traditional chemical processes, providing a more sustainable and efficient approach to cellulose extraction.

3.1.1 Mechanical Grinding and Sieving

The initial mechanical grinding is a vital part of the process of cellulose extraction. It involves preprocessing plant materials wherein they are mechanically ground down into particles to increase the surface for further preparation. The application of sieving then helps to acquire a uniform particle size is important as it ensures homogenous chemical modification and better extraction [17]. To extract the microcrystalline cellulose from the galam wood, the wood fiber is first crushed by water and then sieved to reach a uniform 30-mesh size. This preparation afterward allows a better and more effective chemical and mechanical treatment that results in the advantage of high crystallinity and improved cellulose quality.

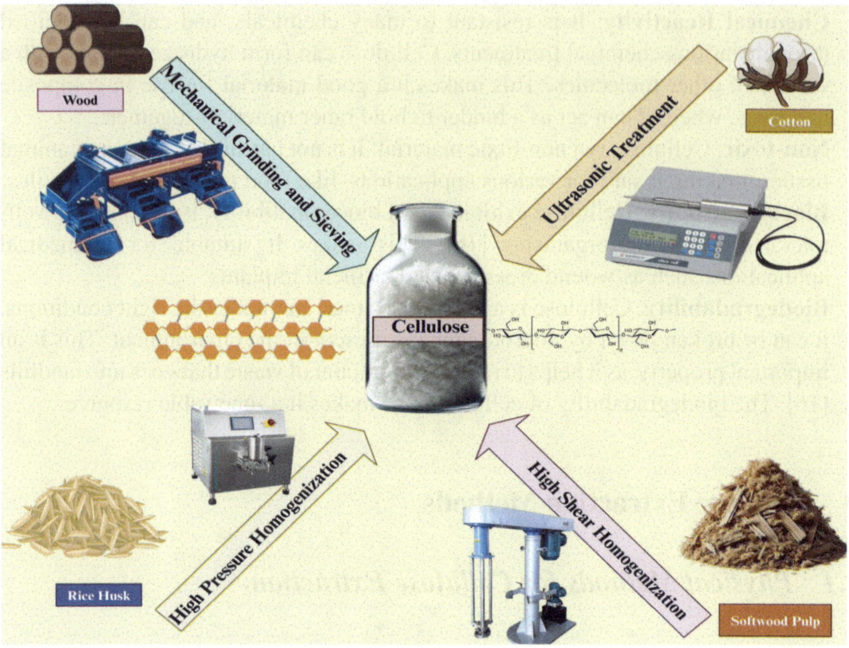

Fig. 3 Physical treatment for cellulose extraction

3.1.2 High Shear Homogenization

A manual application of high pressure to the cellulose that is termed high shear homogenization causes the generation of cutting and so-called impact forces, which allow cellulose fibers to separate into their basic elements; as a result, the cellulose is rubbed, squeezed, and impacted. High shear homogenization is desirable due to its possibility to produce nanocellulose with improved mechanical characteristics [18]. The cellulose nanofibrils were extracted with the application of high-shear homogenization from the dry softwood pulp. The pulp was blended at high speed, and the uniform nanofibers afterward received improved mechanical characteristics.

3.1.3 Ultrasonic Treatment

An application of the ultrasonic treatment implies high-frequency sound waves to be transmitted into a liquid, and their cavitation effect has an effective breakdown over the cellulose fibers. It is particularly beneficial for its fast and simple nature when it comes to cellulose nanocrystals [19]. Using ultrasonic treatment in combination with acid hydrolysis, cellulose nanocrystals were extracted from cotton. The purpose of using ultrasonic waves was that the waves could help in the fragmentation of the cellulose fibers, thereby forming nanocrystals that were characterized by high crystallinity and well-defined morphological features.

3.1.4 High-Pressure Homogenization

High-pressure homogenization allows for exposing cellulose fibers to high pressure and thereby converting the fibers into nanofibrils or nanocrystals. The method helps in the preparation of uniformly sized cellulose objects with a large surface area. Cellulose recovery from rice husk started with the mechanical grinding of the material followed by high-pressure homogenization, which helped in producing cellulose nanofibrils with the required properties for various applications.

3.2 Chemical Methods for Cellulose Extraction

3.2.1 Cellulose Extraction from Wood Chips Using 1-Allyl-3-Methylimidazolium Chloride (AmimCl)

The extraction of cellulose from wood is essential for various industrial applications. Traditional methods, like the Kraft process, are effective but environmentally hazardous due to the use of toxic chemicals. Recently, ionic liquids have emerged as an eco-friendly and more efficient alternative. Pine, poplar, Chinese parasol, and catalpa wood samples were dried, ground into powder, and treated with 1-allyl-3-methylimidazolium chloride (AmimCl) mixed with dimethyl sulfoxide [20]. The dissolution process was optimized at temperatures between 100 and 120 ºC. After dissolution, the mixture was filtered to remove undissolved residues, and cellulose was precipitated using water. Pine was found to be the most suitable wood species, achieving the highest dissolution and regeneration rates. Factors such as dissolution time, temperature, microwave heating, and initial wood concentration were systematically studied to optimize the process. Microwave heating significantly reduced reaction times, and using dimethyl sulfoxide/water as a precipitant yielded higher cellulose content in the extracts. Characterization techniques such as nuclear magnetic resonance spectroscopy, X-ray diffraction analysis, Fourier-transform infrared spectroscopy, and scanning electron microscopy confirmed effective cellulose extraction and transformation into an amorphous structure. The recycled 1-allyl-3-methylimidazolium chloride demonstrated comparable efficiency to the fresh ionic liquid, making the process cost-effective and sustainable.

3.2.2 Alkaline Extraction of Cellulose

Alkaline extraction is a common method used to extract cellulose from various biomass sources. This process involves the use of alkaline solutions, such as NaOH, to remove the non-cellulosic product such as lignin, hemicellulose, and other impurities from the plant product hereby isolating the cellulose. Some of the common approaches are listed.

1. Chemical modification of cellulose extracted from sugarcane bagasse employs the conventional process of alkaline extraction comprehensively. Sugarcane bagasse was treated with NaOH to extract lignin and hemicellulose. This process included soaking the bagasse in an adequate NaOH solution and then heating the product at a temperature range of 80–100 °C for a particular time [21]. The heating steps made the lignin and hemicellulose weak such that it was easily washed away in the next steps. The cellulose was further processed for chemical modification.
2. Extraction, Preparation, and Characterization of Cellulose Fibres and Nanocrystals from Rice Husk cellulose from rice husk was also treated with alkaline. Rice husk was treated with an alkaline solution of NaOH which made it easy to remove lignin and hemicellulose [22]. The rice husk was mixed with NaOH solution and heated at a particular temperature. The resultant product was later washed to obtain the cellulose fibres which were further processed to obtain nanocrystals.
3. Extraction of Cellulose and Preparation of Nanocellulose from Sisal Fibers In the case of sisal fibers, a study was done using two processes each involving alkaline extraction. The primary process involved the pretreatment of the products with NaOH then other treatments were carried out to purify the cellulose [23]. The second process involved treating the sisal fibers with sodium chlorite to remove lignin. NaOH was then used to extract cellulose. Both processes facilitated the extraction of cellulose which was processed to obtain nanocellulose via acid hydrolysis.
4. Extraction of cellulose nanocrystals from Mengkuang leaves the product involved in treatment with alkali to remove this non-cellulosic product. The leave was treated with the solution of NaOH which later facilitated the removal of hemicelluloses [24]. Partial hydrolysis of the lignin occurred during the NaOH treatment which is crucial in breaking down the complex matrix that surrounded the cellulose fibres. The cellulose fibers were later processed to obtain cellulose nanocrystals.

3.2.3 Acid Hydrolysis Mediated Extraction of Cellulose

Acid hydrolysis consists of treating the biomass with concentrated acids, such as sulfuric acids under controlled conditions. This method is used to hydrolyze hemicellulose and lignin, producing purified cellulose. Some of the main stages are pretreatment, hydrolysis, and post-treatment [25].

1. Pretreatment: The biomass is subjected to the initial treatments which enhance the efficiency of the hydrolysis process. This includes both mechanical processing, which is necessary to increase the particle surface, and chemical treatments which are performed to remove surface impurities.

2. Hydrolysis: The pretreated biomass is subjected to strong acid hydrolysis. In most cases, this is the concentrated sulfuric acid of the acid weight percent of 60–70% which is applied for 45–50 °C. This process takes 30–60 min and intends to hydrolyze the non-crystalline regions of cellulose, thus increasing its crystallinity.
3. Post-Treatment: The hydrolyzed material is washed, centrifuged, and dialyzed to remove the excess acid and other impurities. The obtained cellulose is used to produce nitrocellulose or other cellulose derivatives [26].

The efficacy of acid hydrolysis for cellulose extraction has been proved in multiple studies.

1. Sugarcane Bagasse: Using alkaline pretreatment combined with subsequent acid hydrolysis the hemicellulose and lignin could be both easily removed. The process presupposes treating the biomass with 72% sulfuric acids at room temperature [27].
2. Rice Husk: When handling the rice husk with sulfuric acids of 10 mol L^{-1} at 500 °C for 40 min and then washing and dialyzing the material, one can isolate the cellulose fibers, with improved mechanical properties and higher degrees of polymerization [28]. This cellulose is suitable for further application for the production of biodegradable films or nanocomposites.
3. Sisal Fibers: The sisal fibers are chlorinated and bleached and then subjected to sulfuric acid hydrolysis. The acid application rate is 60 wt%, the temperature is 45 °C and the contact time is 30 min [29]. The cellulose nanocrystals obtained can be identified by high crystallinity, mechanical strength, and thermal stability.
4. Mengkuang Leaves: The leaves should be treated with 60 wt% sulfuric acids at 45 °C for 45 min. The method is proven to help in isolating the cellulose nanocrystals, which possess high aspect ratios and high crystallinity and, therefore, can be applied to produce high-performance materials and nanocomposites [30].
5. Formic acid and peroxyformic acid treatment allows for fractionating lignocellulosic objects into cellulose, hemicelluloses, and lignin. The technological process involves treating the biomass with formic acid, followed by the use of peroxy formic acid to delignify the material. Rice straw, wheat straw, corn stalks, and dhaincha were treated with formic acid and peroxy formic acid to extract cellulose microfibrils [31]. First, the formic acid was refluxed with the biomass. In the next step, the material was treated with peroxyformic acid, which helped in obtaining the high-purity cellulose.

3.3 Combined Mechanical and Chemical Treatments

Combined treatment of mechanical and chemical is a robust method that leverages the strengths of both techniques to significantly improve the yield and product quality of cellulose. Popular methods to remove non-cellulosic components are separate treatment which significantly reduces the extraction yield and quality of cellulose. This integrated process can effectively eliminate non-cellulosic, yielding a high-quality product providing greater yield. Due to a large surface area and better access to subsequent chemical treatments, the raw biomass particles are first broken down by mechanical grinding [32]. The complex structure of biomass must be broken down and its chemicals allowed to penetrate more easily. This is why mechanical grinding recently has raised research significance. Mechanical grinding is necessary because it simplifies the complex structure of the biomass and fosters greater access by reagents. They can then operate more efficiently. Following the grinding process, the biomass is subjected to an alkaline soak. The step dissolves and removes lignin and hemicellulose, which hinder the purity and functionality of cellulose, and non-cellulosic components. It also causes the cellulose fibers to swell, thereby facilitating further processing. The alkaline environment breaks down lignin and hemicellulose, removing them. What is left behind is more purified cellulose. Washing alone cannot remove any remaining lignin and other impurities; bleaching is therefore necessary.

Common bleaching agents such as hydrogen peroxide and sodium chlorite not only purify the cellulose effectively but do so without destroying its structure. This step is vital for those applications that require high-purity cellulose, such as biocomposites, biofuels, and biomedical materials. Given the fact that they are high in cellulose and are otherwise wasted, cellulose was successfully extracted from pistachio shells using a combined method of mechanical grinding, acid hydrolysis, and bleaching [33]. It started with grinding the pistachio shells into a fine powder and then acid hydrolysis with a mild acid solution–like sulfuric acid–to break down hemicellulose and partly hydrolyze cellulose. Finally, bleaching agents were used to remove any remaining lignin and impurities. The result was high-purity cellulose with excellent structural properties which is suitable for many industries [34].

4 Cellulose Based Biocomposites

Cellulose-based biocomposites have garnered significant interest due to the rising need for sustainable and environmentally friendly materials in various industries. This interest stems from the unique properties of cellulose, a plentiful and renewable natural polymer, which exhibits exceptional mechanical strength and biodegradability. Because of these attributes, cellulose is emerging as a highly promising material for a wide range of applications, with particular attention being drawn to its potential to enhance the sustainability of food packaging solutions. The eco-friendly nature of cellulose-based biocomposites aligns well with the growing global focus on reducing environmental impact and promoting a more circular economy. By leveraging the inherent properties of cellulose and harnessing its sustainability advantages, the future holds great promise for the widespread adoption of cellulose-based materials in the realm of eco-conscious packaging practices [35]. The growing shift towards utilizing cellulose-based biocomposites signifies a significant step forward in addressing the pressing need for more sustainable material choices in a world increasingly focused on environmental considerations and long-term viability.

4.1 Preparation of Cellulose Biocomposites

Mechanical blending

It is based on mixing cellulose with other materials physically for a composite to be formed. This is one of the classical methods of obtaining composites that is still used because of its simplicity and the opportunity to obtain uniform composites. For instance, the effective formation of all-hemp cellulose composites was reported, in which the alkali-treated hemp fibers were mechanically blended with cellulose solutions [36]. The steps that were followed as part of the process are as follows: Substances such as NaOH that is used to eliminate lignin and hemicellulose in hemp fibers are replaced with impure samples that are difficult to treat. Only NaOH is used and the amount of NaOH used is calculated such that only half of it is soaked in the hemp compared to the cotton. The cellulose solution is then mixed with the already-soaked fiber in a mixer until no clumps of the fiber can be seen. Then the resultant mix can be put into a material like water or ethanol and heated together with a deck top until cellulose starts to precipitate out thereby binding the fiber like a matrix.

4.2 Chemical Treatments

Chemical treatments imply that the surface of cellulose is modified chemically to ensure adequate interaction. The alternative methods that can be implemented are as follows:

- Grafting: Attaching functional groups or polymers on the surface of cellulose is one method. For instance, cellulose fibers can be grafted with polycaprolactone through ring-opening polymerization. This method promotes compatibility with different polymers, gives its materials better mechanical properties, and enables a surface-functionalizable asset which are advantages in biocomposites but also for drug delivery systems or hydrogels [37].
- Crosslinking: It is another method that is used to generate covalent bonds between cellulose chains or cellulose and other materials. Glutaraldehyde or epichlorohydrin are crosslinking agents for cellulose through reaction with the hydroxyl groups in each of these chains resulting in stronger holding [38]. The enhanced mechanical strength, thermal stability, and resistance to moisture as well as solvents of crosslinked cellulose materials make them a perfect prospect for application in textiles.

- Surface modification methods, such as salinization and acetylation, have been developed to change the surface layer of cellulose fibers by efficiently reacting with other materials. This is followed by gas phase silanization to modify the cellulose surface for adhesion and compatibility. It is a process that alters cellulose into ester, forming acetyl groups (CH3CO-), and causes the hydrophilic nature of cellulose to be reduced with an increased resistance towards moisture thus making it compatible with most applications involving cellulosic bonded by either natural or synthetic polymers which are typically hydrophobic. Biomedical devices may be utilized for these adjustments in addition to packaging, textiles, and environmental remediation.

4.3 Structural and Mechanical Properties

The structural properties of cellulose composites were studied as a function of their physical properties: fiber size, surface area, and crystallinity. X-ray diffraction and FTIR spectroscopy methods have been widely employed. For, example, the WAXD profile of resulting composites has shown the transformation of triclinic Iα − D1 crystals of hemp fibers into monoclinic Iβ − D1 at cellulose I + II during the NMMO dissolution process. This evidence suggested that drastic changes took place in the biocomposite crystallinity, which, in turn, would explain the enhanced performance of the processed cellulose biocomposites. Mechanical testing, namely tensile strength and modulus measurements, on all-cellulose composite and cellulose/cellulose undoped or doped with other fillers state a significant impact on fiber reinforcement [39]. Cells-closed composites with a smaller fiber size of 45 μm resulted in superior mechanical properties due to improved fiber dispersion resulting in a higher surface area for enhanced interfacial matrix-fiber adhesion. TGA confirmed that fibers affect thermal stability due to structural order and swelling [40].

5 Food Packaging Applications

The biodegradability and intrinsic mechanical properties of cellulose have resulted in an increased interest in the use of plant fibers/fillers for application to food packaging material, to reduce environmental challenges associated with non-biodegradable polymeric systems. Cellulose-based materials present an interesting green material for food packaging alternative to petroleum-derived plastics, mainly due to their environmental sustainability as one of the main advantages. Cellulose biocomposites have excellent barrier properties, making food packaging one of their most important applications. E.g. Cellulose reinforced with nanocrystalline cellulose increases these properties drastically, and the enhanced mechanical strength and barrier properties of the biocomposites were ascribed to the high aspect ratio, crystalline nature, and biodegradability that nanocrystalline cellulose provided. This is particularly important when it comes to food packaging, as the integrity and quality of what's inside can't be compromised.

A significant improvement demonstrated by cellulose reinforcement is their low oxygen permeability. This feature is especially useful for packaging applications, as it serves to prolong the shelf-life of food by reducing oxidation and spoilage. Moreover, these biocompatibles show a medium water vapor permeability. It is not completely impermeable but it has a low-medium water resistance which will be enough for many food packaging applications, preventing the degradation of material used in humid environments. High extensibility is another important characteristic of cellulose biocomposites. It is this attribute that guarantees that the packaging will be able to withstand external forces during transportation and handling, making it useful in protecting any food. In particular, these mechanical properties combined with the inherent barrier functions make cellulose biocomposites a promising alternative to petroleum-based plastics in food packaging applications.

Therefore, cellulose biocomposites are promising tools using their suitable inherent characteristics like mechanical resistance, impermeability of gases, and others in food packaging as the conclusion. The attributes are further tuned through reinforcement with nanocrystalline cellulose making these materials a viable option over traditional petroleum-based plastics. The successful use of cellulose biocomposites in food packaging is an important contribution to the future sustainability of this broad and complex industry.

6 Environmental Benefits

Sustainability benefits from hemp, all-cellulose, and physically-manufactured biocomposites are robust and fast-growing, hemp is highly productive, which makes it an excellent material to harvest for eco-friendly materials. Due to its high yield and rapid growth, hemp is an appealing choice for biocomposite production, especially within the food industry where there is a dire need for sustainable packaging solutions [36]. It is very environmentally friendly and biodegradable so it is most applicable for the food industry uses. Due to the adverse effects that petroleum-based plastics have on the environment, and given their lack of biodegradability, Hemp takes a big step towards this by replacing these with hemp-based materials. Hemp cultivation has additional environmental benefits such as improved soil health and low chemical inputs. Apart from the sustainability of hemp fibers, a highly environmentally friendly dissolution method N-methylmorpholine N-oxide for all-cellulose biocomposites was also used. This important contribution, will in the future become even more significant as N-methyl morpholine N-oxide is an active solvent for cellulose and a process permitting the dissolution of celluloses without stressing to the environment. N-methyl morpholine N-oxide can return over 99% yields with negligible waste generation due to its ability for solvent recovery, making it one of the most attractive aspects inherent in hot water extraction. Per McGehee, scaling down the other forms of manufacture only cuts costs so much because flexible copper indium gallium selenide can be made in a continuous roll-to-roll process.

The NMMO dissolution process is associated with low waste production and high solvent recovery rates that show the potential of it for some real-world applications [37]. This method is adaptable to industrially produced cellulose biocomposites and facilitates the manufacture of high-performance materials at limited environmental costs. These results indicate that the combination of cellulose and hemp fibers, as well as processing under environmentally friendly conditions can be used to successfully produce sustainable materials for diverse applications from food packaging industries. For example, research has surfaced on the potential use of hemp fibers in bio-fiber composites with promising strength and degradability properties for food packaging. This was done using the N-methyl morpholine N-oxide dissolution, added to improvements in their processing these biocomposites have environmentally friendly impacts.

7 Other Cellulose Based Biocomposites

7.1 Oil Palm Cellulose Polylactic Acid Biocomposite

Polylactic Acid is a sustainable polymer that offers excellent mechanical strength and easy processing, unlike other biopolymers. Thermoplastic and aliphatic polyester produced from renewable resources provides a sustainable alternative to petrochemical-based products. Studies indicate that polylactic acid is an inexpensive packaging material that requires minimal resources. However, modifications can be made that improve the traits of this polylactic acid, which is employed in multiple uses. A different possibility is to bolster polylactic acid through additional polymers or fibers (Table 1).

Renewable cellulose nanofibers are an increasingly common choice for micro-sized reinforcements in composite materials. One issue when employing cellulose nanofibers in composites is due to their non-uniform incidence. Their polar surfaces render it difficult to disperse equally in a nonpolar fluid. The cellulose nanofibers have an extensive surface area and tend to agglomerate when dried. This problem can be avoided by mixing in a suitable medium and later fed into an extruder. Since there are considerably fewer steps required, this process approach has the advantage of yielding cellulose nanofiber composites with greater efficiency. Cellulose from oil palm mesocarp fiber can be nano fibrillated et al. For large-scale nanocomposite manufacture, it would be advantageous to be able to nano-fibrillate cellulose from oil palm mesocarp fiber and then melt compound the generated cellulose nanofibers with cellulose nanofibers in the same extruder, as extrusion will allow for high solid content processing while providing a continuous process.

Table 1 Comparison between various cellulose biocomposites

Type of biocomposite	Constituents	Properties	Applications	Citations
Cellulose/ nanoclay	Cellulose, Nanoclay	High thermal stability, improved mechanical properties	Packaging, automotive parts	[52]
Cellulose/ chitosan	Cellulose, chitosan	Antimicrobial, biodegradable, good mechanical properties	Biomedical, food packaging	[53]
Cellulose/ starch	Cellulose, starch	Biodegradable, renewable, good mechanical properties	Packaging, agriculture	[54]
Cellulose/PLA	Cellulose, polylactic acid (PLA)	Biodegradable, high-strength, renewable	Packaging, biomedical, automotive	[55]
Cellulose/PVA	Cellulose, polyvinyl alcohol (PVA)	Good film-forming ability, high mechanical strength, biodegradable	Packaging, biomedical	[56]

7.1.1 Preparation

After the collection of oil palm mesocarp fibers, they are subjected to disintegration, washing, and subsequently, drying. It is chemically treated to isolate the cellulose fraction. This preparation involves two steps: removing hemicellulose and lignin. A 5% wt Sodium Chlorite solution is used to accomplish the process of delignification with a pH value between 4 anc=d 5. The fiber is immersed in this solution at 70 °C for 1.5 h which is then purified by deionized water until a neutral value of pH is reached. The addition of cellulose nanofiber-oil palm mesocarp fiber in the PLA matrix enhanced the mechanical properties and increased the rigidity of the biocomposite. In addition, the incorporation of cellulose nanofiber-oil palm mesocarp significantly decreased the water contact angle compared with pristine polylactic acid, which could increase surface hydrophilicity in biocomposite material. In packaging applications, superior initial wettability is important in providing hydrophilic properties [41].

7.2 Chitosan Reinforced Nanocellulose Biocomposite

The natural fiber, chitosan is similar to cellulose produced from chitin. Since it is biocompatible, renewable, and advantageous to the environment, its use as a polymer is becoming more and more popular. However, its low mechanical and thermal endurance, dimension of particles, and surface area hinder its industrial usage. Thus, to improve these attributes, cellulose and various nanoparticles can be used. Nanoscale cellulosic fibers optimize the physical properties of chitosan composite films, making them preferable for the environment. Chitosan-based composites have numerous applications, including biomedical, packaging, and water treatment. Cellulose nanowhiskers are beneficial fillers for nanocomposite owing to their superior specific strength and modulus, ease of processing, and substantial reactive surface area. Cellulose nanowhiskers reinforced chitosan biocomposite is a promising future-oriented packaging material due to its sustainability and biodegradability, which are important considerations when selecting polymeric packaging materials. It can be widely used in the food packaging business. For example, Cellulose Nanocomposite incorporation in the chitosan matrix allowed us to obtain active films which resulted in increased chicken meat shelf-life [42–45].

7.2.1 Preparation of Chitosan Cellulose Nanofibers

Numerous pre-treatments are routinely performed before isolating cellulose nanocomposite. Cellulose nanocomposite-reinforced chitosan films treated with alkali or xylanase enzymes exhibit higher wet and dry tensile strength when compared with untreated chitosan. Moreover, treated cellulose nanocomposite-reinforced chitosan nanocomposites demonstrated improved storage modulus and glass transition temperature compared to untreated nanocomposites. This is one of the methods used to prepare chitosan cellulose nanofibers as sustainable packaging films [46].

7.3 Cellulose Alginate Biocomposite Films

Alginate is an additional organic biopolymer, primarily consisting of α-L-guluronic acid units and (1,4)-linked β-D-mannuronic acid units. Food packaging, tissue engineering, biomedicine, and pharmaceuticals have all employed alginate-based biocomposites owing to their unique gel-forming characteristics, non-toxicity, and biodegradability. The application of it is restricted by its limited mechanical and water-resistant traits, especially when there is moisture and humidity prevalent. This problem can be overcome by incorporating cellulose nanofibers in the alginate matrix. In general, nanocomposites have superior qualities when compared to composites reinforced with fillers, either small or large. An increase in the amount of interfacial interaction between the filler and the polymer matrix results in a significant improvement in the mechanical and barrier properties of nanofillers, given that they have a bigger surface area/volume ratio than micro or macro fillers [47, 48].

7.3.1 Preparation

Sodium Alginate can be obtained and cellulose nanofiber can be isolated. Typically, a good aspect ratio is preferred to make biocomposites with impeccable mechanical properties. The solvent-casting approach can be used to produce biocomposite films composed of cellulose and alginate. Alginate is gently dissolved in deionized water and is evenly dispersed. Subsequently, the prepared cellulose nanofiber solution can be appropriately added to the required concentrations for making the nanocomposite films for packaging [48, 49].

7.4 PHA Cellulose Biocomposite

Polyhydroxyalkanoate is a viable biopolymer for food packaging. It is a thermoplastic material. It outperforms some commodities, notably polypropylene, in terms of rigidity. In addition to this, polyhydroxyalkanoate use has the advantage, which is highly sustainable. Physical aging of polyhydroxyalkanoate has been reported to increase its inherent brittleness with time. Thus, it reduces with time due to physical aging. Moreover, the high crystallinity of polyhydroxyalkanoate restricts the material's processing window, which limits its suitability for some food packaging products. Finally, the limited lifetime and expensive nature of this material do not provide the possibility to consider packaging. Cellulose, producing high-performance fillers, is a great opportunity for optimizing resources, and it may be a sustainable solution for developing packaging though it may be difficult to use both on sturdiness and availability of strength, and provides a rigid barrier. Nanocellulose can play a valuable part in packaging applications as an innovative and sustainable oxygen barrier material [50, 51] (Table 1).

8 Various Other Applications of Cellulose-Based Biocomposites

8.1 Medical Application

A biopolymer produced by microorganisms, bacterial cellulose holds great potential for use in diverse industrial applications. Although bacterial cellulose has distinct mechanical, chemical, and physical characteristics, its chemical structure is similar to that of cellulose derived from plants. This exhibits excellent biocompatibility and does not trigger any allergic responses or cytotoxicity. The attributes of bacterial cellulose can be further improved by incorporating other biocomposites such as chitosan, collagen, and alginate. Bacterial cellulose along with chitosan and alginate have excellent antibacterial properties, inhibiting bacterial growth. This inherent property makes this biocomposite viable for treating wounds. Cellulose's surface comprises hydroxyl groups that are capable of adsorbing metal ions and metal oxide nanoparticles, creating hybrid materials. Bacterial cellulose with silver metallic particles can be used to formulate an antibacterial and healing product for wounds. Metallic silver with its components is effective at inhibiting the growth of many types of harmful bacteria. Silver ion reacts to the bacteria with their thiol groups which changes the whole structure of how the bacteria respirates and their substance transfer across the membrane. Thus, they are very suitable for biomedical use such as dressings. The bacterial cellulose-silver nanoparticle composites have distinct physicochemical and mechanical properties that can be tuned to specific applications, such as medical dressings. These antibacterial properties coupled with their biocompatibility make bacterial cellulose biocomposites suitable for their utilization in the medical industry [57–63].

8.2 Textiles

The technological innovation can be used in the textile industry by coating nanofibers on the fibers, yarn, or fabric to improve their surface area and mechanical properties. Adding natural fibers such as hemp and sisal to biopolymers improves their interfacial adhesion dramatically. The use of regenerated cellulose fibers has been a game changer in the textile industry having a stupendous biocompatibility. Microfibers based on oriented bacterial cellulose nanofibers have intriguing applications in smart fabrics and structural reinforcement. Such technological innovation entails developing a renewable, biodegradable, and environmentally friendly textile material. The material is developed by microbes and has several distinctive features, including high crystallinity, specific surface area, water absorption, and mechanical strength [63–65].

8.3 Electronics

Cellulose and chitosan are the most widespread biopolymers on Earth. They are renewable, biocompatible, and biodegradable. Foldable electrical conductors fabricated from bio-based materials are growing popular due to their biocompatibility, modifiable biodegradation rate, and lightweight design. Among biodegradable polymers, cellulose is the most prevalent. In addition to its application in paper-based electronics, cellulose is also becoming a vital material for novel applications such as functional, flexible, and foldable electronics. The biocomposites display metal-like electrical conductivity, which makes it integral for their use in electronic devices. Biocomposites are extremely malleable and foldable, with conductivity equivalent to thin metal foils. Optimized polymer-graphene compositions lead to outstanding electrical conductivity, with resistance values around 10 Ω and conductivities above 1000 $S\ m^{-1}$. Young's modulus could be enhanced to 508 MPa after impregnation of cellulose acetate polymer, indicating that a designed system with this particular biocomposite may sustain appreciable stresses before breaking. Their durability is also very good, exhibiting resilient mechanical and conductive behavior even after over 10k folding-unfolding cycles as well as for abrasion testing. These qualities

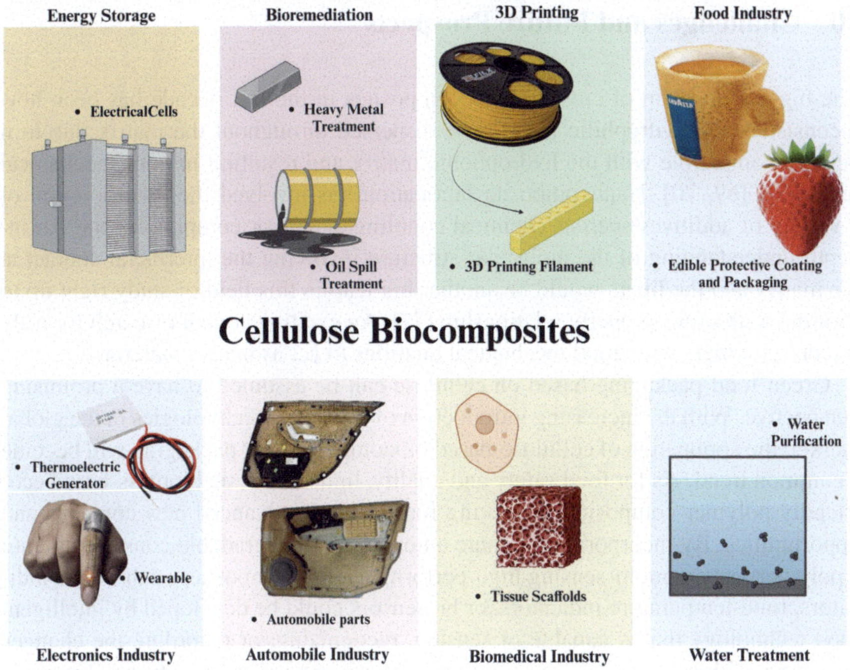

Fig. 4 Other applications of cellulose biocomposites

make biocomposites ideal for high-speed applications, wearable electronics, and electrodiagnostic medicine [66, 67] (Fig. 4).

9 Sustainability of Using Cellulose-Based Biocomposites

With so many environmental and economic advantages, cellulose-based biocomposites mark a major development in sustainable materials science. Conjured from renewable resources like plant fibers, these biocomposites show a low carbon footprint and biodegradability relative to conventional petroleum-based polymers. The disposal of cellulose biocomposites at the end of their life is an integral part of sustainability. Cellulose biocomposites unlike day-to-day plastics decompose into non-toxic products that can be metabolized by natural processes. The effect of this on the environment is notable and also goes a long way to help reduce the spread of plastic. The sustainability of cellulose-based biocomposites is demonstrated through it being renewable, biodegradable, and green manufacturing which is still getting more advanced and less impacting. In any direction, this material is a good example of how the generation must move to make everything better for the environment and their lives [68].

10 Challenges and Future Prospects

The biggest problem of creating nanocomposites in the last decade has been how inconsistent the hydrophilic fibers have dispersed throughout the matrix due to it being incompatible with the hydrophobic matrix and resulting in poor mechanical properties [69, 70]. Nanocomposite fabrication has involved the incorporation of a variety of additives such as chemical coupling agents or compatibilizers. An in-depth understanding of the molecular structure involving the interfacial contact at the matrix and the fibers would be another big leap in this field of study right up to forming a structure-property relationship [71]. An exciting area of research for truly 'green' polymers with good mechanical qualities to act as matrix material.

Green food packaging based on cellulose can be assumed to have a promising perspective. With the increasing introduction of other new technologies in the global market, the application of cellulose-based biocomposites for packaging will become a common trend. As for food safety and quality, lignocellulosic biomass-based eco-friendly polymer composites also bring forth a lot of advanced new concepts and opportunities. By incorporating minute amounts of biodegradable constituents into a polymer matrix, multi-sensing high-performance biocomposites such as gas indicators, time-temperature indicators, or biosensors could be developed by intelligent food technology that is capable of sensing, recognizing, or recording the changes occurring within or outside the product. According to experts, advanced innovative biocomposites such as novel antimicrobials and gas scavengers would also be possible to be incorporated into packaging films through nanoparticles in the next generation of food packaging technology.

11 Conclusions

Synthetic polymeric materials will inevitably be replaced by biodegradable polymers in certain applications. The fact that they are derived from renewable natural resources makes them important. Farmer economics would benefit from the use of cellulose-based products since food packaging and other industries require it in particular. Thus, there would be scope for greater use of cellulose-based mulching films, bags, and other biodegradables in the horticultural and agricultural sectors for short-term application, which could also be attractive to farmers. In this way, non-biodegrading synthetic polymers as well as potential environmental polluting materials can also be reduced. Although synthetic polymers must be phased out as far as packaging is concerned, the customers' concerns regarding environmental pollution due to packaging are far less when compared to the cost of the packaging material. The development of biodegradable polymer materials for use in disposable electrode systems (used in medical applications) and food spoilage detection cheaper packaging materials need to be encouraged also. These two aspects will require interdisciplinary research studies for their development. Finally, the successful implementation and commercialization of biopolymer-based eco-friendly packaging materials will depend on the collective efforts made by scientists working in various fields such as chemistry, polymer technology, microbiology, chemical engineering, and environmental science to mention only a few along with the involvement of bureaucrats for regulation and implementation.

References

1. Wang, W., Liu, X., Guo, F., Yu, Y., Lu, J., Li, Y., Cheng, Q., Peng, J., Yu, G.: Biodegradable cellulose/curcumin films with Janus structure for food packaging and freshness monitoring. Carbohydr. Polym. **324**, 121516 (2024). https://doi.org/10.1016/j.carbpol.2023.121516
2. Gao, Y., Huang, C., Ge, D., Liao, Y., Chen, Y., Li, S., Yu, H.: Highly efficient dissolution and reinforcement mechanism of robust and transparent cellulose films for smart packaging. Int. J. Biol. Macromol. **254**, 128046 (2024). https://doi.org/10.1016/j.ijbiomac.2023.128046
3. Yue, C., Wang, M., Zhou, Z., You, Y., Wang, G., Wu, D.: Cellulose-based intelligent packaging films with antibacterial, UV-blocking, and biodegradable properties for shrimp freshness monitoring. J. Chem. Eng. **488**, 150975 (2024). https://doi.org/10.1016/j.cej.2024.150975
4. Irimia, A., Grigoraş, V.C., Popescu, C.: Active cellulose-based food packaging and its use on foodstuff. Polymers **16**(3), 389 (2024). https://doi.org/10.3390/polym16030389
5. Zhang, X., Fang, C., Cheng, Y., Li, M., Liu, J.: Fine extraction of cellulose from corn straw and the application of eco-friendly packaging films enhanced with polyvinyl alcohol. Int. J. Biol. Macromol. **268**, 131984 (2024). https://doi.org/10.1016/j.ijbiomac.2024.131984
6. Benitez, J.J., Florido-Moreno, P., Porras-Vázquez, J.M., Tedeschi, G., Athanassiou, A., Heredia-Guerrero, J.A., Guzman-Puyol, S.: Transparent, plasticized cellulose glycerol bioplastics for food packaging applications. Int. J. Biol. Macromol. **273**, 132956 (2024). https://doi.org/10.1016/j.ijbiomac.2024.132956
7. Tang, J., Kuo, C., Liao, Y.: Transparent biodegradable composite plastic packaging film from TEMPO-oxidized cellulose nanofibers. Int. J. Biol. Macromol. **260**, 129502 (2024). https://doi.org/10.1016/j.ijbiomac.2024.129502
8. Riahi, Z., Khan, A., Rhim, J., Shin, G.H., Kim, J.T.: Sustainable packaging a film based on cellulose nanofibres/pullulan impregnated with zinc-doped carbon dots derived from avocado peel to extend the shelf life of chicken and tofu. Int. J. Biol. Macromol. **258**, 129302 (2024). https://doi.org/10.1016/j.ijbiomac.2024.129302
9. Mohammed, K., Yu, D., Mahdi, A.A., Zhang, L., Obadi, M., Al-Ansi, W., Xia, W.: Influence of cellulose viscosity on the physical, mechanical, and barrier properties of the chitosan-based films. Int. J. Biol. Macromol. **259**, 129383 (2024). https://doi.org/10.1016/j.ijbiomac.2024.129383
10. Lu, Z., Zhang, H., Toivakka, M., Xu, C.: Current progress in functionalization of cellulose nanofibers (CNFs) for active food packaging. Int. J. Biol. Macromol. **267**, 131490 (2024). https://doi.org/10.1016/j.ijbiomac.2024.131490
11. Zhu, H., Cheng, J., Han, Z.: Construction of a sustainable and hydrophobic high-performance all-green pineapple peel cellulose nanocomposite film for food packaging. Int. J. Biol. Macromol. **256**, 128396 (2024). https://doi.org/10.1016/j.ijbiomac.2023.128396
12. Chen, R., Zhang, Y., Zhang, Z., Yao, L., Liu, L., Wang, J., Wang, R.: Open roads and bridge: preservation of fresh beef by a packaging film constructed from photosensitizing bacterial cellulose. Food Chem. **437**, 137789 (2024). https://doi.org/10.1016/j.foodchem.2023.137789

13. Chen, G., Wang, K., Chen, P., Cai, D., Shao, Y., Xia, R., Li, C., Wang, H., Ren, F., Cheng, X., Yu, Y.: Fully biodegradable packaging films for fresh food storage based on oil-infused bacterial cellulose. Adv. Sci. **11**, e2400826 (2024). https://doi.org/10.1002/advs.202400826
14. Yadav, M., Maurya, A.K., Behera, K., Chiu, F., Rhee, K.: Physical properties of cellulose nanocrystal/magnesium oxide/chitosan transparent composite films for packaging applications. Int. J. Biol. Macromol. **264**, 130560 (2024). https://doi.org/10.1016/j.ijbiomac.2024.130560
15. Zhu, P., Vo, A., Sun, X., Zhang, Y., Mandegari, M., Zargar, S., Tu, Q., Zhu, J., Yu, Z., Sun, H., Zheng, D., Jiang, F.: The water-induced controllable deswelling strategy enabled the rapid fabrication of transparent cellulose film for plastic replacement. J. Chem. Eng. **492**, 152200 (2024). https://doi.org/10.1016/j.cej.2024.152200
16. Hoque, M., Babu, R.P., McDonagh, C., Jaiswal, S., Tiwari, B.K., Kerry, J.P., Pathania, S.: Pectin/sodium alginate-based active film integrated with microcrystalline cellulose and geraniol for food packaging applications. Int. J. Biol. Macromol. **271**, 132414 (2024). https://doi.org/10.1016/j.ijbiomac.2024.132414
17. Momzyakova, K.S., Deberdeev, T.R., Valishina, Z.T., Deberdeev, R.Ya., Ibragimov, A.V., Alexandrov, A.V.: Research of physical and chemical properties of powder cellulose from various types of raw materials. MSF **992**, 791–795 (2020). https://doi.org/10.4028/www.scientific.net/MSF.992.791
18. Zhao, J., Zhang, W., Zhang, X., Zhang, X., Lu, C., Deng, Y.: Extraction of cellulose nanofibrils from dry softwood pulp using high shear homogenization. Carbohydr. Polym. **97**, 695–702 (2013). https://doi.org/10.1016/j.carbpol.2013.05.050
19. Hussin, S.M., Al-Obaidi, Y.M.: Extraction of cellulose nanocrystalline from cotton by ultrasonic and its morphological and structural characterizationInt. J. Mater. Chem. Phys. **1**(2), 99–109 (2015)
20. Wang, X., Li, H., Cao, Y., Tang, Q.: Cellulose extraction from wood chip in an ionic liquid 1-allyl-3-methylimidazolium chloride (AmimCl). Bioresour. Technol. **102**(17), 7959–7965 (2011)
21. Sun, J.X., Sun, X.F., Zhao, H., Sun, R.C.: Isolation and characterization of cellulose from sugarcane bagasse. Polym. Degrad. Stab. **84**(2), 331–339 (2004)
22. Johar, N., Ahmad, I., Dufresne, A.: Extraction, preparation, and characterization of cellulose fibers and nanocrystals from rice husk. Ind. Crops Prod. **37**(1), 93–99 (2012)
23. Morán, J.I., Alvarez, V.A., Cyras, V.P., Vázquez, A.: Extraction of cellulose and preparation of nanocellulose from sisal fibers. Cellulose **15**, 149–159 (2008)
24. Sheltami, R.M., Abdullah, I., Ahmad, I., Dufresne, A., Kargarzadeh, H.: Extraction of cellulose nanocrystals from mengkuang leaves (Pandanus tectorius). Carbohydr. Polym. **88**(2), 772–779 (2012)
25. Nuruddin, M., Chowdhury, A., Haque, S.A., Rahman, M., Farhad, S.F., Sarwar Jahan, M., Quaiyyum, A.: Extraction and characterization of cellulose microfibrils from agricultural wastes in an integrated biorefinery initiative. Biomaterials **3**, 5–6 (2011)
26. Martínez-Sanz, M., Lopez-Rubio, A., Lagaron, J.M.: Optimization of the nanofabrication by acid hydrolysis of bacterial cellulose nanowhiskers. Carbohydr. Polym. **85**(1), 228–236 (2011)
27. Kumar, A., Negi, Y.S., Choudhary, V., Bhardwaj, N.K.: Characterization of cellulose nanocrystals produced by acid-hydrolysis from sugarcane bagasse as agro-waste. J. Mater. Phys. Chem. **2**(1), 1–8 (2014)
28. Hafid, H.S., Omar, F.N., Zhu, J., Wakisaka, M.: Enhanced crystallinity and thermal properties of cellulose from rice husk using acid hydrolysis treatment. Carbohydr. Polym. **260**, 117789 (2021)
29. Lima, C.S.S., Conceição, M.M., Silva, F.L.H., Lima, E.E., Conrado, L.S., Leão, D.A.S.: Characterization of acid hydrolysis of sisal. Appl. Energy **102**, 254–259 (2013)
30. Seltami, R., Ahmad, I.: Acid hydrolysis process of mengkuang cellulose. Malays. NANO-Int. J. **1**(2), 1–12 (2021)
31. Liu, C., Li, B., Du, H., Lv, D., Zhang, Y., Yu, G., Mu, X., Peng, H.: Properties of nanocellulose isolated from corncob residue using sulfuric acid, formic acid, oxidative and mechanical methods. Carbohydr. Polym. **151**, 716–724 (2016)

32. Mantovan, J., Giraldo, G.A.G., Marim, B.M., Kishima, J.O.F., Mali, S.: Valorization of orange bagasse through one-step physical and chemical combined processes to obtain a cellulose-rich material. J. Sci. Food Agric. **101**(6), 2362–2370 (2021)
33. Movva, M., Kommineni, R.: Extraction of cellulose from pistachio shell and physical and mechanical characterization of cellulose-based nanocomposites. Mater. Res. Express **4**(4), 045014 (2017)
34. Ren, H., Shen, J., Pei, J., Wang, Z., Peng, Z., Fu, S., Zheng, Y.: Characteristic microcrystalline cellulose extracted by combined acid and enzyme hydrolysis of sweet sorghum. Cellulose **26**, 8367–8381 (2019)
35. Zhao, G., Lyu, X., Lee, J., Cui, X., Chen, W.-N.: Biodegradable and transparent cellulose film prepared eco-friendly from durian rind for packaging application. Food Packag. Shelf Life **21**, 100345 (2019). https://doi.org/10.1016/j.fpsl.2019.100345
36. Ouajai, S., Shanks, R.A.: Preparation, structure and mechanical properties of all-hemp cellulose biocomposites. Compos. Sci. Technol. **69**(13), 2119–2126 (2009)
37. Dahou, W., Ghemati, D., Oudia, A., Aliouche, D.: Preparation and biological characterization of cellulose graft copolymers. Biochem. Eng. J. **48**(2), 187–194 (2010)
38. Zainal, S.H., Mohd, N.H., Suhaili, N., Anuar, H., Lazim, A.M., Othaman, R.: Preparation of cellulose-based hydrogel: a review. J. Mater. Res. Technol. **10**, 935–952 (2021)
39. Gao, Q., Shen, X., Lu, X.: Regenerated bacterial cellulose fibers prepared by the NMMO· H2O process. Carbohydr. Polym. **83**(3), 1253–1256 (2011)
40. López De Dicastillo, C., Rodríguez, F., Guarda, A., Galotto, M.J.: Antioxidant films based on cross-linked methyl cellulose and native Chilean berry for food packaging applications. Carbohydr. Polym. **136**, 1052–1060 (2016). https://doi.org/10.1016/j.carbpol.2015.10.013
41. Liu, Y., Ahmed, S., Sameen, D.E., Wang, Y., Lu, R., Dai, J., Li, S., Qin, W.: A review of cellulose and its derivatives in biopolymer-based for food packaging application. Trends Food Sci. Technol. **112**, 532–546 (2021). https://doi.org/10.1016/j.tifs.2021.04.016
42. Abdul Khalil, H.P.S., Saurabh, C.K., Adnan, A.S., Fazita, N., Syakir, M., Davoudpour, Y., Rafatullah, M., Abdullah, C., Haafiz, M.M., Dungani, R.: A review on chitosan-cellulose blends and nanocellulose reinforced chitosan biocomposites: properties and their applications. Carbohydr. Polym. **150**, 216–226 (2016). https://doi.org/10.1016/j.carbpol.2016.05.028
43. Di Liberto, E.A., Dintcheva, N.T.: Biobased films based on chitosan and microcrystalline cellulose for sustainable packaging applications. Polym. J. **16**(5), 568 (2024). https://doi.org/10.3390/polym16050568
44. Rong, S.Y., Mubarak, N., Tanjung, F.A.: Structure-property relationship of cellulose nanowhiskers reinforced chitosan biocomposite films. J. Environ. Chem. Eng. **5**(6), 6132–6136 (2017). https://doi.org/10.1016/j.jece.2017.11.054
45. Casalini, S., Baschetti, M.G.: The use of essential oils in chitosan or cellulose-based materials for the production of active food packaging solutions: a review. J. Sci. Food Agric. **103**(3), 1021–1041 (2022). https://doi.org/10.1002/jsfa.11918
46. Hassan, M.L., Hassan, E.A., Oksman, K.N.: Effect of pretreatment of bagasse fibers on the properties of chitosan/microfibrillated cellulose nanocomposites. J. Mater. Sci. **46**(6), 1732–1740 (2010). https://doi.org/10.1007/s10853-010-4992-4
47. Sirviö, J.A., Kolehmainen, A., Liimatainen, H., Niinimäki, J., Hormi, O.E.: Biocomposite cellulose-alginate films: promising packaging materials. Food Chem. **151**, 343–351 (2014). https://doi.org/10.1016/j.foodchem.2013.11.037
48. Deepa, B., Abraham, E., Pothan, L., Cordeiro, N., Faria, M., Thomas, S.: Biodegradable nanocomposite films based on sodium alginate and cellulose nanofibrils. Materials **9**(1), 50 (2016). https://doi.org/10.3390/ma9010050
49. Abdollahi, M., Alboofetileh, M., Behrooz, R., Rezaei, M., Miraki, R.: Reducing water sensitivity of alginate bio-nanocomposite film using cellulose nanoparticles. Int. J. Biol. Macromol. **54**, 166–173 (2013). https://doi.org/10.1016/j.ijbiomac.2012.12.016
50. Bhardwaj, U., Dhar, P., Kumar, A., Katiyar, V.: Polyhydroxyalkanoates (PHA)-cellulose-based nanobiocomposites for food packaging applications. In: ACS Symposium Series, pp. 275–314 (2014). https://doi.org/10.1021/bk-2014-1162.ch019

51. De Carvalho, K.C.C., Montoro, S.R., Cioffi, M.O.H., Voorwald, H.J.C.: Polyhydroxyalkanoates and Their Nano Biocomposites with Cellulose Nanocrystals. Elsevier eBooks, pp. 261–285 (2016). https://doi.org/10.1016/b978-0-323-39408-6.00012-1
52. Smith, J., Doe, J., Johnson, M.: Enhanced properties of cellulose/nanoclay biocomposites. J. Appl. Polym. Sci. **150**(3), 50203 (2020). https://doi.org/10.1002/app.50203
53. Jones, E., Brown, R., Green, S.: Antimicrobial and mechanical properties of cellulose/chitosan biocomposites. Carbohydr. Polym. **215**, 520–528 (2019). https://doi.org/10.1016/j.carbpol.2019.04.052
54. Lee, D., White, A., Harris, K.: Biodegradable cellulose/starch composites for packaging applications. Polym. Compos. **39**(2), 24780 (2018). https://doi.org/10.1002/pc.24780
55. Wang, L., Thompson, W., Martin, A.: Renewable and biodegradable cellulose/PLA composites. Green Chem. **23**(5), 1407–1415 (2021). https://doi.org/10.1039/D1GC01407H
56. Chen, T., Watson, E., Lee, J.: High-strength cellulose/PVA biocomposites for biomedical applications. J. Mater. Sci. **52**(9), 1702–1710 (2017). https://doi.org/10.1007/s10853-017-1702-3
57. Volova, T.G., Shumilova, A.A., Shidlovskiy, I.P., Nikolaeva, E.D., Sukovatiy, A.G., Vasiliev, A.D., Shishatskaya, E.I., Siberian Federal University, Institute of Biophysics SB RAS, & Institute of Physics. L.V. Kirensk SB RAS: Antibacterial properties of films of cellulose composites with silver nanoparticles and antibiotics. Siberian Federal University Digital Repository (2021). https://core.ac.uk/download/pdf/220103896.pdf
58. Saska, S., Barud, H.S., Gaspar, A.M.M., Marchetto, R., Ribeiro, S.J.L., Messaddeq, Y.: Bacterial cellulose-hydroxyapatite nanocomposites for bone regeneration. Int. J. Biomater. **2011**, 1–8 (2011). https://doi.org/10.1155/2011/175362
59. Kwak, M.H., Kim, J.E., Go, J., Koh, E.K., Song, S.H., Son, H.J., Kim, H.S., Yun, Y.H., Jung, Y.J., Hwang, D.Y.: Bacterial cellulose membrane produced by Acetobacter sp. A10 for burn wound dressing applications. Carbohydr. Polym. **122**, 387–398 (2015). https://doi.org/10.1016/j.carbpol.2014.10.049
60. Chang, W., Chen, H.: Physical properties of bacterial cellulose composites for wound dressings. Food Hydrocoll. **53**, 75–83 (2016). https://doi.org/10.1016/j.foodhyd.2014.12.009
61. Shah, N., Ul-Islam, M., Khattak, W.A., Park, J.K.: Overview of bacterial cellulose composites: a multipurpose advanced material. Carbohydr. Polym. **98**(2), 1585–1598 (2013). https://doi.org/10.1016/j.carbpol.2013.08.018
62. Percival, S., Bowler, P., Russell, D.: Bacterial resistance to silver in wound care. J. Hosp. Infect. **60**(1), 1–7 (2005). https://doi.org/10.1016/j.jhin.2004.11.014
63. Da Silva, C.J.G., De Medeiros, A.D.M., De Amorim, J.D.P., Nascimento, H.A.D., Converti, A., Costa, A.F.S., Sarubbo, L.A.: Bacterial cellulose textiles for the future of sustainable fashion: a review. Environ. Chem. Lett. **19**(4), 2967–2980 (2021). https://doi.org/10.1007/s10311-021-01214-x
64. Fernandes, M., Souto, A.P., Dourado, F., Gama, M.: Application of bacterial cellulose in the textile and shoe industry: development of biocomposites. Polysaccharides **2**(3), 566–581 (2021). https://doi.org/10.3390/polysaccharides2030034
65. Klemm, D., Heublein, B., Fink, H., Bohn, A.: Cellulose: fascinating biopolymer and sustainable raw material. Angew. Chem. **44**(22), 3358–3393 (2005). https://doi.org/10.1002/anie.200460587
66. Arantes, A.C.C., Silva, L.E., Wood, D.F., Almeida, C.D.G., Tonoli, G.H.D., De Oliveira, J.E., Da Silva, J.P., Williams, T.G., Orts, W.J., Bianchi, M.L.: Bio-based thin films of cellulose nanofibrils and magnetite for potential application in green electronics. Carbohydr. Polym. **207**, 100–107 (2019). https://doi.org/10.1016/j.carbpol.2018.11.081
67. Cataldi, P., Bonaccorso, F., Del Rio Castillo, A.E., Pellegrini, V., Jiang, Z., Liu, L., Boccardo, N., Canepa, M., Cingolani, R., Athanassiou, A., Bayer, I.S.: Cellulosic graphene biocomposites for versatile high-performance flexible electronic applications. Adv. Electron. Mater. **2**(11) (2016). https://doi.org/10.1002/aelm.201600245
68. Khalil, H.A., Bhat, A., Yusra, A.I.: Green composites from sustainable cellulose nanofibrils: a review. Carbohydr. Polym. **87**(2), 963–979 (2012). https://doi.org/10.1016/j.carbpol.2011.08.078

69. Ashori, A.: Wood–plastic composites as promising green composites for automotive industries! Bioresour. Technol. **99**(11), 4661–4667 (2008). https://doi.org/10.1016/j.biortech.2007.09.043
70. John, M., Thomas, S.: Biofibres and biocomposites. Carbohydr. Polym. **71**(3), 343–364 (2008). https://doi.org/10.1016/j.carbpol.2007.05.040
71. Kim, J., Yoon, T., Mun, S., Rhee, J., Lee, J.: Wood–polyethylene composites using ethylene–vinyl alcohol copolymer as an adhesion promoter. Bioresour. Technol. **97**(3), 494–499 (2006). https://doi.org/10.1016/j.biortech.2005.02.048

Emerging Approaches in Hydrogen Production from Biomass

Niranjan Patra and Shaik Salma Banu

Abstract Hydrogen production from biomass is an appealing prospect as a long-term, eco-friendly energy option. However, there are challenges associated with current methods such as low conversion efficiency and the formation of byproducts. To address these issues, several emerging approaches have been proposed for producing hydrogen from biomass. This chapter provides an overview of these emerging approaches, which includes anaerobic digestion combined pyrolysis, pyrolysis combined with gasification and pyrolysis in the presence of catalyst are revisited and discussed in details. This paper also highlights the technical and economic challenges that need to be resolved in order to have a successful rollout of these emerging approaches. Overall, this chapter aims to provide a critical analysis of the state-of-the-art and future prospects of emerging approaches in hydrogen production from biomass.

Keywords Hydrogen production · Biomass · Emerging approaches

1 Introduction

Petroleum fuels accounted for only around four percent of the world's energy demand in the early 20th century, but today they are a major source of energy, providing about fourty percent. Moreover, roughly 96% of all transportation fuels originate from crude oil [1]. In 2018, 97.6 million barrels of fossil fuels and their derivatives per day was used globally, and this number is expected to rise above 100 million barrels per day by 2035 [2]. Nevertheless, fuels derived from petroleum are both expensive and finite. Carbon dioxide and other pollutants are produced when fossil fuels are burned continuously, posing major problems for the environment. New projections put the annual rate of increase in releases of carbon dioxide from the combustion of fossil fuels and cement production at 1.0% in 2022, for a total of 36.6 billion tonnes of CO_2 [3]. As the world's population approaches 9 billion in 2050, experts predict

N. Patra (✉) · S. S. Banu
Department of Chemistry, Koneru Lakshmaiah Education Foundation, Green Field, Vaddeswaram 522502, Andhra Pradesh, India

K. K. Kesari et al. (eds.), *Functionalized Cellulose Materials*, Engineering Materials,
https://doi.org/10.1007/978-3-031-76953-5_5

that this predicament will only intensify. Because of these issues, it's become urgent to find renewable feedstocks/processes for the production of energy and fuels that won't harm the environment.

The use of hydrogen as a power source is being considered to tackle ever increasing price rise of conventional fuels, and the ever increasing pollution from the fossil fuel energy usage. Hydrogen produced from renewable sources is carbon neutral since the CO_2 released during combustion is taken up by plants during photosynthesis. Biomass has the ability to be a crucial resource for the production of alternative transportation fuels, electricity, heat, and valuable chemicals as a result of its low price and wide availability [1, 4–18]. Using biomass as an alternative energy source has many benefits, including being carbon neutral and producing less sulphur oxides and nitrogen oxides [1, 19–22]. Municipal trash, waste products from farms, forests, and herbaceous biomass processors are all examples of biomass that are typically considered waste and get rid of in different ways depending on the rules of the respective authorities [10, 18]. Thermochemical and biochemical conversion processes can change biomass into fuels and chemicals with significant added value [19, 23–29]. Biomass can be converted into chemicals, fuels, and energy through fungal, bacterial, and enzymatic treatments, while thermochemical conversion uses energy to quickly transforming vast quantities of biomass into useful forms of energy such as liquid, solid, and gaseous fuels or thermal energy [6, 7, 30]. The end product of thermochemical conversion processes is subject to a wide range of variables, biomass variety, reactor size, temperature, and process method are all part of the system.

Based on the temperature, pressure, heating rate, and reaction environment, thermochemical conversion processes can be broken down into four distinct categories: combustion/incineration, pyrolysis, liquefaction and gasification. The decomposition process of pyrolysis, which occurs in an oxygen free environment, has been widely employed to convert biomass into high-value end products [9, 11, 12, 24, 27, 28, 30–33]. Cellulose, hemicellulose, and lignin make up the bulk of biomass, with the cellulose microfibrils sitting in a lignin and hemicellulose-rich matrix. The decomposition of lignin, which influences cellulose degradation by creating a pore in the matrix, is slower between 350 and 500 °C. Moreover, lignin regulates hemicellulose thermal reactivity [29, 31, 34–37]. The pyrolysis regime is dominated by thermal depolymerization of organic molecules between 200 and 700 °C. Depending on the reaction circumstances, a wide variety of products can be produced, each of which has a specific mechanism of thermal cracking and multiple potential uses. Bio-oil, biochar, and gas, for example, have numerous applications, including those in the production of fuels and fuel blends, chemicals, soil amendment, catalysts, adsorbents, hydrogen [10, 14, 21, 38–41]. The development of novel reactor system designs and recovery system has increased interest in the creation of new by-products like methanol.

Despite the fact that numerous has been the subject of numerous reviews using both traditional and cutting-edge methods. Thus, this chapter makes an effort to come out with the recent emerging approaches used in hydrogen production from biomass feedstock. The work also involves highlighting potential challenges with the goal of expanding beyond the present constraints.

2 Prospects of Hydrogen Energy

Hydrogen energy has the potential to become a vital part of the future of sustainable energy. Here are some of the prospects of hydrogen energy:

I. Renewable energy storage: Hydrogen can be used as a means of storing renewable energy from sources such as wind and solar power. This can help to overcome the challenge of intermittency, allowing the energy to be stored and used when needed.
II. Zero-emissions transportation: Hydrogen fuel cell vehicles emit only water and heat, making them a promising alternative to traditional gasoline-powered vehicles. The transportation sector is a major contributor to global greenhouse gas emissions, so this shift to zero-emissions vehicles could have a significant impact on reducing carbon emissions.
III. Industrial processes: Hydrogen is already used in many industrial processes, including refining petroleum, producing ammonia for fertilizer, and manufacturing chemicals. By producing hydrogen from renewable sources, these processes could become more sustainable and reduce their carbon footprint.
IV. Electricity generation: Hydrogen can also be used in fuel cells to generate electricity, either for stationary applications such as backup power for buildings, or for mobile applications such as powering drones.
V. Energy security: Hydrogen can be produced domestically, which eliminates the need to rely on fossil fuels and improves the nation's overall energy security as well as safe guard environment.

However, there are also some challenges to be addressed in realizing the full potential of hydrogen energy, such as the cost of production, storage and transportation, as well as the need for widespread infrastructure development. Nonetheless, the potential benefits of hydrogen energy make it an area of growing interest and investment.

2.1 Hydrogen Production Methods

Hydrogen, a clean and abundant energy carrier, can be produced using a variety of methods. Here are some of the most common methods of hydrogen production:

Steam Reforming: Steam reforming is the most common method used for producing hydrogen. Partial oxidation, auto-thermal reforming, CO_2 reforming, and steam reforming of fossil fuels and biomass are all viable methods for generating hydrogen [3, 18, 23, 36, 42–56]. Since steam reformation can take place at room temperature and atmospheric pressure, it is a practical process for transforming fossil fuels into hydrogen and an attractive strategy for creating hydrogen from glycerol [23, 45, 46, 50, 53, 57]. Glycerol steam reforming (GSR), methane steam reforming (MSR), methanol/ethanol steam reforming, and sorption enhanced-methane steam reforming are the four subcategories of steam reforming [50, 57]. Commercial hydrogen generation utilising MSR from natural gas accounts for over 75% of the world's total hydrogen output annually. Nevertheless, steam reforming of fossil fuels produces considerable CO_2. Around 7 kilogrammes of carbon dioxide are released for every kilogramme of hydrogen, accounting for about 3 percent of all CO_2 emissions.

Electrolysis: Electrolysis is a process that uses an electric current to separate hydrogen and oxygen from water. This method can be powered by renewable sources of electricity such as wind and solar, making it a clean and sustainable method of hydrogen production. Water electrolysis is a commercially viable option because it is abundant in nature, produces no emissions, and has a pure by-product in the form of oxygen [55]. However, due to economic concerns, the proportion of hydrogen produced annually around the world via this approach has dropped to as low as 4 percent in recent years.

Gasification: Gasification is a process that involves the reaction of a solid or liquid fuel with steam and oxygen to produce a syngas, which contains hydrogen, carbon monoxide, and other gases. The syngas can then be separated and purified to produce hydrogen [6, 7, 11, 13, 17, 20, 27, 58–60].

Biomass Gasification: Biomass gasification is similar to gasification, but it uses biomass such as wood chips or agricultural waste as the feedstock instead of fossil fuels [24, 59].

Nuclear Energy: Nuclear energy can also be used to produce hydrogen. The process involves using the heat from a nuclear reactor to split water into hydrogen and oxygen. Photoelectrochemical (PEC) **Water Splitting**: PEC water splitting is a promising method that involves using sunlight to split water into hydrogen and oxygen [55]. The advantages of each of these approaches are distinct. Merits, demerits and the selection of method rely on factors such as availability of resources, cost, and environmental impact.

2.2 Hydrogen from Biomass Based Feedstocks

Since plants use the carbon dioxide (CO_2) released during combustion in the process of photosynthesis to build new biomass, biofuels made from biomass are thought to contain a net quantity of carbon (carbon neutral) [2, 16, 20]. Renewable feedstocks and techniques are separated from their fossil fuel counterparts into separate categories. Coal, oil, and natural gas are examples of fossil feedstock, while hydropower, rainwater, and biomass are examples of renewable feedstock. The renewable feedstock reviewed here comes primarily from biomass. Thermochemical conversion, biological conversion, and steam reforming of glycerol are the three methods of biomass conversion to hydrogen. Second-generation lignocellulosic biomass, agricultural residues, and bio-waste are all examples of low-carbon, cost-effective hydrogen generation using biomass feedstock [24, 31, 36, 37, 61].

Since glycerol has a high volume of output as a by-product of biodiesel production, it holds great stoichiometric promise as a feedstock for the GSR process, which generates hydrogen [62]. The fact that H2 in the amount of 7 mol may be created from 1 molar glycerol makes GSR a promising technique. Based on thermodynamic and experimental findings, the results showed the range of 6 to 6.2% hydrogen by mole of glycerol can be created [63, 64]. Under steam reforming conditions, MSR and MeSR (Methanol/ethanol steam reforming), hydrogen production rates of 4 mol per mole of methane and 3 mol per mole of methanol [65]. Furthermore, the raw materials and technique chosen determine the best process for producing hydrogen.

2.3 Challenges in Hydrogen Production

The high price of hydrogen production is one of the industry's biggest obstacles. Steam reforming is the most frequent technique of creating hydrogen, and it requires both high temperatures and pressures of steam. Steam reforming natural gas, which accounts for 75% of the world's hydrogen output. Steam reforming coal accounts for the remaining 23%. According to the International Energy Agency (IEA) (IEA, 2022), over natural gas equivalent to 230 billion m3, or about 6% of worldwide natural gas, is utilised annually to produce seventy Million Tonnes of hydrogen. Hydrogen is produced using 2% of the world's coal (107 Mt of coal). Hydrogen production using steam reforming of fossil fuels does not appear to be a wholly sustainable method. Nevertheless, dry reforming of methane and dry reforming of biomass can transform the carbon dioxide generated during reforming fuel with steam to produce hydrogen [54, 63, 65]. Carbon dioxide has the potential to be employed widely in chemical synthesis through catalytic conversions to promote mitigation of climate change because as a feedstock, it is cheap, renewable, and safe [66–71]. Hydrogen production also has a high energy footprint, meaning that only a little amount of hydrogen can be produced from a large amount of input energy. This is especially true in electrolysis, electrolysis of water is an example of a low-carbon feedstock that can be used to produce hydrogen, which is an environmentally friendly method due to its zero-emissions status, higher cell efficiency, higher output rates, and more purity; yet, this technology is also the most expensive because it requires a tremendous amount of electricity to run. There is an annual power demand of 3,600 TWh and a water need of roughly 617 million m3 (1.3% of the total world energy sector's water use) (IEA, 2022) which uses electricity to separate hydrogen and oxygen from water molecules. New materials and procedures that can lower the amount of energy needed to manufacture hydrogen are needed to make hydrogen production more energy efficient. Even though the electricity is generated from renewable sources, the price is expected to rise (e.g., nuclear, solar, wind, etc.). However, the European Energy Directive places a cap of roughly 14% on the use of renewable energy for hydrogen production in 2020 [55]. The International Energy Agency (IEA) estimates that between now and 2030, the combined effects of falling renewable energy costs and increased hydrogen production might result in a 30% reduction in price of renewable hydrogen production. Hydrogen generation from low-carbon feedstock is made more economically viable and environmentally friendly.

3 Emerging Techniques in Hydrogen Production via Pyrolysis

Recent studies and developments have focused on the possibility of combining pyrolysis technique with other technologies both in the same area (which is known as thermochemical conversion) and in a other domain (which is known as biochemical conversion). The following part offers an in-depth analysis of various novel methods, such as the combination of pyrolysis with other technologies.

3.1 Anaerobic Digestion Combined Pyrolysis

A promising method, anaerobic digestion (AD) bioenergy recovery from lignocellulosic wastes and biomass [29, 72–80]. Biogas is produced from biodegradable materials by a population of anaerobic bacteria [81]; which is a combined samples of around 60–65% CH_4 and 35–40% CO_2. Hydrolysis is typically considered the rate-limiting phase in AD of organic substrate [82]. AD of organic substrate consists of four primary procedures: acidogenesis, hydrolysis, acetogenesis, and methanogenesis Wet (10% Total Solids), semi-dry (10–20% Total Solids), and dry (20% Total Solids) bioconversion processes can be defined for anaerobic digestion based on the properties of organic substrates [83]. The bioconversion of lignocellulosic biomass can be done using either a wet or dry anaerobic process, each of which has distinct benefits and drawbacks [84, 85]. When different technologies are combined, the negative aspects of each are reduced, which in turn boosts the circular economy and increases the efficiency with which resources are recovered [8, 28, 61, 74]. Many scientists have tried to develop a sustainable integrated system for the systematic degradation of lignocellulosic biomass through a combination of these approaches since the "Supergen Bioenergy Project" first pioneered the integration of anaerobic digestion (AD) and pyrolysis processes [8, 86]. There is a large amount of energy-involved molecules left in the remnant because anaerobic bacteria are unable to degrade the biomass substrate due to recalcitrant components like lignin [18]. Maximising the usefulness of digesting wastes for generation of renewable resources for energy and industrial use is made feasible combining AD and pyrolysis, which also minimises the cost of waste management [8]. As slow pyrolysis accommodates a broad range of particle sizes, it can be used to efficiently remove organic contaminants present in AD-stabilized sewage sludge. Anaerobically stabilised sewage sludge was pyrolyzed in a quartz fixed bed reactor for 2 h (400–800 °C) [60]. It was determined whether or not the AD sludge-chars contained any PCBs, PAHs, medicines, or other chemicals. To get rid of more than 99.8% of the PCBs and PAHs, pyrolysis at temperatures below 600 °C was required [60]. To sum up, when pyrolysis and anaerobic digestion are used together, waste products from one process can be put to better use in the other, such as when remnants of digestion from the AD process are used as the feedstock to the pyrolysis process, when syngas is biomethanized, and when pyrolysis-derived biochar is used as an additive in the AD process [28, 58]. In theory, more of the digestate might be reused if the processes of anaerobic digestion (AD), pyrolysis (pyrolysis), and AD were combined. Pyrolysis liquor anaerobic digestion and pyrolysis of the residue after anaerobic digestion are two methods that are growing in popularity because of their adaptability [8, 18, 28] and better understanding of how the composition of feedstocks affects production. To achieve nutrient-balance and maximise output of methane, anaerobic co-digestion combines multiple substrates to improve the carbon to nitrogen (C/ N) ratio [87, 88]. Therefore, compared to anaerobic digestion of a single substrate, anaerobic co-digestion yields more benefits [89]. Hence, anaerobic digestion (AD) followed by pyrolysis (AD) could be a route for biomass degradation and residue use leading to bioenergy production. Illustration of the combination of AD and pyrolysis represented in Fig. 1 [90].

Aqueous pyrolysis liquid (APL) has been the subject of numerous investigations into its potential use in anaerobic digestion (AD) for biogas production [29, 58]. Hubner et al. [29] found that when APL was used as feed, the AD unit produced essentially no methane and the biogas was primarily composed of carbon dioxide. However, the inhibition that occurs during the AD of APL may be overcome by adding pyrolysis generated biochar as an additive, demonstrating the importance of biochar in AD to increase the methane percentage. Biochar made from digested residues exhibited (LHVs) than biochar made from pristine raw materials, according to an investigation by Song et al. [73]. Energy was converted into syngas when wastes were digested. Pyrolysis of digested leftovers yields biochar, which can be used to fix soil health, preventing their leaching into the soil and groundwater.

Biochar potential use as a soil amendment depends on its properties and the applicable local or national legislation, but it's important to differentiate between biochar leachable and bonded elements [10, 17, 67, 73, 78, 80, 91]. When compared to raw feedstock, digested residues had reduced carbon, hydrogen, and oxygen

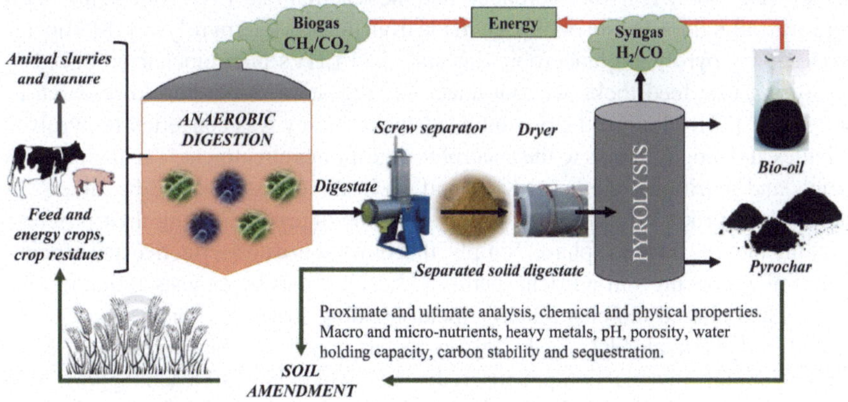

Fig. 1 Anaerobic digestion coupled pyrolysis (Adapted with permission from Elsevier [90])

contents. This suggests that the quality of bio-oil produced during subsequent pyrolysis is enhanced due to the reduction in oxygen concentration [15]. Environmental effect of various AD and pyrolysis configurations are analysed in a recent LCA research on combining the two processes. Righi et al. [79] discovered that as part of the unified procedure, emissions of greenhouse gases were greatly diminished without contributing to a more rapid depletion of abiotic resources. Further potentially synergistic combinations are offered by the combination of anaerobic digestion and pyrolysis, such as the feeding pyrolysis reactors with anaerobic digestate [91], the biomethanation of syngas [92], or adding more anaerobic digestion to the mix with pyrolysis biochars to counteract inhibition [80]. Additionally, research on sewage sludge integration treatments has yielded similarly impressive energy recycling efficiencies [93].

3.1.1 Role of Anaerobic Digestion and Pyrolysis in Synergistic Processes

Anaerobic digestion: Effective anaerobic digestion of the aqueous pyrolysis liquid can increase methane production [28, 29, 61, 78]. Generally, unlike other polymers, Klason lignin was unable to successfully destroyed during the AD process, mostly because of their physicochemical constraints to bacterial invasion [71]. This led to their accumulation in the solid digestate [70, 94]. The digestate residual polymers can serve as a good feedstock for the subsequent process of pyrolysis in its entirety. Reactor drying of solid digestate prior to pyrolysis is a necessary step caused by the need for a lot of energy requirement, and the surplus heat and CH_4 produced from AD can more than fulfil the heat need. That is to say, the AD step enhanced the autonomous power generation in completing the subsequent pyrolysis in an integrated process [51].

Anaerobic digestion showed low performance when exposed to APL. The production of CH_4 ($60 \pm 15\%$ of theoretical) and the reaction rate were both significantly improved after the addition of biochar made from the previous pyrolysis [78]. Biochar produced by pyrolyzing anaerobic digestate had LHVs than biochar produced by pyrolyzing raw feedstocks without anaerobic digestion, according to research by Song et al. [73]. More of the pristine feedstocks energy was converted to pyrolysis volatiles and biogas thanks to the anaerobic digestion pretreatment. The high density metals and nutritional elements K, P, and Ca in the digestate could be "fixed" by the biochars produced during the pyrolysis of the digestate, halting their progress, leaching into the aqueous phase. Simply, this biochar is a cost-effective and efficient means of enhancing soil nutrient quality [73, 91]. It was hypothesised that the AD process had altered the char's heavy metal bonding structure. For instance, if AD treated sludge contains less Fe_2O_3, heavy metals may have a better chance of being bonded into ferromagnet compounds, reducing the likelihood of leaching [95]. Activated carbon (biochar) made by pyrolyzing anaerobic digestate is a cheaper adsorbent for the removal of phosphate from wastewater at a rate of 73% greater than that of the undigested feedstock [96]. In addition, the Fe-modified biochar saw a decrease in efficiency from 73 to 22% in terms of phosphate removal. So, this biochar could be used as an inexpensive adsorbent to get rid of phosphate in water.

Studies showed that O, C, and H levels in anaerobic digestate are lower than those in raw feedstock [69]. Improved bio-oil quality can be expected from the subsequent pyrolysis process if the O content is reduced, since this may result in a reduction of acids and an increase in calorific value [97]. In favour of creating phenolic compounds in pyrolysis was the lignin that persisted in the digestate despite being easily destroyed during the AD process. In addition, FT-IR spectra show that breaking the chemical bonds or groups of the lignocellulose in rice straw through anaerobic digestion treatment can increase the selectivity of pyrolytic products and the efficiency with which bioenergy can be recycled [69].

Pyrolysis: The end results of anaerobic digestion are typically digestate and biogas. Approximately 35–50% of the energy in the feedstock transformed into biogas or syngas during the anaerobic digestion process in an AD plant, while about 30–50% of the refractory part of the feedstock that was unable to be broken down into digestate [93, 94]. In addition, anaerobic digestate contains a lot of useful nutrients like nitrogen and phosphorus [96] and organic molecules [95]. As a result, it's likely that a sizable portion of the energy available in the feed- stock is still being wasted. For instance, organic compounds such as in the final product, Klason lignin was still present following the AD process. As a potential method to get rid of the high-value-recovery digestate and make ecologically benign pyrolysis products [51, 54], Thermal treatment can breakdown these resistant components [60] and boost effectiveness in energy recovery by as much as 85%. In addition, Liang et al. [66] discovered that the Bio-oil's phenolic content can be raised to a greater or lesser amount, depending on the pyrolysis reaction temperature, even while the lignin content remained elevated in the anaerobic digestate. As such, pyrolysis of the valuable digestate is a highly promising approach to AD digestate management. In theory, pyrolyzing anaerobic digestate can help with things like finding new uses for the valuable digestate, cutting down on management expenses, and allowing for the creation of additional products such as gas and bio-oil [63, 64, 71].

Pyrolysis can also be used to get rid of sulphur [69]. Anaerobic digestate contained trace amounts of sulphur. There was no evidence of relative contaminating gases such sulphur dioxide or other sulfocompounds in the pyrolytic products. More study is required to determine the method by which sulphur is eliminated during pyrolysis. Biochar made through pyrolysis has also been said to enhance archaeal development and methanization in anaerobic digestion, as well as prevent moderate ammonia inhibition. Torri et al. and Fabbri et al. [78, 80] discovered that APL inhibited anaerobic microbe digestion in a Py-AD integrated process. However, the inhibition can be greatly reduced by the addition of pyrolysis-derived biochar. It was hypothesised that this biochar may improve the digestive capacity of the anaerobic microbes, increase methane production, and quicken the reaction pace.

3.2 Integrated Pyrolysis and Gasification

Gasification is a multistep process that includes heating, drying, pyrolysis, oxidation, and gasification to turn carbonaceous material into gas using a gasifying agent. Due to their overlap, the individual process steps of a one-stage gasifier cannot be controlled and optimised independently. Char's reactivity and gasification can be negatively affected by volatiles' contact with it [98]. So, to improve gasification efficiency, char gasification should be done without the volatiles. In a multi-stage gasification process, the pyrolysis and gasification steps are separated into their own individually regulated stages. It is possible to achieve high levels of process efficiency and char conversion while simultaneously producing highly pure gas with negligible tar levels.

Two distinct methods have been implemented. The first method employs a two- or three-stage gasification process that combines pyrolysis and gasification directly. The various steps can either be housed in a single unit with dedicated pyrolysis and gasification zones, or in a series of individually operated reactor units. The second method involves separate facilities for pyrolysis and gasification. This approach for producing biofuels by pyrolysis calls for the concentration of biomass at distributed small pyrolysis plants so that the biomass pyrolysis products (liquid and solid) can be transported cheaply to a centralised large gasification plant. The basic layout of a combined pyrolysis and gasification plant is shown in Fig. 2.

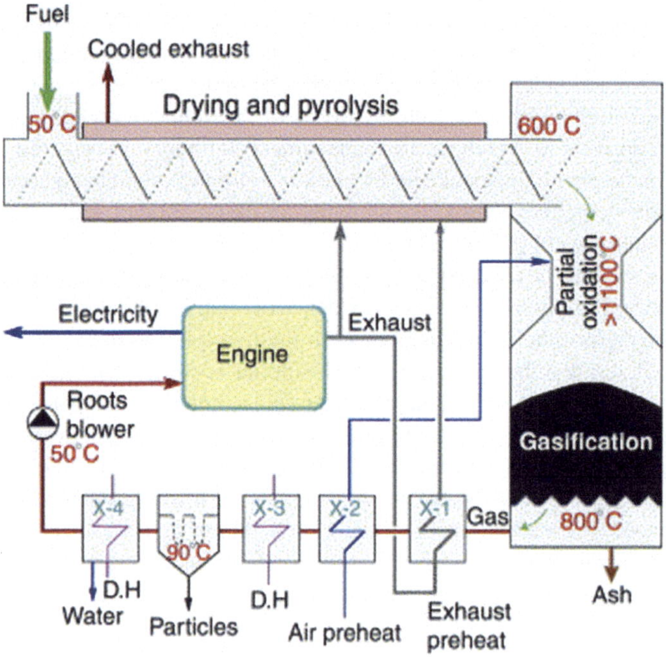

Fig. 2 Schematic of a viking glasifier (Adapted with permission from Elsevier [5]

Some examples of such technology include the Viking gasifier system at the Danish Technical University [5], the FLETGAS process at the University of Seville [11], and the low-temperature circulating fluidized bed gasifier developed by Dong Energy, Denmark [13]. It's doable to produce insignificant tar amounts in the produced gas by performing distinct regions for pyrolysis and gasification and merging them in the subsequent stages. In addition, significant gains in process and char conversion efficiency are possible [13].

Conversely, researchers at the Karlsruhe Institute of Technology (KIT) have proposed a novel method called the Bioliq process, which combines pyrolysis and gasification units in separate facilities [99]. Here, the biomass is processed at multiple independent pyrolysis facilities into slurry of char and oil. The slurry is then transported to a big, centralised gasification plant where biofuels are synthesised. Compared to raw feedstock, transporting char-oil slurry is more cost-effective because its mass energy density is ten times higher. However, a few recurring issues, including storage's impact on slurry composition and characteristics, atomization, and the characteristics of slurry char oil during gasification conditions, must be resolved. Slurries of charred oil derived from various feedstocks and geographical regions must also show steady gasification and atomization [13]. An economic evaluation of the Bioliq method for creating olefins and petrol indicates that it is not financially viable at present market pricing. In addition, the manufacturing of Bioliq cannot contend head-on with fossil fuels in the current market. Yet, the environmental impact can be reduced and the economic viability of pyrolysis increased by optimising this technological method as we transition to a circular economy.

In order to maximise hydrogen yields, a relatively limited amount of studies have been conducted in gasifying the liquid fraction formed during the pyrolysis process. The production of syngas from bio-oil results in a significantly more environmentally friendly product than direct biomass gasification, which is one of the main benefits of bio-oil gasification. Gasifying bio-oil made from rice husks to create ammonia syngas [16]. Raw syngas was produced through gasification of bio-oil obtained via rapid pyrolysis with an H_2/N_2 ratio of 3.0, which can be further used to synthesis ammonia. Syngas quality and tar formation could be enhanced by combining pyrolysis with gasification, for example by employing bio-oil. Bio-oil produced through steam gasification from pyrolysis has the potential to produce high-quality syngas with decreased tar formation [12, 20, 24, 25, 100]. In addition, they showed that it is possible to create hydrogen-rich syngas via efficient tar reformation.

3.3 *Pyrolysis in Bifunctional Catalyst-Sorbent*

Not much is known about the effectiveness of bifunctional metal catalyst in pyrolysis. Using high-temperature sorbents to capture CO_2 on-site is one approach to lowering pyrolysis CO_2 emissions. Catalyst and feedstock continue to interact synergistically during processing. Remains a significant topic of research despite the fact that alternative approaches, such as co-pyrolysis and/or catalytic pyrolysis, might improve product yields and quality [32]. In order to achieve process intensification, bifunctional materials that combine chemical CO_2 sorption and catalytic capabilities have recently attracted a lot of interest [101]. One way to avoid the problems that can arise from having to use two different materials is to combine their catalytic and sorbent properties into a single one. The use of a multipurpose material has been shown to improve mass transfer efficiency, decrease material costs, and alter gas concentration profiles on a particle level [102]. The location of the catalyst, at which the product is formed, can be placed in close proximity to where sorbents are used, the CO_2 is absorbed, using a bifunctional material. Having the catalyst and the chemisorptions site close together can theoretically increase mass transfer and allow for a larger CO_2 partial pressure needed for chemisorptions [22].

When it comes to pyrolysis of biomass, alkaline ceramics-based sorbents stand out among the many others researched since they also provide a supplementary catalytic support [21, 22, 24, 32, 100]. Bifunctional catalyst-sorbent compounds are preferred over traditional mixtures of catalysts and sorbents. CaO has been proposed as a CO_2 sorbent because to its excellent absorbing power (17.8 mmol-CO_2 g sorbent 1) and rapid kinetics [66]. The principal downsides of CaO usage, however, include high-temperature sintering and the reduction in efficiency across several cycles [103]. Despite the widespread application of alkali metal sorbents after combustion carbon dioxide capture can be employed to synthesise bifunctional materials due to insufficient CO_2 absorption. Alkali ceramics such as Li_2ZrO_3, Li_4SiO_4, and Na_2ZrO_3 are suitable alkali metal sorbents that can collect CO_2 at high temperatures (>500 °C). After collecting CO_2, alkaline ceramics form alkaline carbonates (M_2CO_3, M=Li, Na), which are known to perform as a catalyst for several reactions occurring in biomass pyrolysis [100]. During the pyrolysis of cellulose, Memon et al. demonstrated that Na_2ZrO_3 acted as a CO2 sorbent and a catalyst, leading to increased generation of H_2 [22]. Li_2ZrO_3, Li_4SiO_4, and Na_2ZrO_3 were all studied by Zhao et al. [32] under pyrolytic conditions using sawdust as biomass, and identical outcomes were observed. Under laboratory settings, Li_4SiO_4 was more effective than Na_2ZrO_3 and Li_2ZrO_3, respectively, among alkaline ceramics. The removal of CO_2 also increases hydrogen production by stimulating the cracking of tar and the water–gas shift reaction.

4 Conclusions

Hydrogen has been recognized as a promising clean energy carrier due to its high energy density and ability to produce electricity through fuel cells with water being the only by-product. Biomass is seen as a sustainable resource for hydrogen production due to its abundant availability, low cost, and environmental friendliness. Emerging approaches for hydrogen production from biomass include thermochemical, biochemical, and hybrid processes. Hybrid processes (Combination of thermochemical and biochemical) aim to overcome the limitations of individual processes and achieve high hydrogen yield and purity. In conclusion, hydrogen production from biomass is a promising approach for sustainable energy production. The choice of the most suitable process depends on various factors such as feedstock availability, process efficiency, and environmental impact. Thermochemical, biochemical, and hybrid processes offer different advantages and limitations, and the selection of the most appropriate process requires a comprehensive evaluation of their technical and economic feasibility.

References

1. Hu, X., Gholizadeh, M.: Biomass pyrolysis: a review of the process development and challenges from initial researches up to the commercialisation stage. J. Energy Chem. **39**, 109–143 (2019). https://doi.org/10.1016/j.jechem.2019.01.024
2. Mohanty, P., Pant, K.K., Mittal, R., Wiley, J.: Hydrogen generation from biomass materials: challenges and opportunities (2020). https://doi.org/10.1002/wene.111
3. Pierre Friedlingstein, B.Z., O'Sullivan, M., Jones, M.W., Andrew, R.M., Gregor, L., Hauck, J., Quéré, C.L., Luijkx, I.T., Olsen, A., Peters, G.P., Peters, W., Pongratz, J., Schwingshackl, C., Sitch, S., Canadell, J.G.: Global carbon budget 2022 (2022). https://doi.org/10.5194/essd-14-4811-2022
4. Wang, G., Dai, Y., Yang, H., Xiong, Q., Wang, K., Zhou, J., Li, Y., Wang, S.: A review of recent advances in biomass pyrolysis. Energy Fuels **34**, 15557–15558 (2020). https://doi.org/10.1021/acs.energyfuels.0c03107
5. Henriksen, U., Ahrenfeldt, J., Jensen, T.K., Gøbel, B., Bentzen, J.D., Hindsgaul, C., Sørensen, L.H.: The design, construction and operation of a 75 kW two-stage gasifier. Energy **31**, 1542–1553 (2006). https://doi.org/10.1016/j.energy.2005.05.031
6. Zheng, J.L., Zhu, Y.H., Sun, G.T., Dong, Y.Y., Zhu, M.Q.: Bio-oil gasification for production of the raw gas as ammonia syngas. Fuel **327**, 125029 (2022). https://doi.org/10.1016/j.fuel.2022.125029
7. Zheng, J., Zhu, Y., Sun, G., Dong, Y., Zhu, M.: Bio-oil gasification for production of the raw gas as ammonia syngas raw gas heater WGS reactor gasifier bio-oil air CH4 water-1 heater black water water-2 steam. Fuel **327**, 125029 (2022). https://doi.org/10.1016/j.fuel.2022.125029
8. Yang, Z., Liu, Y., Zhang, J., Mao, K., Kurbonova, M., Liu, G., Zhang, R., Wang, W.: Improvement of biofuel recovery from food waste by integration of anaerobic digestion, digestate pyrolysis and syngas biomethanation under mesophilic and thermophilic conditions. J. Clean. Prod. **256**, 120594 (2020). https://doi.org/10.1016/j.jclepro.2020.120594
9. Gao, Z., Li, N., Wang, Y., Niu, W., Yi, W.: Pyrolysis behavior of xylan-based hemicellulose in a fixed bed reactor. J. Anal. Appl. Pyrol. 104772 (2020). https://doi.org/10.1016/j.jaap.2020.104772
10. Antunes, E., Schumann, J., Brodie, G., Jacob, M.V., Schneider, P.A.: Biochar produced from biosolids using a single-mode microwave: characterisation and its potential for phosphorus removal. J. Environ. Manag. **196**, 119–126 (2017). https://doi.org/10.1016/j.jenvman.2017.02.080
11. Gómez-Barea, A., Leckner, B., Villanueva Perales, A., Nilsson, S., Fuentes Cano, D.: Improving the performance of fluidized bed biomass/waste gasifiers for distributed electricity: a new three-stage gasification system. Appl. Therm. Eng. **50**, 1453–1462 (2013). https://doi.org/10.1016/j.applthermaleng.2011.12.025
12. Yoon, S.J., Choi, Y.C., Lee, J.G.: Hydrogen production from biomass tar by catalytic steam reforming. Energy Convers. Manag. **51**, 42–47 (2010). https://doi.org/10.1016/j.enconman.2009.08.017

13. Heidenreich, S., Foscolo, P.U.: New concepts in biomass gasification. Prog. Energy Combust. Sci. **46**, 72–95 (2015). https://doi.org/10.1016/j.pecs.2014.06.002
14. Vuppaladadiyam, A.K., Vuppaladadiyam, S.S.V., Awasthi, A., Sahoo, A., Rehman, S., Pant, K.K., Murugavelh, S., Huang, Q., Anthony, E., Fennel, P., Bhattacharya, S., Leu, S.Y.: Biomass pyrolysis: a review on recent advancements and green hydrogen production. Bioresour. Technol. **364**, 128087 (2022). https://doi.org/10.1016/j.biortech.2022.128087
15. Alvarez, J., Amutio, M., Lopez, G., Santamaria, L., Bilbao, J., Olazar, M.: Improving bio-oil properties through the fast co-pyrolysis of lignocellulosic biomass and waste tyres. Waste Manag. **85**, 385–395 (2019). https://doi.org/10.1016/j.wasman.2019.01.003
16. Zheng, J.L., Zhu, Y.H., Zhu, M.Q., Kang, K., Sun, R.C.: A review of gasification of bio-oil for gas production. Sustain. Energy Fuels **3**, 1600–1622 (2019). https://doi.org/10.1039/c8se00553b
17. Yang, C., Li, R., Zhang, B., Qiu, Q., Wang, B., Yang, H., Ding, Y., Wang, C.: Pyrolysis of microalgae: a critical review. Fuel Process. Technol. **186**, 53–72 (2019). https://doi.org/10.1016/j.fuproc.2018.12.012
18. Varsha, S.S.V., Vuppaladadiyam, A.K., Shehzad, F., Ghaedi, H., Murugavelh, S., Dong, W., Antunes, E.: Co-pyrolysis of microalgae and municipal solid waste: a thermogravimetric study to discern synergy during co-pyrolysis process. J. Energy Inst. **94**, 29–38 (2021). https://doi.org/10.1016/j.joei.2020.10.010
19. Singh, V., Poho, M., Meers, E., Skoblia, S., Jeremi, M.: Potential of coupling anaerobic digestion with thermochemical technologies for waste valorization **294** (2021). https://doi.org/10.1016/j.fuel.2021.120533
20. Hwang, J.G., Choi, M.K., Choi, D.H., Choi, H.S.: Quality improvement and tar reduction of syngas produced by bio-oil gasification. Energy **236**, 121473 (2021). https://doi.org/10.1016/j.energy.2021.121473
21. Shan, R., Han, J., Gu, J., Yuan, H., Luo, B., Chen, Y.: Resources, conservation & recycling a review of recent developments in catalytic applications of biochar-based materials. Resour. Conserv. Recycl. **162**, 105036 (2020). https://doi.org/10.1016/j.resconrec.2020.105036
22. Memon, M.Z., Ji, G., Li, J., Zhao, M.: Na2ZrO3 as an effective bifunctional catalyst-sorbent during cellulose pyrolysis. Ind. Eng. Chem. Res. **56**, 3223–3230 (2017). https://doi.org/10.1021/acs.iecr.7b00309
23. Fasolini, A., Cespi, D., Tabanelli, T., Cucciniello, R., Cavani, F.: Hydrogen from renewables: a case study of glycerol reforming. Catalyst **9**, 722 (2019). https://doi.org/10.3390/catal9090722
24. Dou, B., Zhang, H., Song, Y., Zhao, L., Jiang, B., He, M., Ruan, C., Chen, H., Xu, Y.: Hydrogen production from the thermochemical conversion of biomass: issues and challenges. Sustain. Energy Fuels **3**, 314–342 (2019). https://doi.org/10.1039/c8se00535d
25. Hossain, M.A., Jewaratnam, J., Ganesan, P.: Prospect of hydrogen production from oil palm biomass by thermochemical process – a review. Int. J. Hydrog. Energy **41**, 16637–16655 (2016). https://doi.org/10.1016/j.ijhydene.2016.07.104
26. Nai, P., Yek, Y., Adibah, W., Mahari, W., Kong, S.H., Ying, S.: Pilot-scale co-processing of lignocellulosic biomass, algae, shellfish waste via thermochemical approach: recent progress and future directions. Bioresour. Technol. 126687 (2022). https://doi.org/10.1016/j.biortech.2022.126687
27. Lepage, T., Kammoun, M., Schmetz, Q., Richel, A.: Biomass-to-hydrogen: a review of main routes production, processes evaluation and techno-economical assessment. Biomass Bioenergy **144**, 105920 (2021). https://doi.org/10.1016/j.biombioe.2020.105920
28. Feng, Q., Lin, Y.: Integrated processes of anaerobic digestion and pyrolysis for higher bioenergy recovery from lignocellulosic biomass: a brief review. Renew. Sustain. Energy Rev. **77**, 1272–1287 (2017). https://doi.org/10.1016/j.rser.2017.03.022
29. Hübner, T., Mumme, J.: Integration of pyrolysis and anaerobic digestion - Use of aqueous liquor from digestate pyrolysis for biogas production. Bioresour. Technol. **183**, 86–92 (2015). https://doi.org/10.1016/j.biortech.2015.02.037. Elsevier Ltd.

30. Liu, P.K.T., Sahimi, M., Tsotsis, T.T.: Process intensification in hydrogen production from coal and biomass via the use of membrane-based reactive separations. Curr. Opin. Chem. Eng. **1**, 342–351 (2012). https://doi.org/10.1016/j.coche.2012.06.001
31. Wang, J., Yin, Y.: Fermentative hydrogen production using various biomass-based materials as feedstock. Renew. Sustain. Energy Rev. **92**, 284–306 (2018). https://doi.org/10.1016/j.rser.2018.04.033
32. Zhao, M., Memon, M.Z., Ji, G., Yang, X., Vuppaladadiyam, A.K., Song, Y., Raheem, A., Li, J., Wang, W., Zhou, H.: Alkali metal bifunctional catalyst-sorbents enabled biomass pyrolysis for enhanced hydrogen production. Renew. Energy **148**, 168–175 (2020). https://doi.org/10.1016/j.renene.2019.12.006
33. Sołowski, G., Shalaby, M.S., Abdallah, H., Shaban, A.M., Cenian, A.: Production of hydrogen from biomass and its separation using membrane technology. Renew. Sustain. Energy Rev. **82**, 3152–3167 (2018). https://doi.org/10.1016/j.rser.2017.10.027
34. Novaes, E., Kirst, M., Chiang, V., Winter-sederoff, H., Sederoff, R.: Lignin and biomass: a negative correlation for wood formation and lignin content in trees. Plant Physiol. **1**(154), 555–561 (2010). https://doi.org/10.1104/pp.110.161281
35. Ansari, K.B., Arora, J.S., Chew, J.W., Dauenhauer, P.J., Mushrif, S.H.: Fast pyrolysis of cellulose, hemicellulose, and lignin: effect of operating temperature on bio-oil yield and composition and insights into the intrinsic pyrolysis chemistry (2019). https://doi.org/10.1021/acs.iecr.9b00920
36. Wood, J., Wang, J., Minami, E., Kawamoto, H.: Thermal reactivity of hemicellulose and cellulose in cedar and beech wood cell walls. J. Wood Sci. **66**, 1–10 (2020). https://doi.org/10.1186/s10086-020-01888-x
37. Liu, H., Kong, T.H.: Toward a fundamental understanding of cellulase-lignin interactions in the whole slurry enzymatic saccharification process. Biofuels Bioprod. Biorefin. **10**, 648–663 (2016). https://doi.org/10.1002/bbb.1670
38. Hashim, H., Narayanasamy, M., Ala, N.: A cleaner and greener fuel : Biofuel blend formulation and emission assessment 1–10 (2016). https://doi.org/10.1016/j.jclepro.2016.06.021
39. Chaiwat, W., Gunawan, R., Gholizadeh, M., Li, X., Lievens, C., Hu, X., Wang, Y., Mourant, D., Rossiter, A., Bromly, J., Li, C.: Upgrading of bio-oil into advanced biofuels and chemicals. Part III. Importance of holdup of heavy species during the hydrotreatment of bio-oil in a continuous packed-bed catalytic reactor. Fuel 1–9 (2013). https://doi.org/10.1016/j.fuel.2013.05.004
40. Raghavan, V.: Soil biochar amendment as a climate change mitigation tool: key parameters and mechanisms involved. J. Environ. Manag. **181**, 484–497 (2016). https://doi.org/10.1016/j.jenvman.2016.06.063
41. Tursunov, O.: A comparison of catalysts zeolite and calcined dolomite for gas production from pyrolysis of municipal solid waste (MSW). Ecol. Eng. **69**, 237–243 (2014). https://doi.org/10.1016/j.ecoleng.2014.04.004
42. Cipriani, G., Di, V., Genduso, F., La, D., Liga, R., Miceli, R., Ricco, G.: ScienceDirect perspective on hydrogen energy carrier and its automotive applications. Int. J. Hydrog. Energy. 1–13 (2014). https://doi.org/10.1016/j.ijhydene.2014.03.174
43. Dou, B., Wang, C., Song, Y., Chen, H., Xu, Y.: Activity of Ni – Cu – Al based catalyst for renewable hydrogen production from steam reforming of glycerol. Energy Convers. Manag. **78**, 253–259 (2014). https://doi.org/10.1016/j.enconman.2013.10.067
44. Szima, S., Cormos, C.-C.: CO2 utilization technologies: a techno-economic analysis for synthetic natural gas production. Energies MDPI **14**, 1258 (2021)
45. Blasi, A., Fiorenza, G., Freda, C., Calabrò, V.: Steam reforming of biofuels for the production of hydrogen-rich gas (2013). https://doi.org/10.1533/9780857098658.3.145
46. Pompeo, F., Santori, G.F., Nichio, N.N.: Hydrogen production by glycerol steam reforming with Pt/SiO2 and Ni/SiO2 catalysts. Catal. Today **172**, 183–188 (2011). https://doi.org/10.1016/j.cattod.2011.05.001
47. Marda, J.R., DiBenedetto, J., McKibben, S., Evans, R.J., Czernik, S., French, R.J., Dean, A.M.: Non-catalytic partial oxidation of bio-oil to synthesis gas for distributed hydrogen

production. Int. J. Hydrog. Energy **34**, 8519–8534 (2009). https://doi.org/10.1016/j.ijhydene.2009.07.099
48. Cai, W., Wang, F., Zhan, E., Van Veen, A.C., Mirodatos, C., Shen, W.: Hydrogen production from ethanol over Ir/CeO2 catalysts: a comparative study of steam reforming, partial oxidation and oxidative steam reforming. J. Catal. **257**, 96–107 (2008). https://doi.org/10.1016/j.jcat.2008.04.009
49. Romano, M.C., Cassotti, E.N., Chiesa, P., Meyer, J., Mastin, J.: Application of the sorption enhanced-steam reforming process in combined cycle-based power plants. Energy Procedia **4**, 1125–1132 (2011). https://doi.org/10.1016/j.egypro.2011.01.164
50. Mehrpooya, M., Ghorbani, B., Abedi, H.: Biodiesel production integrated with glycerol steam reforming process, solid oxide fuel cell (SOFC) power plant. Energy Convers. Manag. **206**, 112467 (2020). https://doi.org/10.1016/j.enconman.2020.112467
51. Papalas, T., Antzaras, A.N., Lemonidou, A.A.: Intensified steam methane reforming coupled with Ca-Ni looping in a dual fluidized bed reactor system: a conceptual design. Chem. Eng. J. **382**, 122993 (2020). https://doi.org/10.1016/j.cej.2019.122993
52. Jastrzębska, M., Janik, H., Paukszta, D.: The investigation of polyester composites filled by modified bentonite. Polimery/Polym. **59**, 656–661 (2014). https://doi.org/10.14314/polimery.2014.656
53. Seadira, T., Sadanandam, G., Ntho, T.A., Lu, X., Masuku, C.M., Scurrell, M.: Hydrogen production from glycerol reforming: conventional and green production (2017)
54. Soltani, R., Rosen, M.A., Dincer, I.: Assessment of CO 2 capture options from various points in steam methane reforming for hydrogen production. Int. J. Hydrog. Energy. 1–10 (2014). https://doi.org/10.1016/j.ijhydene.2014.09.161
55. Kumar, S.S., Himabindu, V.: Hydrogen production by PEM water electrolysis – a review. Mater. Sci. Energy Technol. **2**, 442–454 (2019). https://doi.org/10.1016/j.mset.2019.03.002
56. Kumar, S.S., Himabindu, V.: Materials science for energy technologies hydrogen production by PEM water electrolysis – a review. Mater. Sci. Energy Technol. **2**, 442–454 (2019). https://doi.org/10.1016/j.mset.2019.03.002
57. Dou, B., Zhao, L., Zhang, H., Wu, K., Zhang, H.: Renewable hydrogen production from chemical looping steam reforming of biodiesel byproduct glycerol by mesoporous oxygen carriers. Chem. Eng. J. **416**, 127612 (2021). https://doi.org/10.1016/j.cej.2020.127612
58. Pecchi, M., Baratieri, M.: Coupling anaerobic digestion with gasification, pyrolysis or hydrothermal carbonization: a review. Renew. Sustain. Energy Rev. **105**, 462–475 (2019). https://doi.org/10.1016/j.rser.2019.02.003
59. Parthasarathy, P., Narayanan, K.S.: Hydrogen production from steam gasification of biomass: Influence of process parameters on hydrogen yield - a review. Renew. Energy **66**, 570–579 (2014). https://doi.org/10.1016/j.renene.2013.12.025
60. Moško, J., Pohořelý, M., Cajthaml, T., Jeremiáš, M., Robles-Aguilar, A.A., Skoblia, S., Beňo, Z., Innemanová, P., Linhartová, L., Michalíková, K., Meers, E.: Effect of pyrolysis temperature on removal of organic pollutants present in anaerobically stabilized sewage sludge. Chemosphere **265**, 129082 (2021). https://doi.org/10.1016/j.chemosphere.2020.129082
61. Singh, R., Paritosh, K., Pareek, N., Vivekanand, V.: Integrated system of anaerobic digestion and pyrolysis for valorization of agricultural and food waste towards circular bioeconomy: review. Bioresour. Technol. **360**, 127596 (2022). https://doi.org/10.1016/j.biortech.2022.127596
62. Ikhsan, M., Viswanathan, K., Wu, W., Hattu, N., Atabani, A.E.: A critical review of the hydrogen production from biomass-based feedstocks: challenge, solution, and future prospect. Process. Saf. Environ. Prot. **164**, 384–407 (2022). https://doi.org/10.1016/j.psep.2022.06.006
63. Dahdah, E., Aouad, S., Gennequin, C., Estephane, J., Nsouli, B., Aboukaı, A.: Glycerol steam reforming over Ru-Mg-Al hydrotalcite-derived mixed oxides: role of the preparation method in catalytic activity 1–9 (2018). https://doi.org/10.1016/j.ijhydene.2018.09.042
64. Chen, M., Zhou, Z., Wang, Y., Liang, T., Li, X., Yang, Z., Chen, M., Wang, J.: Effects of attapulgite-supported transition metals catalysts on glycerol steam reforming for hydrogen production. Int. J. Hydrog. Energy. 1–14 (2018). https://doi.org/10.1016/j.ijhydene.2018.09.122

65. Kim, C., Han, J., Kim, S., Lee, B., Lim, H., Lee, K., Ryi, S.: Hydrogen production by steam methane reforming in a membrane reactor equipped with a Pd composite membrane deposited on a porous stainless steel. Int. J. Hydrog. Energy. 1–9 (2017). https://doi.org/10.1016/j.ijhydene.2017.11.176
66. Dewoolkar, K.D., Vaidya, P.D.: Improved hydrogen production by sorption-enhanced steam methane reforming over hydrotalcite- and calcium-based hybrid materials. Energy Fuels **29**, 3870–3878 (2015). https://doi.org/10.1021/acs.energyfuels.5b00584
67. Troy, S.M., Nolan, T., Leahy, J.J., Lawlor, P.G., Healy, M.G., Kwapinski, W.: Effect of sawdust addition and composting of feedstock on renewable energy and biochar production from pyrolysis of anaerobically digested pig manure. Biomass Bioenergy **49**, 1–9 (2012). https://doi.org/10.1016/j.biombioe.2012.12.014
68. Tambone, F., Scaglia, B., Imporzano, G.D., Schievano, A., Orzi, V., Salati, S., Adani, F.: Chemosphere assessing amendment and fertilizing properties of digestates from anaerobic digestion through a comparative study with digested sludge and compost. Chemosphere **81**, 577–583 (2010). https://doi.org/10.1016/j.chemosphere.2010.08.034
69. Liang, J., Lin, Y., Wu, S., Liu, C., Lei, M., Zeng, C.: Enhancing the quality of bio-oil and selectivity of phenols compounds from pyrolysis of anaerobic digested rice straw. Bioresour. Technol. **181**, 220–223 (2015). https://doi.org/10.1016/j.biortech.2015.01.056
70. Monlau, F., Kaparaju, P., Trably, E., Steyer, J.P., Carrere, H.: Alkaline pretreatment to enhance one-stage CH4 and two-stage H2/CH4 production from sunflower stalks: mass, energy and economical balances. Chem. Eng. J. **260**, 377–385 (2015). https://doi.org/10.1016/j.cej.2014.08.108
71. Monlau, F., Latrille, E., Carvalho, A., Costa, D., Steyer, J., Carrère, H.: Enhancement of methane production from sunflower oil cakes by dilute acid pretreatment. Appl. Energy **102**, 1105–1113 (2013). https://doi.org/10.1016/j.apenergy.2012.06.042
72. Zhang, T., Yang, Y., Xie, D.: Insights into the production potential and trends of China's rural biogas. Int. J. Energy Res. **39**, 1068–1082 (2015). https://doi.org/10.1002/er.3311
73. Song, X.D., Chen, D.Z., Zhang, J., Dai, X.H., Qi, Y.Y.: Anaerobic digestion combined pyrolysis for paper mill sludge disposal and its influence on char characteristics. J. Mater. Cycles Waste Manag. **19**, 332–341 (2017). https://doi.org/10.1007/s10163-015-0428-1
74. Tayibi, S., Monlau, F., Marias, F., Cazaudehore, G., Fayoud, N.E., Oukarroum, A., Zeroual, Y., Barakat, A.: Coupling anaerobic digestion and pyrolysis processes for maximizing energy recovery and soil preservation according to the circular economy concept. J. Environ. Manag. **279**, 111632 (2021). https://doi.org/10.1016/j.jenvman.2020.111632
75. Varsha, S.S.V., Soomro, A.F., Baig, Z.T., Vuppaladadiyam, A.K., Murugavelh, S., Antunes, E.: Methane production from anaerobic mono- and co-digestion of kitchen waste and sewage sludge: synergy study on cumulative methane production and biodegradability. Biomass Convers. Biorefin. **12**, 3911–3919 (2022). https://doi.org/10.1007/s13399-020-00884-x
76. Zhang, R., El-mashad, H.M., Hartman, K., Wang, F., Liu, G., Choate, C., Gamble, P.: Characterization of food waste as feedstock for anaerobic digestion. Bioresour. Technol. **98**, 929–935 (2007). https://doi.org/10.1016/j.biortech.2006.02.039
77. Monlau, F., Sambusiti, C., Antoniou, N., Barakat, A., Zabaniotou, A.: A new concept for enhancing energy recovery from agricultural residues by coupling anaerobic digestion and pyrolysis process. Appl. Energy **148**, 32–38 (2015). https://doi.org/10.1016/j.apenergy.2015.03.024
78. Torri, C., Fabbri, D.: Biochar enables anaerobic digestion of aqueous phase from intermediate pyrolysis of biomass. Bioresour. Technol. **172**, 335–341 (2014). https://doi.org/10.1016/j.biortech.2014.09.021
79. Righi, S., Bandini, V., Marazza, D., Baioli, F., Torri, C., Contin, A.: Bioresource technology life cycle assessment of high ligno-cellulosic biomass pyrolysis coupled with anaerobic digestion. Bioresour. Technol. **212**, 245–253 (2016). https://doi.org/10.1016/j.biortech.2016.04.052
80. Mumme, J., Srocke, F., Heeg, K., Werner, M.: Bioresource technology use of biochars in anaerobic digestion. Bioresour. Technol. **164**, 189–197 (2014). https://doi.org/10.1016/j.biortech.2014.05.008

81. Fabbri, D., Torri, C.: Linking pyrolysis and anaerobic digestion (Py-AD) for the conversion of lignocellulosic biomass. Curr. Opin. Biotechnol. **38**, 167–173 (2016). https://doi.org/10.1016/j.copbio.2016.02.004
82. Aquino, F., Stuckey, D.C.: Integrated model of the production of soluble microbial products (SMP) and extracellular polymeric substances (EPS) in anaerobic chemostats during transient conditions **38**, 138–146 (2008). https://doi.org/10.1016/j.bej.2007.06.010
83. Hartmann, H., Ahring, B.K.: Strategies for the anaerobic digestion of the organic fraction of municipal solid waste: an overview. Water Sci. Technol. **53**, 7–22 (2006). https://doi.org/10.2166/wst.2006.231
84. Colazo, A., Sánchez, A., Font, X., Colón, J.: Environmental impact of rejected materials generated in organic fraction of municipal solid waste anaerobic digestion plants: comparison of wet and dry process layout. Waste Manag. **43**, 84–97 (2015). https://doi.org/10.1016/j.wasman.2015.06.028
85. Kothari, R., Pandey, A.K., Kumar, S., Tyagi, V.V., Tyagi, S.K.: Different aspects of dry anaerobic digestion for bio-energy: an overview. Renew. Sustain. Energy Rev. **39**, 174–195 (2014). https://doi.org/10.1016/j.rser.2014.07.011
86. Supergen Bioenergy (n.d.) 3–6
87. Wang, X., Yang, G., Feng, Y., Ren, G., Han, X.: Bioresource technology optimizing feeding composition and carbon – nitrogen ratios for improved methane yield during anaerobic co-digestion of dairy, chicken manure and wheat straw. Bioresour. Technol. **120**, 78–83 (2012). https://doi.org/10.1016/j.biortech.2012.06.058
88. Hamawand, I.: Anaerobic digestion process and bio-energy in meat industry: a review and a potential. Renew. Sustain. Energy Rev. **44**, 37–51 (2015). https://doi.org/10.1016/j.rser.2014.12.009
89. Callaghan, F.J., Wase, D.A.J., Thayanithy, K., Forster, C.F.: Continuous co-digestion of cattle slurry with fruit and vegetable wastes and chicken manure. Biomass Bioenergy **27**, 71–77 (2002)
90. Monlau, F., Francavilla, M., Sambusiti, C., Antoniou, N., Solhy, A., Libutti, A., Zabaniotou, A.: Toward a functional integration of anaerobic digestion and pyrolysis for a sustainable resource management. Comparison between solid-digestate and its derived pyrochar as soil amendment. Appl. Energy **169**, 652–662 (2016). https://doi.org/10.1016/j.apenergy.2016.02.084
91. Inyang, M., Gao, B., Pullammanappallil, P., Ding, W., Zimmerman, A.R.: Bioresource technology biochar from anaerobically digested sugarcane bagasse. Bioresour. Technol. **101**, 8868–8872 (2010). https://doi.org/10.1016/j.biortech.2010.06.088
92. Guiot, S.R., Cimpoia, R.: Potential of wastewater-treating anaerobic granules for biomethanation of synthesis gas. Environ. Sci. Technol. **45**, 2006–2012 (2012)
93. Fonts, I., Azuara, M., Gea, G., Murillo, M.B.: Study of the pyrolysis liquids obtained from different sewage sludge. J. Anal. Appl. Pyrol. **85**, 184–191 (2009). https://doi.org/10.1016/j.jaap.2008.11.003
94. Menardo, S., Gioelli, F., Balsari, P.: Bioresource technology the methane yield of digestate: effect of organic loading rate, hydraulic retention time, and plant feeding. Bioresour. Technol. **102**, 2348–2351 (2011). https://doi.org/10.1016/j.biortech.2010.10.094
95. Folgueras, M.B., Alonso, M., Díaz, R.M.: Influence of sewage sludge treatment on pyrolysis and combustion of dry sludge 1–10 (2013). https://doi.org/10.1016/j.energy.2013.03.063
96. Yao, Y., Gao, B., Inyang, M., Zimmerman, A.R., Cao, X., Pullammanappallil, P., Yang, L.: Bioresource technology biochar derived from anaerobically digested sugar beet tailings: characterization and phosphate removal potential. Bioresour. Technol. **102**, 6273–6278 (2011). https://doi.org/10.1016/j.biortech.2011.03.006
97. Liang, J., Shan, G., Sun, Y., Al, S.: Catalytic fast pyrolysis of lignocellulosic biomass: critical role of zeolite catalysts. Renew. Sustain. Energy Rev. **139**, 110707 (2021). https://doi.org/10.1016/j.rser.2021.110707
98. Li, C.: Importance of volatile – char interactions during the pyrolysis and gasification of low-rank fuels – a review. Fuel **112**, 609–623 (2013). https://doi.org/10.1016/j.fuel.2013.01.031

99. Nicolaus Dahmen, R.S., Dinjus, E., Kolb, T., Arnold, U., Leibold, H.: State of the art of the bioliq® process for synthetic biofuels production. Environ. Prog. Sustain. Energy **31**, 676–680 (2012). https://doi.org/10.1002/ep.10624
100. Memon, M.Z., Zhao, X., Sikarwar, V.S., Vuppaladadiyam, A.K., Milne, S.J., Brown, A.P., Li, J., Zhao, M.: Alkali metal CO2 sorbents and the resulting metal carbonates: potential for process intensification of sorption-enhanced steam reforming. Environ. Sci. Technol. **51**, 12–27 (2017). https://doi.org/10.1021/acs.est.6b04992
101. Satrio, J.A., Shanks, B.H., Wheelock, T.D.: Development of a novel combined catalyst and sorbent for hydrocarbon reforming 3901–3911 (2005)
102. Dietrich, W., Lawrence, P.S., Grünewald, M., Agar, D.W.: Theoretical studies on multifunctional catalysts with integrated adsorption sites. Chem. Eng. J. **107**, 103–111 (2005). https://doi.org/10.1016/j.cej.2004.12.016
103. Blamey, J., Anthony, E.J., Wang, J., Fennell, P.S.: The calcium looping cycle for large-scale CO2 capture. Prog. Energy Combust. Sci. **36**, 260–279 (2010). https://doi.org/10.1016/j.pecs.2009.10.001

Comparative Analysis of Nanomaterials and Artificial Intelligence for Sustainable Nutrient Management in Soil

Madhu Bala and Virat Khanna

Abstract Nanomaterials, with their large surface area and reactivity, offer unique benefits in nutrient delivery and retention whereas artificial intelligence approaches offer significant tools for optimising nutrition application strategies and decision-making processes. For sustainable farming techniques to be implemented and ecosystem health to be maintained, soil nutrient management is essential. This chapter presents a comparative analysis of nanomaterials and artificial intelligence (AI) approaches for sustainable nutrient management in soil. Evaluation of both the strategies based on effectiveness, environmental implications and scalability is also done. The results of this study will help to clarify the benefits and drawbacks of using AI and nanomaterials, and it will also provide the guidance for the creation of integrated plans that maximise nutrient availability and sustainably increase soil fertility.

Keywords Nanomaterials · Artificial intelligence · Soil nutrients · Sustainability · Agriculture · Sensors · Nano fertilizers · Machine learning · Neural networks · Precision farming · Environment · Surface area · Macro nutrients · Pore structure · Nanotechnology · Soil pH · Degradation · Pollutants · Adsorption

---

The original version of the chapter has been revised. A correction to this chapter can be found at https://doi.org/10.1007/978-3-031-76953-5_13

M. Bala
Department of Computer Science and Engineering, Lovely Professional University, Phagwara 144411, Punjab, India

V. Khanna (✉)
University Centre for Research and Development, Chandigarh University, Gharuan, Mohali 140413, Punjab, India
e-mail: Khanna.virat@gmail.com

K. K. Kesari et al. (eds.), *Functionalized Cellulose Materials*, Engineering Materials, https://doi.org/10.1007/978-3-031-76953-5_6

1 Introduction

The global population is projected to reach 9.6 billion by 2050, placing immense pressure on ability of humans to produce enough food [1]. Natural resources including water, minerals, and forests are under more stress. The overuse of these resources for agriculture, building, and fuel leads to ecosystem disruption, habitat loss, and biodiversity decline. To be specific, the overuse of water can result in soil salinization, reducing agricultural productivity and degrading land. Overuse of mining and mineral extraction destroys habitats and erodes soil, upsetting ecosystems and landscapes. Heavy metals and other pollutants are frequently released into the soil and water by mining operations, polluting natural resources and endangering the health of both people and wildlife [2–4]. Plant growth and soil fertility may be impacted by nitrogen imbalances caused by mineral loss from the soil. This necessitates a significant increase in agricultural productivity while minimizing environmental impact. One of the critical challenges in achieving this goal is sustainable nutrient management in soil [5, 6].

Traditionally, farmers have relied on conventional fertilizers to replenish soil nutrients and boost crop yields. However, these practices often lead to inefficient nutrient use, with significant portions leaching into groundwater or volatilizing into the atmosphere. This not only reduces fertilizer efficacy but also contributes to environmental pollution, eutrophication of water bodies, and greenhouse gas emissions [7–9]. Precipitation patterns, variations in temperature, and extreme weather events are all impacted by climate change, and these factors have an impact on soil erosion rates, nutrient cycling, and general soil health. Additionally, productive agricultural land is turned into urban districts or industrial zones as a result of the fast urbanisation that is occurring to accommodate a growing population. As a result, there is less arable land available overall, which puts pressure on the soil that is still there to produce more food. Urbanisation can worsen soil quality and reduce its capacity to sustain plant development by causing pollution and compaction.

In recent years, two emerging technologies offer promising solutions for sustainable nutrient management: nanomaterials and artificial intelligence (AI). Nanomaterials, with their unique properties at the nanoscale, hold immense potential for targeted nutrient delivery and improved soil health. AI, with its ability to analyze vast amounts of data, can optimize fertilizer application strategies and promote precision agriculture [10–12]. This chapter focuses onto the current state of nutrient management challenges, explores the exciting possibilities offered by nanomaterials and AI, and highlights the potential of their combined application for a more sustainable agricultural future.

1.1 *Importance of Soil Nutrients for Plant Growth*

Soil acts as the foundation for terrestrial ecosystems, providing essential nutrients (depicted in Fig. 1) for plant growth and development. These nutrients, including macronutrients like nitrogen (N), phosphorus (P), and potassium (K), and micronutrients like iron (Fe), zinc (Zn), and copper (Cu), play crucial roles in various plant physiological processes. Nitrogen is a key component of proteins, nucleic acids, and chlorophyll, contributing to plant growth, photosynthesis, and overall yield. Phosphorus is vital for energy transfer, root development, and seed formation. Potassium plays a role in regulating water balance, enzyme activity, and plant stress tolerance [13]. Micronutrients, though required in smaller quantities, are essential for various metabolic functions, and their deficiencies can lead to stunted growth and reduced yield.

The availability of these nutrients in soil depends on various factors, including natural weathering processes, organic matter decomposition, and microbial activity. A number of growth problems, such as chlorosis (yellowing of the leaves), necrosis (death of tissue), stunted growth, reduced yield, and increased susceptibility to pests and diseases, can be brought on by deficiencies or imbalances. For example, a lack in phosphorus can result in weak root systems and dark green or purplish foliage, while a deficiency in nitrogen typically causes pale green or yellow leaves and limited growth. Further, agricultural practices like intensive cultivation, monoculture, and excessive use of conventional fertilizers can deplete soil nutrients over time. This necessitates the replenishment of nutrients to maintain soil fertility and support healthy plant growth. Soil management techniques are essential for assuring plants receive a sufficient amount of nutrients. These methods include applying fertilisers, using cover

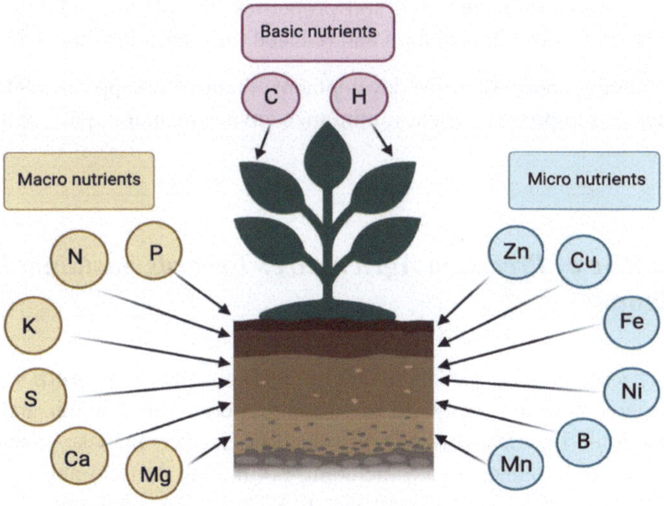

Fig. 1 Essential nutrients for plant growth

crops, rotating crops, and adding organic amendments like manure and compost. Both organic and inorganic fertilisers offer the required nutrients in easily obtainable forms. By boosting microbial activity, soil structure, and water-holding ability, organic additions increase nutrient availability. While cover crops stop soil erosion, increase the amount of organic matter in the soil, and fix nitrogen from the atmosphere, crop rotation aids in preserving soil fertility and lowering the growth of pests and diseases.

1.2 Challenges of Conventional Nutrient Management Practices

Conventional fertilizers have played a significant role in boosting agricultural productivity throughout the 20th century. However, their reliance on broad-acre application methods often leads to several challenges like:

- Inefficient Nutrient Use: A significant portion of applied fertilizers, estimated to be around 30–50% for nitrogen, remains unused by plants and is lost through various pathways. This includes leaching into groundwater, volatilization into the atmosphere, and immobilization in the soil [14].
- Environmental Pollution: Fertilizer runoff contaminates water bodies, leading to eutrophication, algal blooms, and hypoxia (oxygen depletion). Excess nitrogen contributes to ammonia emissions, a potent greenhouse gas, while phosphorus runoff can stimulate harmful algal growth in freshwater and coastal ecosystems [15, 16].
- Soil Degradation: Overuse of fertilizers can disrupt the natural balance of soil microbial communities and alter soil properties like pH and salinity. This can ultimately lead to soil degradation and reduced long-term fertility [17].

These challenges necessitate the development of innovative approaches to nutrient management that improve efficiency, minimize environmental impact, and promote soil health.

1.3 The Rise of Precision Agriculture: Towards Sustainable Solutions

Precision agriculture (PA) [18] is a rapidly evolving field that utilizes advanced technologies to collect and analyze data on soil properties, crop health, and environmental conditions. This data-driven approach allows farmers to make informed decisions about resource management, including optimizing nutrient application rates and timing. Several technologies contribute to PA practices, including:

- Remote Sensing: Satellite imagery and aerial drones provide high-resolution data on crop health and variations within fields. This enables targeted fertilizer application based on specific needs of different areas within a field [19].
- Soil Sensors: In-situ sensors can continuously monitor soil moisture, nutrient levels, and other parameters, providing real-time data for optimizing irrigation and fertilizer application [20].
- Yield Monitors: These tools collect data on crop yield variations within a field, helping identify areas requiring adjustments in nutrient management strategies [20–23].

While PA offers a significant step towards sustainable nutrient management, it still relies on conventional fertilizers with their inherent limitations. This paves the way for the exploration of new technologies like nanomaterials and AI for precise and targeted nutrient delivery. The control of nutrients is a crucial component of sustainable agriculture [24–26]. It is recognized that approximately twenty nutrients are necessary for healthy crop growth. Many of these nutrients can be found in soil that has a pH of 6.0–7.0 and enough soil organic matter. However, additional nitrogen, phosphorus, and potassium are typically required for optimum yields. It is also observed that excess amount of these nutrients may harm the health of soil. For example when fruiting plants receive too much nitrogen, they generate comparatively more foliage, which lowers fruit output and delays fruit maturity. Likewise, huge amount of phosphorous can result in iron and zinc micronutrient deficiencies. Proper management of soil nutrients ensures that plants have access to the necessary nutrients in the right quantities and at the right times, which ultimately maximizes crop yield and quality. It minimize nutrient losses to the environment, reducing the risk of pollution and protecting natural ecosystems.

By means of precise engineering, formulations based on nanomaterials demonstrate improved nutrient uptake efficiency, extended release kinetics, and focused delivery to plant roots. This results in the optimisation of nutrient utilisation and a decrease in the requirement for traditional fertilisers. Additionally, by providing a sustainable method of soil regeneration and rehabilitation, nanomaterials contribute in the remediation of contaminated soil [27]. AI platforms provide specific solutions for improving crop production while minimising resource inputs by utilising complex algorithms and sensor networks to offer insights into soil health parameters, nutrient requirements, and crop-specific demands [28]. Predictive models powered by AI also enable farmers to plan ahead for nutrient shortages, optimise fertilisation schedules, and implement adaptive techniques in response to changing environmental circumstances [29].

2 Nanomaterials for Sustainable Nutrient Management

Nanomaterials, particles with at least one dimension in the range of 1–100 nm, exhibit unique physical and chemical properties compared to their bulk counterparts [30]. These properties offer exciting possibilities for targeted nutrient delivery in soil:

- Enhanced Nutrient Availability: Nanofertilizers can be engineered to release nutrients slowly and directly to plant roots, minimizing waste and environmental pollution [31]. Nutrients like potassium, phosphorus, and nitrogen can be encapsulated by nanomaterials to prevent runoff, volatilization, and leaching. Nanofertilizers can minimise environmental pollution, lower fertiliser application rates, and enhance plant uptake of nutrients [32].
- Targeted Delivery: Nanoparticles can be designed to target specific deficiencies in soil or even individual plant needs based on their surface modifications and coatings. By increasing aeration, nutrient retention, and water retention, nanomaterials can improve the qualities of soil [33]. Particles the size of nanoparticles can alter the fertility and structure of soil, increasing crop yields while lowering the requirement for artificial fertilisers.
- Improved Soil Health: Some nanomaterials can promote beneficial soil microbes, leading to better nutrient cycling and overall soil health. Nanomaterials can be used for recovering nutrients from organic waste streams, such as sewage sludge and agricultural residues. Nanotechnology-based approaches enable the efficient extraction and recycling of valuable nutrients, closing the nutrient loop and reducing reliance on mineral fertilizers [34].

2.1 Characteristics and Properties of Nanomaterials Relevant to Soil Nutrient Management

Although there are many potential advantages of using nanomaterials in soil nutrient management, their safe and sustainable application necessitates rigorous assessment of the effects on the environment and careful consideration of how they interact with soil organisms and components. In this regard, realising the potential of nanomaterials while maintaining agricultural and environmental sustainability requires an understanding of their features (shown in Fig. 2) and attributes related to soil nutrient management.

Because of their large surface area, they can interact more effectively with soil components and nutrients. More active sites for chemical reactions, such as the adsorption of nutrients like potassium, phosphorus, or nitrogen, are made possible by this increased surface area. Increased interaction with soil particles and organic matter due to the larger surface area improves the soil's ability to retain and release nutrients [17]. Greater surface reactivity is encouraged by nanomaterials' large surface areas. This implies that the surface of the nanomaterial can facilitate chemical reactions like ion exchange or dissolution more easily. For instance, ions (such as cations like

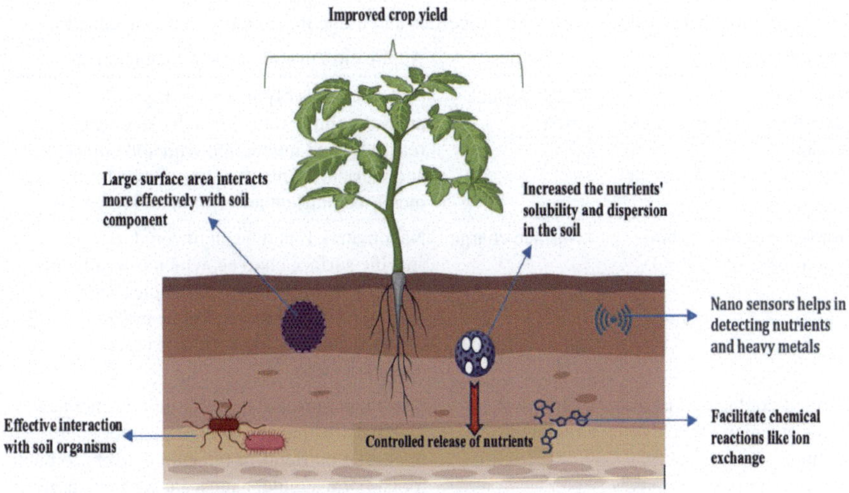

Fig. 2 Characteristics of nanomaterials

calcium, magnesium, or ammonium) can be more effectively adsorbed and released by nanoparticles.

Targeted nutrition delivery, controlled release, and effective adsorption in soil conditions can be facilitated by these materials by taking advantage of their nanoscale properties, such as small particle size, surface charge, and porosity. Nutrient availability can be maximised by engineering nanoparticles to release nutrients in response to particular environmental triggers, like as variations in soil moisture or pH levels.

Nanomaterials can be designed to release nutrients slowly over time, ensuring a steady supply to plants and reducing the need for frequent fertilizer applications. By increasing the nutrients' solubility and dispersion in the soil, nanomaterials can increase their availability to plants for uptake. Their capacity to stay uniformly dispersed throughout the soil avoids hotspots for nutrients and guarantees a consistent supply to plant roots. Table 1 is representing few nanomaterials along with their properties and usage in soil nutrient management. Precise nutrient management is possible with the integration of certain nanomaterials into smart fertiliser systems that have sensors to track soil nutrient levels and ambient factors [35].

2.2 *Applications of Nanomaterials in Enhancing Nutrient Uptake and Delivery in Soil*

The possible application of nanomaterials to improve soil nutrient uptake and distribution generated a lot of attention in agriculture. By preventing leaching and volatilization and guaranteeing a steady delivery of nutrients to plants, nanofertilizers lower nutrient loss and improve crop nutrient availability. Plant cells have

Table 1 Nanomaterials along with their properties and usage in soil nutrient management

Nano material	Properties	Application in soil nutrient management
Nano clays, nano-hydroxyapatite nano zeolites	High surface area	Nanomaterials often possess high surface area-to-volume ratios, which can enhance their reactivity and interaction with soil components and nutrients. This property allows for increased nutrient adsorption and retention
Graphene oxide, carbon nano tubes, TiO2, ZnO	Surface charge	Nanomaterials can be engineered to have specific surface charges, which can influence their interaction with soil particles and nutrients. This property can be utilized to enhance nutrient adsorption or repel contaminants
Layered double hydroxides, chitosan nanoparticles, zeolites	Chemical reactivity	Some nanomaterials exhibit unique chemical reactivity, which can be exploited for controlled release of nutrients or targeted delivery to plant roots. For example, nanoscale zero-valent iron (nZVI) can be used for remediation of contaminated soils while simultaneously supplying nutrient
Mesoporous silica nnaoparticles, metal organic frameworks, carbon nanotubes, nano clay	Pore structure	Certain nanomaterials possess inherent porosity and well-defined pore structures, which can serve as reservoirs for nutrient storage and slow release. This property enables sustained nutrient availability to plants over time
Titanium dioxide nanoparticles	Photocatalytic activity	Some nanomaterials exhibit photocatalytic activity under light irradiation. This property can be utilized for soil remediation and nutrient transformation processes

the ability to internalise some nanoparticles through a process called endocytosis (presented in Fig. 3), in which molecules or particles from the surrounding environment are taken up by the cell membrane. These nanoparticles have the ability to distribute nutrients straight into the intracellular region once they are inside the cell [36]. Additionally, by improving nutrient mobility inside the plant, they can improve nutrient distribution and utilisation across the entire tissue of the plant.

The capacity of nanoparticles to interact with transport proteins embedded in cell membranes affects how important substances, like nutrients, are taken up and transported by plant cells. Several processes, such as direct binding to certain locations on transport proteins or changing the shape and activity of these proteins, might cause this interaction. Through the manipulation of transport protein behaviour, nanoparticles can improve nutrition uptake efficiency by making it easier for nutrients to cross cell membranes. Plant cells may be better able to absorb nutrients as a result of this focused interaction, which may encourage healthier growth and greater resistance to environmental challenges. To guarantee the safe and efficient use of nanoparticles in agricultural applications, more research is necessary to determine the specifics of these interactions and how they affect biological processes.

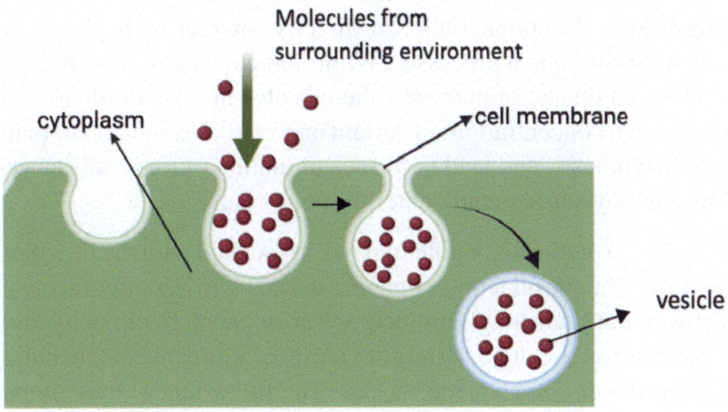

Fig. 3 Nanoparticles distributing the nutrients into intracellular region

To improve the structure of soil, certain nanomaterials can be employed, such as nano-silica or nano clays [37]. They can enhance the aeration, water retention, and porosity of the soil, all of which can help plants absorb nutrients and expand their roots. Besides nutrients, nano particles can be utilised to more efficiently administer other advantageous chemicals to plants, such as growth-promoting agents, insecticides, or herbicides.

2.3 Environmental Considerations and Potential Risks Associated with Nanomaterial Use in Agriculture

The use of nanomaterials in agriculture has gained significant attention due to their potential to revolutionize farming practices, improving crop yields, pest control, and soil health. However, alongside these benefits come environmental considerations and potential risks that need to be carefully assessed.

- **Soil Health and Microbial Activity**: Nanomaterials, such as nanoparticles, can significantly affect soil properties and microbial communities. While some nanomaterials might improve nutrient availability and promote plant growth, others could disrupt the delicate balance of soil ecosystems. For example, silver nanoparticles, known for their antimicrobial properties, might reduce the beneficial microbial populations essential for soil health and nutrient cycling.
- **Water Quality**: The application of nanomaterials in agriculture may lead to their runoff into water bodies. This poses a risk of contamination, affecting aquatic life and potentially entering human water supplies [38]. Nanoparticles can persist in water systems, and their small size allows them to pass through conventional water filtration systems, making it challenging to control and mitigate their spread [16].

- **Plant Physiology**: Nanomaterials can directly interact with plants, potentially altering their physiological processes. While some interactions may be beneficial, such as enhanced uptake of nutrients, there is also the risk of phytotoxicity [39]. For instance, high concentrations of certain nanoparticles can impair plant growth, disrupt photosynthesis, and lead to the accumulation of toxic substances in plant tissues, which may subsequently enter the food chain.

Ecotoxicity is one of the major potential risk associated with the use of nano material in agriculture. The introduction of nanomaterials into agricultural environments raises concerns about their long-term ecological impacts. Ecotoxicity studies have shown that nanoparticles can be harmful to a range of organisms, including beneficial insects, earthworms, and aquatic species. The full extent of these impacts is not yet fully understood, necessitating comprehensive risk assessments and long-term monitoring.

Bioaccumulation is the another risk. Nanomaterials have the potential to bioaccumulate in the food chain. This occurs when nanoparticles taken up by plants are ingested by herbivores, and then by predators higher up the food chain. The cumulative effect of this bioaccumulation could pose health risks to wildlife and humans, including potential toxicity and the disruption of biological processes.

2.4 Nano-Soil Amendments

The application of nanoparticles to soil to improve its qualities and increase crop yields is known as nano-soil amendments (presented in Fig. 4), and it is a class of innovative agricultural technology. Several soil and plant-related issues are addressed by these amendments by utilising the special qualities of nanoparticles.

Nanofertilizers are a form of advanced agricultural input that utilizes nanotechnology to enhance the delivery and efficiency of nutrients to plants. They are designed to overcome the limitations of conventional fertilizers by providing a more controlled

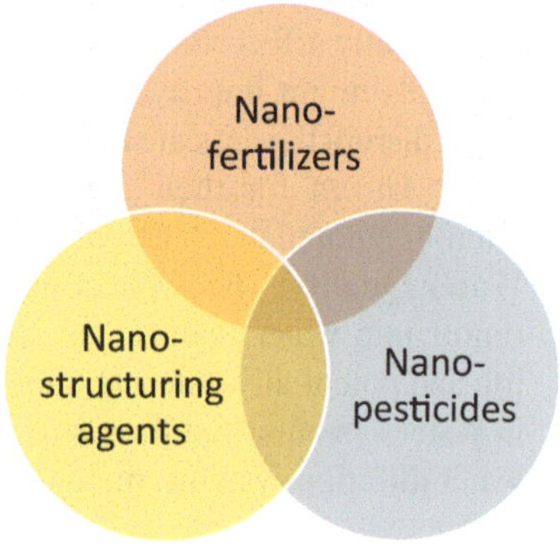

Fig. 4 Nano soil Amendments

and efficient release of nutrients, thereby improving plant growth and reducing environmental impact. Utilising nanotechnology, nano-pesticides are a sophisticated class of agricultural compounds that improve pest control agents' specificity, effectiveness, and safety for the environment. By using nanoparticles to more efficiently carry active ingredients to their target pests, these insecticides reduce the total amount of chemicals needed and lessen their environmental impact. Nanocapsules, nanosuspensions, nano-emulsions and nanogels are examples of nano pesticides.

Nano clays, nano silica and nano polymers are nano structuring agents. They play a crucial role in enhancing soil properties, promoting plant growth, and improving the efficiency of agrochemicals. These agents can enhance the aggregation of soil particles, increasing porosity and improving aeration. This leads to better root growth and soil health. The water-holding capacity of the soil can be greatly increased by using nano-hydrogels and nano-clays, which will provide a consistent supply of water for plants and lessen the need for regular watering.

3 Artificial Intelligence in Soil Nutrient Management

Agriculture's use of artificial intelligence (AI) is transforming the field of soil nutrient management. AI is assisting farmers in making better decisions regarding agricultural production, which will increase crop quality, added value, and productivity. This is achieved by utilising cutting-edge technology like robotics, the Internet of Things (IoT), and data analysis [40, 41]. As part of its smart agriculture promotion project, the Ministry of Agriculture, Forestry and Fisheries in Japan is pushing the adoption of Wide Area Generalised Agricultural Data Infrastructure (WAGRI), with the goal of having "almost all of the leaders in agriculture practise data-driven agriculture" by 2025.

Using sensors to gather data on soil conditions and then analysing it using AI systems to provide farmers ideas on how to maximise crop development is one of the main uses of AI in soil nutrient management [42]. AI-driven systems, for instance, are able to identify patches of a field where the soil is excessively dry or wet and may advise on the best times and amounts of water to apply. In a similar vein, AI is able to detect nutrient deficits in the soil and recommend the right kinds and dosages of fertiliser.

AI is also being used to identify plant illnesses and insects faster and more precisely than humans, allowing farmers to respond rapidly to avoid crop damage. AI-powered solutions can also automate insect or weed control, which lowers the demand for pesticides and increases crop yields [43]. It can assist farmers in implementing regenerative agriculture techniques like cover crops, decreased tillage, and nutrient management to improve the health of their soil. Farmers may make better judgements by using AI technologies to assess how these activities affect soil carbon and financial performance [44].

3.1 Predictive Analytics

In soil nutrient management, predictive analytics is the use of advanced statistical and machine learning methods to data analysis in order to forecast future occurrences or results pertaining to crop yield and soil health. This requires applying models and algorithms to massive dataset which frequently contain data on crop varieties, soil characteristics, climate, and other variables in order to spot patterns and trends that might guide decision-making [45].

Predictive analytics can be used in various aspects of soil nutrient management, including:

- **Crop selection and planning**: By analysing soil data, historical weather patterns, and other factors, farmers can use predictive analytics to select the most suitable crops for their fields and plan their planting schedules accordingly.
- **Optimizing irrigation and fertilizer application**: Predictive models can help farmers determine the optimal amounts of water and nutrients to apply to their crops, reducing waste and minimizing environmental impacts.
- **Soil health monitoring**: Predictive analytics can be used to identify potential nutrient deficiencies or other issues in soil health, enabling farmers to take proactive measures to address these problems before they affect crop yields [46].
- **Pest and disease management**: By analysing data on weather patterns, soil conditions, and other factors, predictive models can help farmers anticipate and prepare for potential pest and disease outbreaks, reducing the need for pesticides and other chemicals.
- **Yield prediction and supply chain management**: Predictive analytics can be used to forecast crop yields, enabling farmers and agribusinesses to better manage their supply chains and make informed decisions about pricing, storage, and distribution.

3.2 Machine Learning Models

Agriculture is being revolutionised by machine learning (ML) models for soil nutrient management, which offer accurate, data-driven insights for maximising nutrient utilisation. Large-scale data analysis is done by these models to produce well-informed predictions and suggestions that improve crop yields, cut costs, and have the least negative effects on the environment [47, 48]. The key components of machine learning models are depicted in Fig. 5.

Data Collection: Data can be collected from various sources to train the machine learning models [49]. Soil Sampling is one of the method which is a collection of soil samples for laboratory analysis to determine nutrient levels and other properties. Use of satellite imagery and drones to gather data on soil properties and crop health is called remote sensing. Further, sensors can be deployed in soil to monitor soil moisture, temperature, pH, and nutrient levels in real-time.

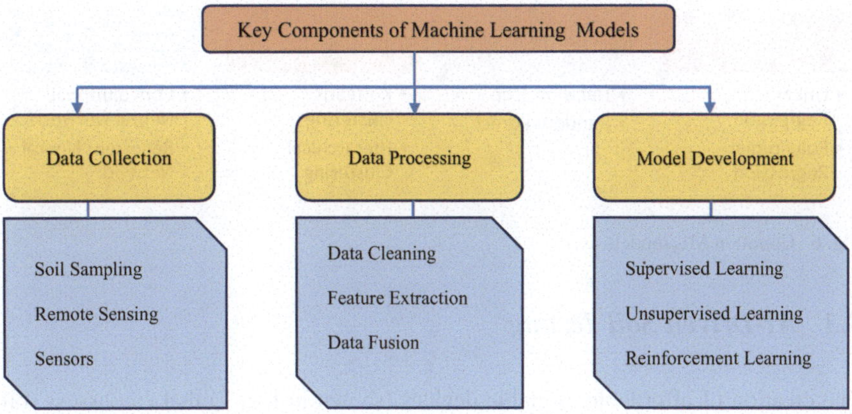

Fig. 5 Key components of machine learning models for soil nutrient management

Data Processing and Integration: Once the data is collected from any of the above mentioned methods, it needs to be processed. The first step includes data cleaning where removal of noise and errors from raw data is done to ensure accuracy [50]. Identification of key variables (features) that influence soil nutrient levels and crop growth are predicted after data cleaning. Data fusion is another step that comes under data integration where integration of data from multiple sources (e.g., soil samples, remote sensing, weather data) for comprehensive analysis is done.

Model Development: A model can learn in different ways i.e. supervised learning, un supervised learning and Reinforcement Learning. Training models using labelled data (e.g., known soil nutrient levels and corresponding crop yields) to predict nutrient needs is called supervised learning whereas identifying patterns and clusters in unlabelled data (e.g., grouping fields with similar soil properties) refers to unsupervised learning. Reinforcement Learning includes developing models that learn optimal nutrient management strategies through trial and error.

A wide range of models (depicted in Fig. 6) are included in machine learning (ML), and each is appropriate for a particular set of tasks and data. Linear Regression model predicts soil nutrient levels based on input variables. On the other hand, polynomial regression model handles non-linear relationships between variables. Decision trees help to classify soil samples into categories (e.g., nutrient-rich, nutrient-poor). Random Forest is an ensemble method that improves accuracy by combining multiple decision trees. K-Means Clustering is used to group soil samples with similar nutrient profiles for targeted management. Hierarchical Clustering builds a hierarchy of clusters to understand the relationships between different soil samples.

A neural network is a computational model that is based on the information processing mechanisms of biological neural networks found in the human brain. Convolutional Neural Networks (CNNs) analyzes spatial data from remote sensing images to assess soil and crop health [51]. Recurrent Neural Network models temporal changes in soil properties and nutrient levels over time.

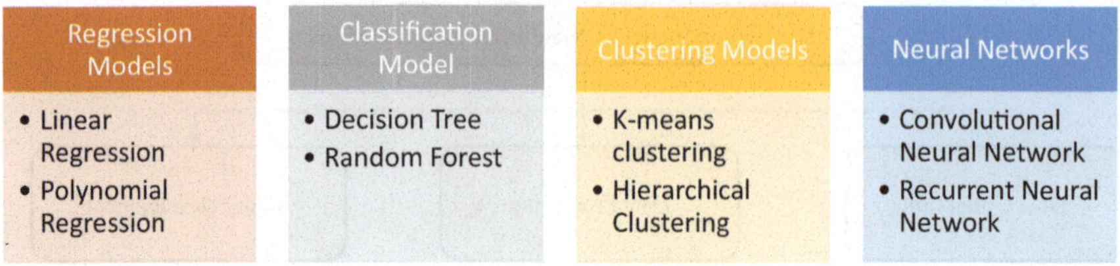

Fig. 6 Common ML models used

3.3 AI-Driven Soil Testing

The creation of affordable, portable devices (shown in Fig. 7) that can assess soil fertility and health in real-time and give farmers data-driven insights to improve their crop management techniques is one novel component of AI-driven soil testing. These gadgets use AI-powered sensors to measure a variety of soil factors, such as pH levels, nutrient content, humidity, radiation, temperature, and soil moisture. Examples of these devices are Omishtu-Joy in Ethiopia and Farmspace in Nigeria. After data collection, it is sent to a smartphone app for analysis. The software tells farmers which crops to plant for their particular soil type and suggests how much fertiliser to use. These illustrations show how AI-driven soil testing has the potential to transform agriculture by giving farmers quick, precise, and data-driven insights regarding the fertility and health of their soil.

Fig. 7 AI based applications for nutrient management in soil

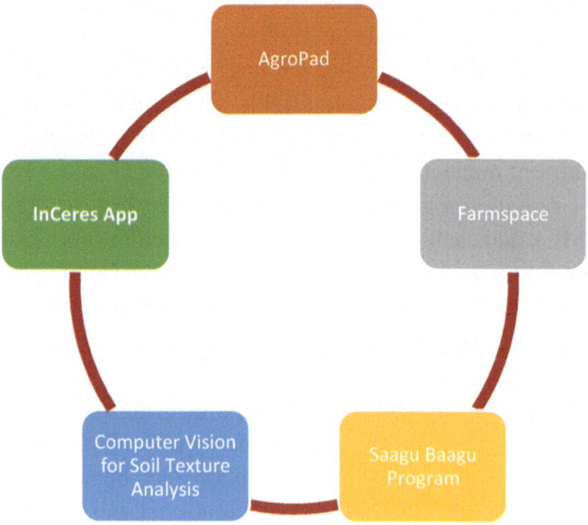

3.4 *AI for Nutrient Management Recommendations*

AI is being increasingly used in nutrient management to provide personalized recommendations for optimal nutrient intake. Some examples are mentioned below:

DRI Calculator: The Dietary Reference Intake (DRI) Calculator is a tool designed for healthcare professionals to calculate daily nutrient recommendations based on the DRIs established by the National Academies of Sciences, Engineering, and Medicine. This calculator takes into account factors such as height, weight, age, and activity level to generate a report on recommended intakes of macronutrients, water, vitamins, and minerals.

Machine Learning Models for Soil Nutrient Properties Prediction: A systematic review explored the use of machine learning models for predicting soil nutrient properties, highlighting the potential of AI in smart agriculture and digital soil management. The review examined the current state of AI models for soil property management and soil fertility prediction, including machine learning and deep learning algorithms.

Unilever's Nutrient Profiling Model: Unilever uses its own nutrient profiling model to assess the levels of nutrients and healthiness of its products. The company has been using this model for over a decade and has recently started using it alongside six other models to improve the nutritional quality of its products. This approach demonstrates the use of AI in the food industry to optimize product formulations and improve nutritional quality.

Bayer's Artificial Intelligence in Healthcare: Bayer is harnessing cutting-edge artificial intelligence (AI) technologies to transform healthcare, although the specific application to nutrient management is not detailed.

Remote Sensing for Nutrient Monitoring: Deep Planet discusses the potential of remote sensing from satellite imagery to monitor nutrient levels within soil and tissue, identifying nutrient deficiencies in a cost-effective and timely manner. This approach uses machine learning technologies to analyze the spatial distribution of nutrients in crop plants, providing recommendations for farmers on fertilizer application.

4 Comparative Analysis

Enhancing soil nutrient management sustainably is essential for good agricultural yield and protecting the environment. Artificial intelligence (AI) and nanomaterials both provide creative answers to this problem. Table 2 is representing the comparative analysis of nanomaterials and AI for sustainable nutrient management in soil.

AI and nanomaterials both have potential for long-term soil nutrient management. Improved nutrition uptake and delivery are provided by nanomaterials, which may lessen environmental effect. However, the costs and regulatory obstacles prevent them from being widely used. Although it is also expensive, artificial intelligence (AI) can provide scalable solutions with less impact on the environment by optimising nutrient

Table 2 Comparative analysis of nanomaterials and AI for sustainable nutrient management in soil

Aspect	Nanomaterials	Artificial intelligence
Nutrient delivery	Enhanced, slow, and controlled release	Optimized based on precise data analysis
Nutrient uptake	Increased bioavailability for plants	Predictive analytics to identify nutrient deficiencies
Environmental impact	Reduced runoff and pollution, but uncertain long-term effects	Minimized waste and over-application, no new materials added
Pest and disease control	Can serve as carriers for pesticides and fungicides	N/A
Safety concerns	Potential toxicity and long-term environmental persistence	Low, focused on data and software
Cost	High production cost, challenging scalability	High initial investment, but potential for cost savings
Scalability	Limited by production and regulatory hurdles	Easier global scalability with digital infrastructure
Data requirements	Not applicable	High-quality, granular data needed for effective implementation
Technical complexity	Moderate, requires material engineering	High, requires expertise in AI and data management
Integration	Physical integration with soil and plants	Digital integration with farming systems
Real-time monitoring	Not applicable	Provides real-time monitoring and dynamic adjustments
Adoption barriers	Regulatory scrutiny, cost, and production scale	Technical expertise, initial cost, and farmer resistance
Long-term viability	Dependent on overcoming safety and regulatory concerns	Strong potential with continuous advancements in AI

management techniques using data driven insights and predictive analytics [46]. Combining the two technologies- using AI to identify exact nutrient requirements and using nanomaterials to effectively distribute those nutrient might be the best course of action. By maximising the advantages of each technology, this synergy could result in more productive and sustainable agriculture practices.

5 Integration of Nanotechnology and AI

Table 3 highlights the integration of nanomaterials and AI in various research papers for sustainable nutrient management in soil. The application methods include precision agriculture, nanosensors for monitoring soil conditions, and smart nano formulations and delivery systems. The nutrient management techniques include soil health monitoring, fertilizer efficiency, and soil remediation. The AI techniques employed include AI and machine learning for safety assessment of nanomaterials, and AI for sustainable agriculture.

Table 3 Integration of nanomaterials and AI for sustainable nutrient management

Research paper	Nanomaterials used	Application methods	Nutrient management techniques	AI techniques employed	Performance metrics evaluated
Nanotech and AI could hold key to unlocking global food security	Not specified	Precision agriculture	Soil health monitoring, fertilizer efficiency	AI and machine learning for safety assessment of nanomaterials	Crop yields, resource waste reduction
Eco-friendly nanomaterial synthesis for soil–plant systems	Not specified	Nanosensors for monitoring soil moisture, pH, herbicide and pesticide concentrations, and nutrient levels	Soil health monitoring, fertilizer efficiency	Not mentioned	Not mentioned
Nano enabled agriculture for sustainable soil	Biogenic silicon nanoparticles, iron oxide-functionalized magnetic imogolite nanocomposites	Soil remediation, plant nutrition	Soil microbiome enhancement, nutrient management	Not mentioned	Soil fertility, microbial abundance, diversity, activity
Nano-enable materials promoting sustainability and resilience in agriculture	Smart nanoformulations and delivery systems	Improving crop protection and plant nutrition	Soil health monitoring, fertilizer efficiency	Not mentioned	Crop yields, resource waste reduction
Integration of hybrid nanomaterials and artificial intelligence for sustainable agriculture	Hybrid nanomaterials	Not specified	Not specified	AI for sustainable agriculture	Not specified

5.1 Potential Benefits of Integrating Nanomaterials and AI

The integration of nanomaterials and AI in agriculture has the potential to revolutionize the way we manage nutrients in soil, leading to more efficient and sustainable practices.

- Improved Nutrient Management: Nanomaterials can enhance the efficiency of fertilizers, reducing waste and environmental pollution. AI can optimize fertilizer application by analyzing soil conditions, crop growth, and weather patterns, ensuring that nutrients are delivered precisely when and where they are needed.
- Enhanced Crop Yields: Nanomaterials can increase crop yields by improving nutrient uptake and reducing stress. AI can analyze data from sensors and drones to identify areas where crops are under stress, allowing for targeted interventions to improve yields.
- Reduced Environmental Impact: Nanomaterials can reduce the environmental impact of agriculture by minimizing the use of chemical fertilizers and pesticides. AI can help optimize irrigation systems, reducing water waste and conserving this valuable resource.
- Increased Food Security: The integration of nanomaterials and AI can help address global food security challenges by increasing crop yields, reducing waste, and improving the efficiency of agricultural systems.
- Precision Agriculture: Nanomaterials and AI can enable precision agriculture, where farmers respond in real-time to changes in crop growth and soil conditions. This approach can reduce waste, improve yields, and promote sustainable agriculture practices [52].
- Sustainable Agriculture: The integration of nanomaterials and AI can support sustainable agriculture by reducing the environmental impact of farming, improving crop yields, and promoting more efficient use of resources.

6 Conclusion

The comparative analysis of nanomaterials and artificial intelligence for sustainable nutrient management in soil highlights the potential of these technologies to transform agriculture towards a more sustainable and resilient future. By enhancing crop yields, decreasing waste, and encouraging better resource efficiency, the combination of nanomaterials and AI can help accomplish the Sustainable Development Goals (SDGs) for food security, agriculture, and the environment. It has been demonstrated that nanomaterials improve fertiliser efficiency, lowering waste and contamination to the environment. Through the analysis of weather patterns, crop development, and soil conditions, artificial intelligence (AI) can optimise the application of fertiliser, ensuring that nutrients are provided precisely when and where they are needed.

Precision farming and less environmental effect are made possible by the use of nanosensors in agriculture, which can measure a variety of parameters such as soil moisture, pH, concentrations of pesticides and herbicides, and nutrient levels. By enhancing crop yields, encouraging more effective resource use, and lessening the negative environmental effects of farming, the integration of nanomaterials and AI can promote sustainable agriculture. This strategy can assist address issues with global food security by boosting crop yields, cutting waste, and enhancing agricultural systems' effectiveness. Investigating the effects of engineered nanomaterials on soil microbial communities and potential risks to human and environmental health, developing intelligent nano formulations and delivery systems, investigating nano remediation strategies for contaminated soils, developing nanosensors for plant health and food quality and safety monitoring, and developing nanomaterial-related solutions for smart food packaging are some of the research gaps and areas for innovation that still need to be filled. In summary, the combination of AI and nanomaterials has the power to completely change how we manage soil nutrients, resulting in more effective and sustainable agriculture methods. In order to fully reap the rewards of these technologies and tackle the problems associated with sustainable agriculture, more study and development are required.

References

1. Tripathi, A.D., Mishra, R., Maurya, K.K., Singh, R.B., Wilson, D.W.: Estimates for world population and global food availability for global health. In: The Role of Functional Food Security in Global Health, pp. 3–24. Elsevier (2018). https://doi.org/10.1016/B978-0-12-813148-0.00001-3
2. Bhattacharjee, T., Konwar, A., Boruah, J.S., Chowdhury, D., Majumdar, G.: A sustainable approach for heavy metal remediation from water using carbon dot based composites: a review. J. Hazard. Mater. Adv. **10**, 100295 (2023). https://doi.org/10.1016/j.hazadv.2023.100295
3. Thakur, A., Khanna, V., Murtaza, Q.: Characterization, strength and wear optimization of aluminium based boron carbide/graphite hybrid composite by TOPSIS for dry condition. ECS J. Solid State Sci. Technol. **12**(12), 127005 (2023). https://doi.org/10.1149/2162-8777/ad1209
4. Singh, K., Khanna, V., Chaudhary, V.: Effect of hybrid reinforcements on the mechanical properties of copper nanocomposites. ECS J. Solid State Sci. Technol. **11**, 097001 (2022). https://doi.org/10.1149/2162-8777/ac8bf9
5. Chowdhury, S., et al.: Role of cultural and nutrient management practices in carbon sequestration in agricultural soil. In: Advances in Agronomy, vol. 166, pp. 131–196. Academic Press Inc. (2021). https://doi.org/10.1016/bs.agron.2020.10.001
6. Powlson, D.S., et al.: Soil management in relation to sustainable agriculture and ecosystem services. Food Policy **36**(SUPPL 1), S72–S87 (2011). https://doi.org/10.1016/j.foodpol.2010.11.025
7. Selim, M.M.: Introduction to the integrated nutrient management strategies and their contribution to yield and soil properties. Int. J. Agron. Hindawi Limited (2020). https://doi.org/10.1155/2020/2821678
8. Khanna, V., Kumar, V., Bansal, S.A.: Mechanical properties of aluminium-graphene/carbon nanotubes (CNTs) metal matrix composites: advancement, opportunities and perspective. Mater. Res. Bull. **138**, 111224 (2021). https://doi.org/10.1016/j.materresbull.2021.111224
9. Khanna, V., Kumar, V., Bansal, S.A.: Effect of carbonaceous nanomaterials' reinforcement on mechanical properties of aluminium metal-based nanocomposite: a review. Mater. Today Proc. **38**, 289–295 (2021). https://doi.org/10.1016/j.matpr.2020.07.221

10. Bala, M., Sharma, R., Gupta, S.: Integration of hybrid nanomaterials and artificial intelligence for sustainable agriculture, pp. 97–118 (2024). https://doi.org/10.4018/979-8-3693-1261-2.ch004
11. Dahiya, M., Khanna, V., Bansal, S.A.: Finite element analysis of the mechanical properties of graphene aluminium nanocomposite: varying weight fractions, sizes and orientation. Carbon Lett. **33**, 1601–1613 (2023). https://doi.org/10.1007/s42823-023-00543-x
12. Nagarajan, T., Sridewi, N., Wong, W.P., Walvekar, R., Khanna, V., Khalid, M.: Synergistic performance evaluation of MoS2–hBN hybrid nanoparticles as a tribological additive in diesel-based engine oil. Sci. Rep. **13**(1), 12559 (2023). https://doi.org/10.1038/s41598-023-39216-0
13. Shepherd, R.M., Oliverio, A.M.: Micronutrients modulate the structure and function of soil bacterial communities. Soil Biol. Biochem. **192** (2024). https://doi.org/10.1016/j.soilbio.2024.109384
14. Solanki, P., Bhargava, A., Chhipa, H., Jain, N., Panwar, J.: Nano-fertilizers and their smart delivery system. In: Nanotechnologies in Food and Agriculture, pp. 81–101. Springer International Publishing (2015). https://doi.org/10.1007/978-3-319-14024-7_4
15. Ji, M., AL-Huqail, A.A., Majdi, A., Liu, H., Ali, I., Marzouki, R.: Remediation of mine polluted soil with nano-enhanced materials: Development of extreme learning machine approaches. Chemosphere **307** (2022). https://doi.org/10.1016/j.chemosphere.2022.135772
16. Bin Khalid, Z., Nasrullah, M., Nayeem, A., Wahid, Z.A., Singh, L., Krishnan, S.: Application of 2D graphene-based nanomaterials for pollutant removal from advanced water and wastewater treatment processes. In: ACS Symposium Series, vol. 1353, pp. 191–217. American Chemical Society (2020). https://doi.org/10.1021/bk-2020-1353.ch009
17. Raja, M.A., Husen, A.: Role of nanomaterials in soil and water quality management. In: Nanomaterials for Agriculture and Forestry Applications, pp. 491–503. Elsevier (2020). https://doi.org/10.1016/B978-0-12-817852-2.00020-2
18. Duhan, J.S., Kumar, R., Kumar, N., Kaur, P., Nehra, K., Duhan, S.: Nanotechnology: the new perspective in precision agriculture. Biotechnol. Reports **15**, 11–23 (2017). Elsevier B.V. https://doi.org/10.1016/j.btre.2017.03.002
19. Gontijo da Cunha, V.A., Hariharan, J., Ampatzidis, Y., Roberts, P.D.: Early detection of tomato bacterial spot disease in transplant tomato seedlings utilising remote sensing and artificial intelligence. Biosyst. Eng. **234**, 172–186 (2023). https://doi.org/10.1016/j.biosystemseng.2023.09.002
20. Lutz, É., Coradi, P.C.: Applications of new technologies for monitoring and predicting grains quality stored: sensors, internet of things, and artificial intelligence. Meas.: J. Int. Meas. Confed. **188** (2022). Elsevier B.V. https://doi.org/10.1016/j.measurement.2021.110609
21. Singh, K., et al.: Paradigm of state-of-the-art CNT reinforced copper metal matrix composites: processing, characterizations, and applications. J. Market. Res. **24**, 8572–8605 (2023). https://doi.org/10.1016/j.jmrt.2023.05.083
22. Gupta, P., et al.: Synergetic effect of CeO2 doping on structural and tribological behavior of Fe-Al_2O_3 metal matrix nanocomposites. ECS J. Solid State Sci. Technol. **11**(11), 117001 (2022). https://doi.org/10.1149/2162-8777/ac9c92
23. Singh, K., et al.: Panorama of physico-mechanical engineering of Graphene-reinforced copper composites for sustainable applications. Mater. Today Sustain. 100560 (2023). https://doi.org/10.1016/j.mtsust.2023.100560
24. Ramachandran, R.P., Erkinbaev, C., Vellaichamy, C.: Smart nano-biosensors in sustainable agriculture and environmental applications. In: Food, Medical, and Environmental Applications of Nanomaterials, pp. 527–542. Elsevier (2022). https://doi.org/10.1016/B978-0-12-822858-6.00019-4
25. Agrawal, S., Kumar, V., Kumar, S., Shahi, S.K.: Plant development and crop protection using phytonanotechnology: a new window for sustainable agriculture. Chemosphere **299** (2022). Elsevier Ltd. https://doi.org/10.1016/j.chemosphere.2022.134465
26. Zhang, P., Guo, Z., Ullah, S., Melagraki, G., Afantitis, A., Lynch, I.: Nanotechnology and artificial intelligence to enable sustainable and precision agriculture. Nat. Plants **7**(7), 864–876 (2021). Nature Research. https://doi.org/10.1038/s41477-021-00946-6

27. Danai-Varsou, D., Zhang, P., Afantitis, A., Guo, Z., Lynch, I., Melagraki, G.: Nanoinformatics and artificial intelligence for nano-enabled sustainable agriculture. In: Nano-Enabled Sustainable and Precision Agriculture, pp. 503–531. Elsevier (2023). https://doi.org/10.1016/B978-0-323-91233-4.00015-6
28. Martinelli, F., et al.: Advanced methods of plant disease detection. A review. Agron. Sustain. Dev. **35**(1), 1–25 (2015). https://doi.org/10.1007/s13593-014-0246-1
29. Institute of Electrical and Electronics Engineers and Hindusthan Institute of Technology. In: Proceedings of the International Conference on Electronics and Sustainable Communication Systems (ICESC 2020), 02–04 July 2020 (2020)
30. Yin, Z., Cui, C., Chen, H., Duoni, Yu, X., Qian, W.: The application of carbon nanotube/graphene-based nanomaterials in wastewater treatment. Small **16**(15) (2020). Wiley-VCH Verlag. https://doi.org/10.1002/smll.201902301
31. Babu, S., et al.: Nanofertilizers for agricultural and environmental sustainability. Chemosphere **292** (2022). https://doi.org/10.1016/j.chemosphere.2021.133451
32. Machado, T.O., Grabow, J., Sayer, C., de Araújo, P.H.H., Ehrenhard, M.L., Wurm, F.R.: Biopolymer-based nanocarriers for sustained release of agrochemicals: a review on materials and social science perspectives for a sustainable future of agri- and horticulture. Adv. Colloid Interface Sci. **303** (2022). Elsevier B.V. https://doi.org/10.1016/j.cis.2022.102645
33. Khin, M.M., Nair, A.S., Babu, V.J., Murugan, R., Ramakrishna, S.: A review on nanomaterials for environmental remediation. Energy Environ. Sci. **5**(8), 8075–8109. Royal Society of Chemistry (2012). https://doi.org/10.1039/c2ee21818f
34. Bala, M., Kumar Bansal, S., Fatima, F.: Nanotechnology: a boon for agriculture. Mater. Today Proc. **73**, 267–270 (2023). https://doi.org/10.1016/j.matpr.2022.09.498
35. Nishu, Kumar, S.: Smart and innovative nanotechnology applications for water purification. Hybrid Adv. **3**, 100044 (2023). https://doi.org/10.1016/j.hybadv.2023.100044
36. Saha, S., Bharadwaj, A.: A step towards smart agriculture using metallic nanostructures. Plant Stress **10** (2023). Elsevier B.V. https://doi.org/10.1016/j.stress.2023.100216
37. Merino, D., Tomadoni, B., Salcedo, M.F., Mansilla, A.Y., Casalongué, C.A., Alvarez, V.A.: Nanoclay as carriers of bioactive molecules applied to agriculture. In: Handbook of Nanomaterials and Nanocomposites for Energy and Environmental Applications, pp. 1–22. Springer International Publishing (2020). https://doi.org/10.1007/978-3-030-11155-7_62-1
38. Tohamy, H.A.S., El-Sakhawy, M., Kamel, S.: Development of magnetite/graphene oxide hydrogels from agricultural wastes for water treatment. J. Renew. Mater. **10**(7), 1889–1909 (2022). https://doi.org/10.32604/jrm.2022.019211
39. Tewari, C., et al.: Can graphene-based composites and membranes solve current water purification challenges - a comprehensive review. Desalination **567** (2023). Elsevier B.V. https://doi.org/10.1016/j.desal.2023.116952
40. Thangaraj, R., Anandamurugan, S., Pandiyan, P., Kaliappan, V.K.: Artificial intelligence in tomato leaf disease detection: a comprehensive review and discussion. J. Plant Dis. Prot. **129**(3), 469–488 (2022). https://doi.org/10.1007/s41348-021-00500-8
41. Nagasubramanian, K., Jones, S., Singh, A.K., Sarkar, S., Singh, A., Ganapathysubramanian, B.: Plant disease identification using explainable 3D deep learning on hyperspectral images. Plant Methods **15**(1), 1–10 (2019). https://doi.org/10.1186/s13007-019-0479-8
42. Divya, S., et al.: Smart data processing for energy harvesting systems using artificial intelligence. Nano Energy **106** (2023). Elsevier Ltd. https://doi.org/10.1016/j.nanoen.2022.108084
43. Bedi, P., Gole, P.: Plant disease detection using hybrid model based on convolutional autoencoder and convolutional neural network. Artif. Intell. Agric. **5**, 90–101 (2021). https://doi.org/10.1016/j.aiia.2021.05.002
44. Zürner, S., Deutschländer, L.P., Schieck, M., Franczyk, B.: Sustainable development of AI applications in agriculture: a review. Procedia Comput. Sci. **225**, 3546–3553 (2023). https://doi.org/10.1016/j.procs.2023.10.350
45. Nishad, M.A.R., Mitu, M.A., Jahan, N.: Predicting and classifying potato leaf disease using K-means segmentation techniques and deep learning networks. Procedia Comput. Sci. **212**(C), 220–229 (2022). https://doi.org/10.1016/j.procs.2022.11.006

46. Ahmed, A., He, P., He, P., Wu, Y., He, Y., Munir, S.: Environmental effect of agriculture-related manufactured nano-objects on soil microbial communities. Environ. Int. **173** (2023). Elsevier Ltd. https://doi.org/10.1016/j.envint.2023.107819
47. Iqbal, M.A., Talukder, K.H.: Detection of potato disease using image segmentation and machine learning. In: 2020 international conference on wireless communications, signal processing and networking, WiSPNET 2020, pp. 43–47 (2020). https://doi.org/10.1109/WiSPNET48689.2020.9198563
48. Kiranmai, B., Venu Vasantha, S., Rama Krishna, S.: Techniques for rice leaf disease detection using machine learning algorithms. Int. J. Eng. Techn. Res. **9**(8), 162–166 (2021). www.ijert.org
49. Bera, T., Das, A., Sil, J., Das, A.K.: A Survey on Rice Plant Disease Identification using Image Processing and Data Mining Techniques, vol. 814. Springer Singapore (2019). https://doi.org/10.1007/978-981-13-1501-5_31
50. Pal, A., Kumar, V.: AgriDet: plant leaf disease severity classification using agriculture detection framework. Eng. Appl. Artif. Intell. **119**, 105754 (2023). https://doi.org/10.1016/j.engappai.2022.105754
51. Agarwal, M., Singh, A., Arjaria, S., Sinha, A., Gupta, S.: ToLeD: tomato leaf disease detection using convolution neural network. In: Procedia Computer Science, pp. 293–301. Elsevier B.V. (2020). https://doi.org/10.1016/j.procs.2020.03.225
52. Eltohamy, K.M., Khan, S., He, S., Li, J., Liu, C., Liang, X.: Prediction of nano, fine, and medium colloidal phosphorus in agricultural soils with machine learning. Environ. Res. **220** (2023). https://doi.org/10.1016/j.envres.2023.115222

Biocompatible Cellulose Derivatives: Green Chemistry and Its Sustainable Applications

Debangana Das, Anuvansh Singh, Rithika Adari, Gaurav Mudgal, and Panjanathan Radha

Abstract In material science, cellulose the main building block of plant cell walls is becoming known as a practical and affordable alternative. Cellulose, which is derived from a variety of materials including wood and agricultural waste, has the potential to replace petroleum-based resources with renewable ones. Because of its distinct structure, biocompatibility, and effective qualities, biopolymer nanocellulose which comes from a variety of plants and agricultural waste is gaining attention in the field of advanced nanotechnology for use in biosensing, catalysis, and drug delivery. Chiral chromatographic separation is being used for cellulose derivatives, which offer viable substitutes. This chapter addresses the environmental effects and economic feasibility of cellulose manufacturing processes, including the use of organic acids, wood pulp, and ionic liquids. Difficulties are addressed in detail, including competition from established products and restricted usefulness. Applications for cellulose and its derivatives are found in a wide range of industries, such as biomedicine, sensors, water purification, and microbiological activity. The chapter also highlights the critical role that continuous research plays in improving cellulose-based materials for sustainable and functional reasons by projecting future developments in green solubilization systems, smart electronics, and sustainable alterations.

Keywords Functionalized cellulose · Mineralization · Chemical functionalization · Nanotechnology · Sustainable applications

---

D. Das · A. Singh · R. Adari · P. Radha (✉)
Bioprocess Principles Laboratory, Department of Biotechnology, Faculty of Engineering and Technology, SRM Institute of Science and Technology, Kattankulathur, Chengalpattu District, Chennai 603203, Tamil Nadu, India
e-mail: radhap@srmist.edu.in

G. Mudgal (✉)
University Institute of Biotechnology, Chandigarh University, Mohali 140413, Punjab, India
e-mail: gauravmdgl@gmail.com

Center for Waste Management and Renewable Energy, Saveetha Dental College and Hospitals, Saveetha Institute of Medical and Technical Sciences, Saveetha University, Chennai 600077, India

R&D, Vanproz Agrovet LLP, Bagroda Industrial Area, Bhopal 462026, Madhya Pradesh, India

K. K. Kesari et al. (eds.), *Functionalized Cellulose Materials*, Engineering Materials,
https://doi.org/10.1007/978-3-031-76953-5_7

1 Introduction

Cellulose has always been an asset to humanity in its natural form. However, even though it was widely used and turned out to be extremely suitable for various applications, its chemical structure was unknown for centuries. Modern cellulose chemistry was first discovered by Anselme Payen, who isolated cellulose from plants and defined it chemically in 1837 [1]. In the early 1920s, that before, a superior number of researchers and scientists assumed that cellulose is made up of one or another limited number of very tiny glucose or cellobiose molecules. Nevertheless, as this material was a polymer, it could neither be produced nor recognized by the scientific world. At that time, the concept of cellulose's polymer nature was resisted by the scientists largely. As a result, for over ten years, the debates about the structure of cellulose continued. In 1920, Hermann Staudinger, finally, ended the opposing arguments and showed that cellulose was a polymer. In its unmodified and natural form, cellulose is commonly used in the paper and textile industry, construction, and personal care. Its rigidity, capability to regenerate, structural properties, resistance, and good thermal insulation features make cellulose vital to these industries. Natural cellulose is broadly used in the human food industry and the manufacture of medication as it is high in fibre, which is good for food and patient health and nutrition. Besides, it is a non-hazardous way to improve drug formulations, enabling them to be released better and formulated into tablets. Moreover, the cellulose has a porous rigid structure, making it suitable as a wire filter medium for air, water, and beverage filtration [2].

1.1 Attributes of Cellulose

Cellulose is a nucleic polymer of glucose. It has a unique crystal structure mediated by intermolecular and intramolecular interactions. Cellulose gets its stability from the conformation called the 'chair' of the glucopyranose cycles (C1). The understanding of these structures is crucial to the interpretation of cellulose's physical properties. Rotary isomerism, which follows rotations around glycosidic bonds, affects cellulose's macromolecule forms. Such forms include intramolecular hydrogen bonds, which play a vital role in glucopyranose residue's stability. Hydrogen bonds also greatly contribute to cellulose's physical structure. The rigidity and stability, tensile strength, elastic properties and solubility of cellulose are all outcomes of hydrogen bonds. A successful cooperation between hydrogen and Van der Waals forces is responsible for the stability of the rod-like structure of the cellulose molecule.

As for cellulose's chemical properties, the majority of its chemical reactivity can be explained by the cellulose's hydroxyl groups. Glucose hydroxyl groups cause fluctuations that give rise to the formation of rotamers that can be fabricated into desired polymers' chemical reactivity. The most common reactions include esterification, etherification or oxidation. Cellulose also can be found in various structural forms (I, II and III) with various conformational characteristics that affect its chemical reactivity. Thus, cellulose can undergo various chemical processes or interact with other molecules. Therefore, its applicability in various areas of technology or pharmacy is defined. [3].

1.2 Limitations of Natural Cellulose

Although natural cellulose is abundant, cheap, versatile and easily obtainable, there are many limitations to its applicability to reactions as it is in many industries. One of the most important is its restricted chemical reactivity, which results from the presence of only hydroxyl groups. In the natural state, other groups enhancing cellulose's hydrophobicity, antibacterial properties or activity against selected substances cannot be inserted [4, 5]. The reactivity of these substances to manufactured forms of cellulose is therefore limited. Another great disadvantage of natural cellulose is its low solubility in solvents. This affects its applicability in high-performing industries like aerospace or automotive. Its hydrophilic nature makes it prone to water. Natural cellulose swells when exposed to water causing dimension instability. Its biodegradability can also be perceived as a disadvantage when applied in an industry that requires microbial resistance and durability.

1.3 Functionalized Cellulose

Cellulose's chemical reactivity is mainly determined by the crystallinity degree and hydroxyl group arrangement. This could be suggestive of compounds such as cellulose acetate, carboxymethyl cellulose and methylcellulose. Cellulose functionalization is a renewable tool for non-renewable sources. Advanced applications include the use of nanocomposites, 3D printing and in the biomedical field [6]. It should also be added that functionalized nanocelluloses improve the hydrophobic interaction with the polymeric matrices. They could be used for environmental protection and sustainability. Figure 1 represents the consolidated characteristics of functionalized cellulose compared with the natural cellulose [4, 7].

2 Functionalization and Manufacturing of Cellulose Materials

Cellulose is the most prevalent polymer in the world, and it has the property of providing physical robustness to plant materials. Cellulose can be obtained from a plethora of natural sources like plants or bacteria [8]. The past few years have seen a

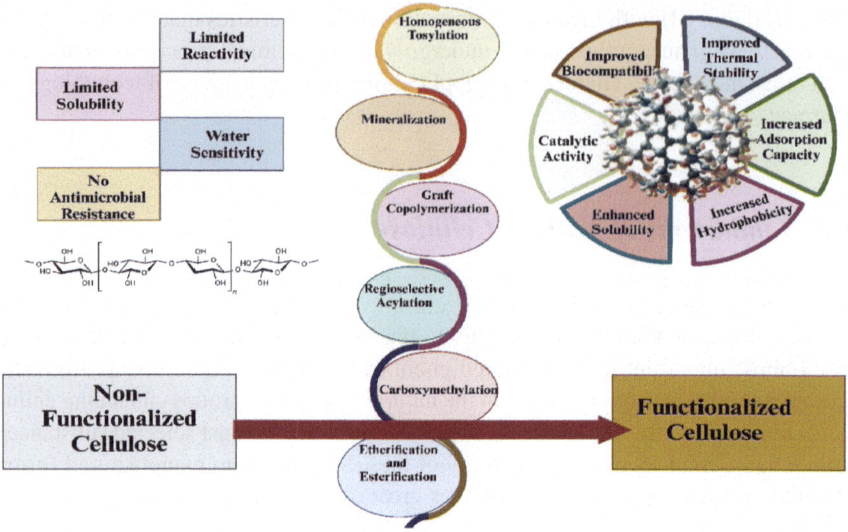

Fig. 1 Attributes of functionalized cellulose

spike in the exploration and research of cellulose and its functionalization for various applications. Many scientists and researchers have devoted their time and effort to the examination of cellulose and its derivatives bringing about an extensive number of research articles. The idea of altering or functionalization of cellulose or its derivatives is enticing as it brings out the possibility of forming new or advanced materials with enhanced properties. Cellulose also has a lot of derivatives depending on its compatibility and application. By integrating functional materials into the system, synergistic outcomes can be generated, offering electronic, magnetic, catalytic, fluorescence, and optical attributes. Consequently, this will elevate their capabilities and widen the range of potential applications [7]. In the subsequent sub-topics, the various functionalization processes of cellulose derivatives and the suitability of conventional or unconventional methods are emphasized.

2.1 Mineralization of Cellulose

The cellulose's hierarchical arrangement and presence of abundant hydroxyl groups, make it amenable to diverse treatments that aim at enhancing its functionalities, properties, and overall application. These objectives are often accomplished through a plethora of chemical modifications. Mineralization processes can proceed at room temperature using eco-friendly or sustainable methods, which are scalable for implementation in the paper and textile sectors.

2.1.1 Methods of Mineralization

There exist primarily three distinct mineralization procedures for cellulose as specified in Fig. 2, each characterized by its unique underlying principles.

Graft-Induced Mineralization

This method, primarily suitable for bonding polymeric chains to cellulose molecules, is occasionally employed for grafting polydimethylsiloxane. For reference, research was conducted where they covalently attached silica nanoparticles containing amino groups to cotton fabric. These NH_3 groups acted as sites for grafting polydimethylsiloxane chains terminated with monoglycidyl ether. Consequently, the hydrophilic surface of the fabric was transformed into a superhydrophobic one, evidenced by a water contact angle of 155°.

Inclusion Mineralization

The strong hydrogen bonding among cellulose macromolecules renders nano/microfibrils impermeable to solvents, necessitating specific solubilization methods for biocomposite fabrication. Cellulose can be dissolved under harsh conditions or in ionic liquids, often with additives or chemical modifications during dissolution.

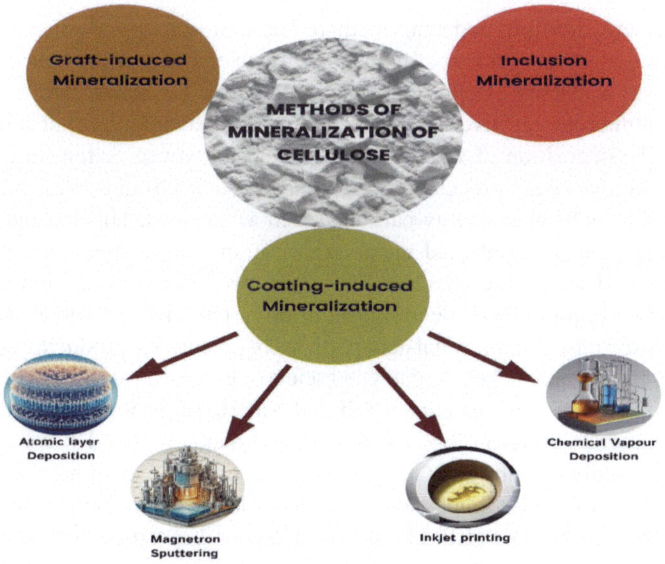

Fig. 2 Different methods of mineralization of cellulose and further classification

For instance, lithium chloride was used to dissolve microcrystalline cellulose in anhydrous N, N-dimethylacetamide, followed by hexamethyldisilazane treatment to produce trimethylsilyl cellulose for film production [11]. To dissolve cellulose, a titania precursor was added to the ionic liquid 1-butyl-3-methylimidavolium chloride, and after solvent exchange, obtained regenerated cellulose with entrapped titania in fibril form [9]. To create films, titania nanoparticles in solution were combined with soluble cellulose acetate, later deacetylating them to obtain cellulose. While mineralization offers performance advantages, it can involve toxic solvents and inconvenient conditions [10].

Coating-Induced Mineralization

In the functionalization of cellulose, deposits of inorganics are the most common as it portrays its simplicity and flexibility. These inorganic substances can be applied either via vapour deposition or through solutions. Vapour deposition often leads to uneven coatings due to phenomena known as "shadowing" or "step coverage," where the substrate is attacked unidirectionally by the evaporated material. Wet chemistry is a subject which involves a material submerged in a liquid and this material is immersed or absorbed by cellulose. As a result, the interior and surface undergo treatment, leading to a material coating with improved uniformity.

Coating Through Vapour

There are several divisions under this method. The atomic layer deposition talks about the formation of an inorganic coating on the surface of a fibre using a metal–organic precursor and a reactant. Cellulose is particularly apt for atomic layer deposition due to its numerous reactive hydroxyl groups. An exemplary application of this method is the deposition of zinc oxide on paper and woven cotton. In magnetron sputtering, samples were processed by expelling particles from a solid source under the bombardment of high-energy particles, such as gas ions. This technique allows for the deposition of metals and metal oxides at ambient temperature. The inkjet printing method is gaining attention due to its cost-effectiveness, minimal material and power requirements, flexibility in printing patterns, and direct-write capability. Inks used may contain metal nanoparticles or precursors, producing conductive stripes like copper and nickel. A renowned scientist employed a diverse colour printhead to sediment an aqueous copper salt and $NaBH_4$ sequentially, yielding highly conductive copper stripes (1.8×10^6 S m^{-1}) and nickel (2.2×10^4 S m^{-1}). Techniques were combined, printing Cu patterns on paper using a commercial Pd/Sn ink as a catalyst, followed by immersion in an electroless bath for copper precipitation through catalytic reduction [12]. The chemical vapour disposition method used silica to render textiles, paper, and wood hydrophobic. An illustrative contemporary application involves hydrophobization nanofibrillated cellulose aerogel using 1H, 1H, 2H, and 2H-perfluorodecyltrichlorosilane. Multiple alkyl silanes are used, and they are mentioned in Table 1.

2.2 *Chemical Functionalization of Cellulose*

There are various methods for the chemical functionalization of cellulose.

2.2.1 Dip-Pad-Dry-Cure Method

This term originated with the infusion of cotton with titania nanoparticles and expanded to include other types like silver and zinc oxide. To create crystalline titania, the fibres are submerged in a solution of artificial TiO_2 nanoparticles, then pressed, dried, and cured. Various research teams use similar methods. However, a drawback of this mineralization process is that nanoparticles may not adhere well to the cellulose surface during immersion. To address this, the oxidation of cellulose is done to introduce negatively charged groups using methods such as plasma treatments, UV irradiation, or chemical oxidants. However, these treatments can lead to fibre degradation and reduced performance. To mitigate polysaccharide degradation, cross-linking fibres with fatty acids containing multiple carboxylic groups have been proposed [12].

2.2.2 Electroless Plating

Typically, this method involves a two-stage synthesis and is commonly employed for silver deposition on paper or aerogels. Numerous researchers have investigated the metals applied through this process. Electroless plating offers advantages such as scalability using existing commercial technologies, minimal need for costly equipment, and adherence to principles of green chemistry.

2.2.3 Sol–Gel-Chemistry

This method stands out as the most prominent among its counterparts due to a multitude of advantages. It doesn't necessitate complex equipment, making it notably simple and convenient to employ. The availability of pure precursors, such as metal salts and alkoxides, has increased, allowing for precise control over precursor stoichiometry and functional groups. It operates at minimal temperatures close to ambient conditions, leading to reduced toxicity, minimal evaporation losses, and environmental impact, particularly beneficial for biopolymers and biological materials. Although sol–gel technology has its drawbacks, including the costliness of certain precursors, prolonged processing times for silica, limited control over the formation

Table 1 Shows various silanes, their functional groups, and their applications

Silane name	Functional group(s)	Applications
Tetramethylsilane (TMS)	$Si(CH_3)_4$	Used as a standard in NMR spectroscopy (as a chemical shift reference)
Trimethylchlorosilane (TMCS)	$Si(CH_3)_3Cl$	Surface modification of silica and glass, protection of hydroxyl groups in organic synthesis
Vinyltrimethoxysilane (VTMS)	$Si(OCH_3)_3\text{-}CH{=}CH_2$	Used as a coupling agent, adhesion promoter in sealants and paints, cross-linking in silicone rubbers
3-Aminopropyltriethoxysilane	$Si(OCH_2CH_3)_3\text{-}(CH_2)_3\text{-}NH_2$	Surface modification of nanoparticles, adhesion promoter in coatings, and primer for bonding metal/glass
Methyltrichlorosilane (MTCS)	$Si(CH_3)Cl_3$	Waterproofing agent for textiles and construction materials, precursor in silicone polymer synthesis
Tetramethoxysilane (TMOS)	$Si(OCH_3)_4$	Used in sol–gel processes, precursor for silica materials, optical coatings
Phenyltrimethoxysilane (PTMS)	$Si(OCH_3)_3\text{-}C_6H_5$	Used as an adhesion promoter in paints, coatings, and plastics
Bis(triethoxysilylpropyl)tetrasulfide (TESPT)	$Si(OCH_2CH_3)_3\text{-}(CH_2)_3\text{-}S_4\text{-}(CH_2)_3\text{-}Si(OCH_2CH_3)_3$	Used as a coupling agent in rubber industry to enhance silica reinforcement in tires
Isocyanatopropyltriethoxysilane	$Si(OCH_2CH_3)_3\text{-}(CH_2)_3\text{-}NCO$	Used as a cross-linker in coatings, adhesives, and surface modification of fillers and pigments
Chloromethyltrimethoxysilane (CMTMS)	$Si(OCH_3)_3\text{-}CH_2Cl$	Used in the functionalization of surfaces, adhesion of polymers, and chemical intermediates

of certain metal oxides due to rapid processes, and considerable shrinkage and brittleness during preparation, these are overshadowed by the numerous benefits outlined earlier, rendering sol–gel technology attractive and well-suited for industrial use [13].

2.3 Unconventional Methods in Cellulose Functionalization

Unconventional methods for preparing biopolymers typically indicate that they are not commercially available. While cellulose nitrate remains the sole commercially significant inorganic cellulose ester, current research focuses on developing novel cellulose-based polymers. These innovations include polymers with multiple functional groups, compilation of various functional groups (such as lipophilic-hydrophilic), integration of new functionalities like chromophores or fluorophores, addition of redox-active groups, and incorporation of substituents with unique magnetic, optical, or biological properties, all within a controlled functionalization framework. Interest in crafting new cellulosic materials arises from cellulose's renewable nature, being the primary component of woody plant cell walls [14].

2.3.1 Homogenous Cellulose Esterification

Non-derivatizing solvents undergoing acylation

Solvents like N, N-dimethylacetamide (DMA) or N-methylpyrrolidone combined with lithium chloride (LiCl) play a significant role without introducing derivatives. The initial exploration of cellulose chemical modification under homogeneous conditions involved esterification in DMA/LiCl using carbonic acid anhydrides and chlorides. Dicarbonic acid anhydrides were used to produce water-soluble carbonic acid half-esters of cellulose. Subsequently, more effective and efficient homogeneous acylation methods were developed. Among these, the shift of cellulose in DMA/LiCl with diketene emerged as notably efficient.

Derivatizing solvents undergoing acylation

Various solvents were explored as the reaction medium for the homogeneous conversion of cellulose, including Dimethyl sulfoxide (DMSO)/paraformaldehyde, chloral/Dimethylformamide (DMF)/pyridine, Trifluoroacetic acid (TFA), and DMF/N_2O_4. One method involves polymer conversion with multiple carbonic acid anhydrides at low temperatures in the presence of pyridine. It was noted that DMSO present in the solvent system can be replaced with DMF/DMA. The cellulose solvent DMF/N_2O_4 is highly toxic.

2.3.2 Heterogenous Esterification

In addition to homogenous esterification, novel approaches were devised for heterogeneous processes. It is to be noted that heterogeneous processes are still frequently used. Most large-scale processes typically occur heterogeneously. Among the most effective techniques is the impeller method, where specific carbonic acids are transformed into reactive mixed anhydrides. Impellers such as chloroacetyl-, methoxyacetyl-, and trifluoroacetyl moieties are utilized. Concurrently, numerous new catalysts have been developed, including titanium-(IV)-alkoxide compounds like titanium-(IV)-isopropoxide. These catalysts can be employed for partially esterified cellulose derivatives. Sulfonic acid chlorides also serve as activating agents for cellulose conversion when suspended in pyrimidine or DMF.

2.3.3 Regioselective Acylation

Selective deacylation and acylation after protection are a few special methods for regioselective functionalization. Selective removal of acetyl groups from secondary functional groups was achieved by hydrolyzing cellulose triacetate with amines. The resulting selectively substituted cellulose acetates were utilized in producing cellulosic sulfuric acid half-esters with heparin-like properties. Typically, the regioselective esterification of the secondary hydroxyl group involves protecting the primary function with the bulky triphenyl methyl group. A more recent approach shields the primary hydroxyl groups, allowing for the creation of regioselectively substituted cellulose esters.

3 Sustainable Production of Cellulose, and Its Derivatives

In the forestry sector, using logging wastes that were previously unexplored as a form of woody biomass can have positive environmental effects like improving energy efficiency and addressing global warming concerns. One approach to optimize the utilization of woody biomass involves producing cellulose nanofibers (CNFs), which are considered high-value products. CNFs have found applications across various industries including paper, electronics, pharmaceuticals, medicine production, cosmetics, food, and automotive parts manufacturing.

3.1 Methodology

The cellulose life cycle process boundaries were segmented into four stages: (1) planting and harvesting; (2) chipping; (3) transportation; and (4) CNF manufacturing.

The mechanical-chemical method can be subdivided into three stages: (1) initial grinding, (2) fine grinding, and (3) classification.

3.1.1 Production

Two primary methods exist for producing All-Cellulose-Composites (ACCs). The first is the one-step method, where cellulosic materials undergo partial dissolution, with the same source material acting as both reinforcement and matrix. This yields what's known as a self-reinforcing composite. Here, a cellulose solvent system partially dissolves the surface or weaker regions of the fibers, while the dissolved cellulose forms a matrix around the initial fiber structure upon precipitation. The core of the reinforcing fibers retains its original structure and chemical crystallinity, reinforcing the composite. One advantage of this method is the creation of interfaceless composites, where boundaries between reinforcement and matrix are not clearly defined. Additionally, adjusting the proportion of undissolved fibers in the composite is easily achieved by varying dissolution time [15].

Alternatively, ACCs can be produced through the two-step technique. In this method, a robust reinforcing fiber is soaked in a solution containing dissolved cellulose, which then envelops the fiber structure as a matrix during the regeneration phase. This results in a composite comprising solid cellulose and dissolved cellulose, chemically bonded under specific temperature, pressure, and curing conditions. This method offers the advantage of easy control over the fiber content by adjusting the mixing ratio, and it is particularly beneficial for large-scale composite production [16].

4 Challenges Faced

We have observed that cellulose materials can be of great importance to us and have been one of the most widely used biopolymers. But now comes the challenges faced during its implication and functionalization in the industry. There are numerous chemical, economical, and sustainable reasons why cellulose functionalization has its challenges.

4.1 Limited Functionality Due to the Chemical Nature

There are a lot of hindrances when analyzing the chemical composition of cellulose and its derivatives. The constrained number of reactions that can occur at the hydroxyl groups of cellulose derivatives to functionalize them is one of the main obstacles to their utilization as functional derivatives. They have a lot of inferior mechanical and thermal properties.

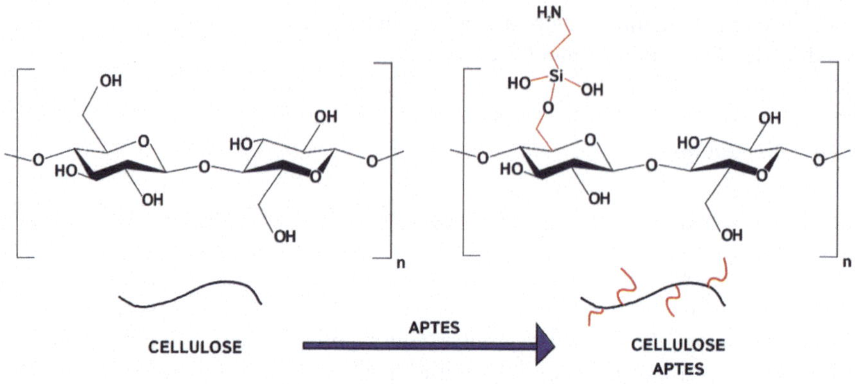

Fig. 3 This is the conversion reaction of Cellulose APTES from Cellulose

4.1.1 Aminopropyltriethoxysilane as a Functionalizing Molecule for Cellulose and Its Derivatives

Aminopropyltriethoxysilane (APTES) is being utilized primarily as a binding agent to immobilize diverse species serving various functions. Cellulose microfibrils underwent modification with APTES to enhance the adhesion of mesoporous silica nanoparticles. We can see the conversion in Fig. 3. Polyglutamic acid was utilized as an intermediary molecule linking mesoporous silica with APTES for this objective.

Since the alteration was done in an aqueous solution of urea and sodium hydroxide to stop APTES from polymerizing, it was very noteworthy. Images captured by SEM microscopy showed a noticeable crosslinking impact on the changed microfibrils' surface, which was explained by many ether bond formation processes. The membranes thus functionalized exhibited favourable outcomes when compared to untreated cellulose acetate membranes. Several molecules were examined for attachment onto bacterial cellulose surfaces. Utilizing acids can improve membrane hydrophobicity, holding promise in the food industry for packaging materials. Nevertheless, for more advanced functionalization, APTES is important because of the versatility of its amino group, surpassing the reactivity of vinyl-triethoxysilane and enabling subsequent reactions [4].

4.1.2 Inherent Hydrophilicity of Cellulose Nanocrystals CNCs

Cellulose nanocrystals (CNCs) represent a burgeoning nanomaterial applicable in various fields such as coatings, construction, adhesives, and biomedical devices. With their notable characteristics including high aspect ratio, stiffness, and reinforcement capability, CNCs hold significant potential for integration into polymer nanocomposites. Nonetheless, their inherent hydrophilicity and affinity for polar environments have constrained their utilization in hydrophobic polymer matrices or formulations based on organic solvents.

The cellulose chains in CNCs possess numerous hydroxyl groups, rendering them predominantly hydrophilic. Nonetheless, the arrangement of these chains within the individual CNC particles creates hydrophobic crystal plane edges, imparting amphiphilic properties. These properties, contingent upon the surface charge of CNCs and the isolation method used, enable CNCs to stabilize various emulsions and foams [17].

4.2 Technical, Economic, and Sustainability Issues

The difficulties posed by the main technical and financial variables influencing the cost of cellulose and its derivatives will be covered further.

4.2.1 Xylan-Active Enzymes

Apart from cellulose, fibers isolated through chemical treatments often retain residual hemicelluloses. Hemicelluloses are intricate polymers comprising various sugar units bonded by (1–4) glycosidic bonds, constituting both the primary core and branching structures. Xylan stands as the most common hemicellulose present in hardwood and glass fibers. Due to the heterogeneous nature of natural xylans, their enzymatic breakdown is complex, necessitating a diverse range of enzymes to collaborate for full hydrolysis. This complexity poses hindrances in comprehending how enzymes interact with cellulose fibers at a molecular level.

4.2.2 Enzymatic Treatment of Cellulosic Fibers

The utilization of cellulases in enzymatic hydrolysis of cellulosic materials under high solids loading has garnered significant interest and research funding. Employing high solids loading enables the production of products at higher concentrations, thus driving research efforts in this area. Another challenge arises when fermenting both pentoses and hexoses (C5/C6) simultaneously, especially when the sugars released during enzymatic hydrolysis come from cellulose pulp containing substantial hemicellulose content.

4.2.3 Technological and Economic Challenges

The most preferred method used in the CNC plants is the acid hydrolysis method using mineral acids. Nevertheless, the acid pathway poses techno-economic challenges. An economic disadvantage is that a highly concentrated acid reaction demands significant capital expenditure due to the need for sturdy reactors, acid-resistant equipment, and pipelines.

Additional challenges include the inability to efficiently recover and recycle acids, and also to retrieve the solubilized sugars from the spent liquor [18].

5 Diverse Applications of Functionalized Cellulose

The multifaceted application of functionalized cellulose is comprehensively presented in Fig. 4.

5.1 *Utilization of Cellulose in Biomedical Applications*

Significant growth has been seen in the pharmaceutical industry for increasingly turning to polymers as a sustainable alternative to fossil fuels, as research on biomaterials using unsustainable resources such as fossil fuels results in severe global pollution and changing the climate [19]. Assessing the material's biocompatibility and confirming how it interacts with cells is crucial for biomedical applications, particularly when the material must stay in proximity to living tissues and shouldn't cause any cytotoxic or other unfavourable consequences [20]. Regarding this, cellulose is the most abundant, affordable, and environmentally friendly biopolymer available. The special qualities of cellulose include its nontoxicity, abundance, sustainable resource availability, low manufacturing cost compared to synthetic biopolymers, biodegradability, biocompatibility, and superior mechanical capabilities. Nanocellulose materials are made of cellulose and have no negative impacts on the natural world or human health. They are recyclable, biodegradable, and non-toxic [21].

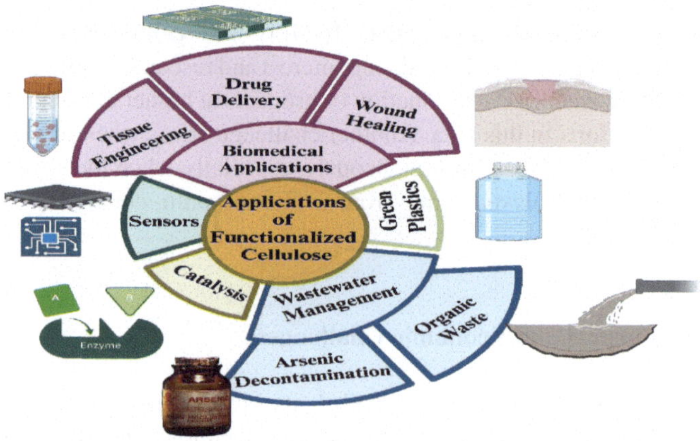

Fig. 4 Multifaceted applications of functionalized cellulose

In the realm of biomedical applications, cellulose finds widespread use in the following areas:

- **Tissue Engineering**

Tissue engineering is a multidisciplinary discipline that uses life sciences and engineering ideas to create intelligent biological replacements that could potentially improve, preserve, and restore tissue characteristics that have malfunctioned. This area of study focuses on the coordinated growth of three components: biomaterial scaffolds; stem cells and cells interacting with three-dimensional scaffolds; and biochemical molecules that transmit signals. The biodegradable and biocompatible polymers used to create 3D scaffolds have enough mechanical strength to closely resemble native tissues while also favorably resembling the extracellular matrix (ECM) of native tissues [22]. To produce tissue scaffolds, nanocellulosic materials can have their permeability, thickness, and interconnectivity adjusted without sacrificing their mechanical qualities [23]. Because of its advantageous characteristics, which include its stability, flexibility, ease of chemical modification, and sensitivity to pH changes, CMC has been hailed as a potential scaffold biomaterial with potential uses in the fields of tissue engineering, 3D printing, delivery of drugs, cosmetic fillers, and cancer therapy. A combination of cells, scaffolds, and growth factors are used in conventional tissue engineering techniques. Here, a framework supporting regeneration is supplemented with cells and substances that encourage growth. These techniques have successfully created functional scaffolds for the trachea, heart, liver, esophagus, and bone [24].

- **Wound dressing**

An intricate web of biological processes is involved in wound healing to maintain barrier function, avoid dehydration, and lower the chance of bacterial infection. New materials for wound dressings must be developed to treat burn injuries and skin grafting. There are various stages to the healing process of wounds, including granulation, collagenization, scar maturation, and collagen maturation. By 2024, it is anticipated that the wound treatment market will have attracted over USD 22 billion in investment due to the introduction of new products, rising rates of chronic wounds, and government assistance [25].

Ways in which wound dressing can be proceeded are:

(a) Based on its ability to absorb and retain water, nanocellulose offers significant promise for applications related to wound healing. These applications can be applied directly to the wound, helping to reduce inflammatory reactions and encourage fibroblast growth in the healing process cascade [26].
(b) It has been established that using human adipose mesenchymal stem cells is an efficient way to encourage tissue regeneration in wounds from surgery [27].
(c) In these situations, dressings made of films, foams, and hydrogels are frequently used to cover wounds; however, they are prone to bacterial infections [28].
(d) The three-dimensional, flexible scaffold that the bacterial cellulose microfibrillar structure provides can both encourage tissue granulation and cell adhesion

while functioning as an external barrier against pathogens. Because of the way it is constructed, it can hold a lot of water, which helps to maintain gas exchange, keep the wound moist, and absorb exudates from the wounded tissue. Furthermore, it is non-toxic, non-allergenic, and biocompatible due to its exceptionally high level of inherent purity [29].

- **Drug Delivery**

The release of medications in a predetermined quantity and at the right time to particular organs is known as the 'drug delivery system'. The transport mechanism of drug delivery systems can be adjusted to react to ambient influences like temperature, sunlight, chemical reactions, pH, and magnetic and electrical fields. Next, the drugs are administered to the specific tissues, cells, or organs. As a drug delivery method, soluble and adjustable diffusive properties should be promoted using cellulosic materials. For instance, in oral dosage forms, certain drug delivery patterns, such as rapid, controlled, or delayed release, have been found for cellulose and its derivatives. [30]. CNFs possess advantageous flexibility, elastic characteristics, lightweight nature, limited toxicity, and reactive surface, making them promising carriers for regulated drug administration in biomedical applications. They may be attached to particular functional groups more easily on this surface, which gives them versatility for different applications. In addition, CNFs are economical and renewable, which makes them more appealing for use in biomedicine.

Some of the recent drug deliveries are as follows:

(a) Transdermal drug delivery
(b) Anticancer drug delivery
(c) Antimicrobial drug delivery

5.2 Utilization of Functionalized Cellulose Materials to Sensors

Sensors play an important role in our modern-day world. To obtain a secure statistically correct and accurate model of our world, it is in dire emergence that we place wired or wireless sensors that are cost-efficient at prominent places. IoT called the Internet of Things, is the interconnection via the internet of computing devices incorporated in daily usable objects, which enable them to receive as well as send data. For instance, these networks might incorporate gas sensors capable of detecting environmental pollutants. Similarly, the deployment of temperature, pressure, and humidity sensors across various locations globally can enhance our ability to forecast weather patterns.

Functionalized cellulose materials promise a major prospect for sensors and detection. Rull-Barrull et al. [31] employed cellulose grafted with thioglycolic acid for selectively detecting copper in water utilizing optical and colour-based techniques.

Other metals such as mercury (Hg) or nickel (Ni) combined with modified cellulose can be detected using colourimetric methods [31]. Weishaupt et al. successfully developed sensors capable of detecting low levels of copper in aqueous solutions [32]. The material for their sensor comprises red fluorescent phycobiliprotein fixed onto 2, 2, 6, 6-tetramethyl-1-piperidinyloxy (TEMPO)-oxidized nanofibrillated cellulose. Zhang et al. [33] documented the creation of a cellulose-derived fluorescent chemosensor specifically designed for detecting Fe (III) ions. In 2017, Zhang et al. [34] presented strain sensors employing carbonized plain-weave cotton fabric. These sensors can sense small changes in strain of 0.02% and have a vast strain range of up to 140%. This technology shows great potential in areas such as robotics and wearable electronics [34].

A captivating realm of contemporary investigation revolves around altering cellulose with conductive materials derived from carbon, including diverse carbon forms like nanotubes and graphene, as well as conjugated polymers such as poly(3,4-ethylenedioxythiophene): poly(styrene sulfonate) (PEDOT: PSS), polypyrrole, and polyaniline. When the percolation threshold is attained, these modifications yield flexible conductive materials capable of detecting temperature variations. Additionally, they can utilize temperature variances to generate low power, such as in the case of remote self-sustaining sensors.

5.3 Green Synthesis Routes for Nanoparticles and Catalysis

In contemporary times, there has been a notable curiosity in adopting green synthesis methods for nanoparticles. Functionalized cellulose, due to its outstanding coordination and adsorption characteristics, rises as a potential environmentally friendly base and stabilizing agent predominantly for nanoparticle production. For instance, carboxylated cellulose shows proficiency in binding metallic ions and inhibiting nanoparticle clustering [35]. Moreover, it functions as a reducing agent in eco-friendly synthesis, as demonstrated in the research conducted by Li et al. [35], where carboxymethyl cellulose serves as both a reducing agent and a stabilizer in the production of palladium nanoparticles. Different types of modified cellulose present innovative routes for synthesizing a diverse array of metallic and metal oxide-based nanoparticles, philately titanium dioxide (TiO_2), gold (Au), silver (Ag), platinum (Pt), and nickel (Ni).

Cellulose-bound nanoparticles have a range of applications, including catalysis. Khan et al. [36] described a catalytic reduction process using copper (Cu) and silver (Ag) nanoparticles enamoured with cellulose acetate, which transformed 4-nitrophenol into 4-aminophenol [36]. The in-depth detail of this extensively examined reaction model has recently been clarified. Additionally, Keshipour et al. [37] developed copper(I) nanoparticles supported on ethylenediamine-functionalized cellulose, introducing a new catalyst for 1,3-dipolar cycloaddition in direct arylation sequences [37]. These studies provided a comprehensive understanding of the properties of cellulose-CuNP catalysts, with the added advantage of catalyst recovery through filtration and subsequent reuse up to five times with satisfactory yields. On the other hand, cellulose might function as a sacrificial source of electrons towards photocatalytic production of hydrogen operational on the CdS/CdOx nanoparticles, as evidenced by the research of Wakerley et al. [39]. Bacterial cellulose has been utilized as a base for adjustable positive electrodes in supercapacitors. In this context, anodes made of bismuth oxide/graphene were erected, capitalizing on the complex structure offered by the substrate material [38].

5.4 Wastewater Management Using Functionalized Cellulose

Wastewater management is a leading question of modern environmental issues practising the removal of noxious waters in reducing the effect of waters contaminated on the flora, fauna, and humans. Being utilized effectively for water treatment and sustainability, renewable materials have experienced significant advancement. Cellulose is a cheap, available, and easily functionalized biopolymer that has been utilized for the removal of heavy metals and arsenic from industrial waters. Cellulose is one of the greenest and most effective techniques to achieve water remediation sustainability, taking advantage of its biopolymer's availability. Cellulose-based adsorbents are ideal in wastewater treatment due to their potential to eliminate harmful substances including heavy metals, arsenic, and some organic contaminants from water.

The recent achievements in water management where adsorption of arsenic facilitated by chemically functionalized cellulose-based materials. As the researcher's state, functionalization of cellulose promotes an interaction with arsenic. Thus, the removal of arsenic carried by the adsorbent increased as the functionalized products facilitated the recovery and regeneration of the adsorbent. Thus, a waste product was directed to a second use alongside the removal of arsenic in waters. The reviewed literature proves the development and implementation of functionalized cellulose and its potential to remove toxic ions such as heavy metals and arsenic from waters. Moreover, the global approach to waste and environmental management is suspicious but insists that the use of functionalized cellulose in waters should be exercised as one of the progressive issues inclusive of reducing carbon emission and its replenishment with other energy sources. Meanwhile, the integration of this cellulose remains on a case-to-case basis and requires further research and attention.

5.5 Development of Polymers Using Functionalized Cellulose

The problem of a need for sustainable materials has led to outstanding progress in polymer science, and a relevant example is an improvement in the development of green composites and polymers. Functionalized cellulose, being renewable, is one of the key components. The distinctive characteristics of this material, including biodegradability, high mechanical strength, and chemical modification, make it suitable and effective for eco-friendly polymers. Research indicates that surface functionalization of bacterial cellulose enhances interfacial adhesion between cellulose fibres and polylactic acid (PLA) matrix, resulting in stronger and more durable composites.

The development of anisotropic materials through the use of functionalized colloidal cellulose and cellulose derivatives, and their the outcomes illustrate efficiency of the concerned methodology in the augmenting of complex materials with certain characteristics. Meanwhile, recent developments in the creation of sustainable polymeric materials made of lignocellulosic biomass and point to the application of functionalized cellulose as an opportunity To generate lasting alternatives for conventional synthetic materials. The technique shows considerable pledge, yielding materials that can serve several sectors such as packaging, automotive, construction, and biomedical applications. For instance, cellulose-based composites are renewable and biodegradable, which makes them suitable for lightweight and sustainable packaging, and the material can also be applied as biocompatible scaffolds for polymer tissue engineering, and as an alternative to synthetic polyactides. Functionalized cellulose serves an integral part in developing sustainable polymers by enabling the production of environmentally friendly materials with superior attributes. Simultaneously, advancements in current research and applications provide further academic studies with the potential to generate innovative material solutions for significant social and environmental challenges.

6 Environmental Impact

Given the depletion and environmental issues caused by the consumption of fossil fuels, there is a growing adoption of sustainable development using environmentally safe materials. One effective solution to the issue is substituting conventional petroleum-based plastics with biodegradable alternatives like cellulose [40]. The environment is filled with different types of contaminants, both inorganic and organic, which affect the ecology in ways that are unanticipated and can negatively impact the health of humans. To address rising worries about the usage of fossil fuels and pollution, many cellulose-based materials have been developed to protect the environment. Researchers are now exploring the potential of using these cellulose materials for treating a wide range of contaminants [41].

The study compared the environmental efficacy of a green approach using life cycle assessment (LCA) to a traditional approach that looks out for typical acetylation performed after an alkali-bleach pretreatment. This study aims to produce cellulose acetate from corncob using possible acetylation and pretreatment methods that are ecologically friendly. Additionally, the environmental influence of the applied procedure will be calculated [42].

This research aims to examine the environmental effects of recycling textile fibres to new fibres utilizing cellulose carbamate technology, with particular attention to land use, water scarcity, energy consumption, and climate. Cellulose carbamate technology is a cutting-edge technique that regenerates cotton-like textile fibre by converting cellulose from available alternative materials or textile waste. This fibre has certain characteristics similar to cotton and has comparisons to other artificially regenerated cellulosic fibres. This recycling technology provides significant potential to reduce environmental influence compared to conventional fibres like viscose and cotton [43]. The mass-based practical unit "kilograms of cellulose carbamate stable fibre at a moisture content of 13%" was used to evaluate the effects on the environment. [44]. To overcome the drawbacks of these environmental impacts, Cellulose is commonly subjected to modifications resulting in derivative forms.

6.1 Derivation of Cellulose Materials: Selective Oxidation

Oxidation catalysis of industry is widely used in the pulp, paper, and laundry industries for cellulose bleaching materials for its high-volume application. Cellulose oxidation is very important. The industrial bleaching of cellulose-containing materials commonly utilizes hydrogen peroxide or hypochlorite as oxidizing agents to remove coloured substances. The usage of hypochlorite Generates chlorides in stoichiometric quantities, indicating a preference for more environmentally friendly and atom-efficient methods [45].

Recently, natural polysaccharides like cellulose, alginate, starch, gum, chitin, pectin, and chitosan have been in the spotlight due to their distinct structures and potential applications among all polymers. Oxidation modifications have garnered significant interest due to their functional properties and potential implementation in various industries, including chemical material, and industrial sectors. The transformation of alcohols, CO, alkenes, olefins, and other molecules into useful oxygenated chemicals is a well-known and important process within the context of oxidation processes. Ecologically benign and cost-effective catalytic systems are mediated by polysaccharides. They are well-known for being appealing options for catalytic systems that can be afforded for a variety of chemical transformations, including oxidation processes. They exhibit excellent efficacy in oxidative transformations and can be largely recycled and reused for several catalytic reactions [46].

To alter the surface charge density of cellulose a necessary step for delaminating fibres into nanometric fibrils using a low-energy procedure and to produce a variety of products with different structures and attributes that can be processed further, cellulose has undergone several derivatization reactions. One of the most significant techniques for chemical modification, selective oxidation of cellulose, has proven to be a multipurpose platform for the production of new, highly effective, versatile cellulose-based materials. These materials have many uses outside of the environmental field, including energy storage, food packaging, healthcare and biomedical engineering, sensing applications, and energy storage [47]. Two main difficulties have been brought to light by advancements in both theory and practice regarding the production process using cellulose and its derivatives of a novel, high-performance substances: (a) It makes sense to use nanostructured cellulosic (nanocrystals, nanofibers, and nano-whiskers) to very effectively add new or significantly better bulk qualities to materials. (b) It has shown to be beneficial to functionalize cellulose through chemical reactions, such as oxidation, to increase its variety of applications, improve its specificity, and lessen its adverse effects on the environment. [48, 49]. Agro-industrial waste and raw cellulose are low-cost and effective for environmental applications like industrial wastewater decontamination [50–54].

Because it produces products with a variety of structures and features based on the substrate, chemicals, reaction conditions, and medium, oxidation is a dependable approach for functionalizing cellulose. Additionally, it was utilized to alter the surface charge density of cellulose, which is necessary to separate fibres into nanometric fibrils using a low-energy method [55]. The selective oxidation of cellulosic material is a crucial chemical modification method that has proven to be a useful platform for creating high-performance, and versatile cellulose-based substances. These materials have a wide range of applications beyond just environmental ones. They can be used in fields such as biomedical engineering and healthcare [56, 57], barrier applications [58], and active packaging of food [59].

6.2 Hazards Due to Poor Thermal Stability

Aside from their propensity to degrade biologically, the choice of biomaterials for a given application may depend on their ability to exhibit predictable behaviours in the face of changing environmental factors like temperature and pressure. During this session, we looked at one of the features of biocompatible cellulose (BC) that have been researched for use in biomedicine: its heat stability [60]. Natural fibres and thermoset matrices are frequently used in the automotive industry for both heavy and light applications, including dashboards, package trays, headliners, door panels, and seat backs [61]. Commercially, composite materials are particularly successful for non-structural (cosmetics) and semi-structural applications; the main limitations are their thermal and physical stability [62]. One major obstacle to the practical application of natural fibre-reinforced composites is their heat characteristics. Higher temperatures cause cellulose, hemicellulose, and lignin that make fibres to begin to degrade, changing the composite's main characteristics. The suitability of natural fibres with polymers and elevated temperatures stability has been studied for several decades, even though comparative studies are necessary to obtain information regarding the firmness of natural fibres with different polymers, two distinct organic fibre-reinforced polymers composites, or natural fibre strengthened blended with two distinct polymers. Researchers can choose reinforced materials and matrices with the necessary thermal stability with the aid of this study [63].

The decomposition temperature (Td) of natural fibres varies, including sisal (*Agave sisalana Perr.*), water hyacinth, roselle, and reed (*Phragmites vulgaris*). To assess how thermal deterioration behaved, a temperature range of 290–490 °C was investigated [64]. The weight loss of a substance was measured at different temperature ranges. The initial loss of weight was caused by water molecules evaporating between 50 and 100 °C. Following this, weight loss was noted in the 200–350 °C temperature range as a result of hemicellulose breakdown. The lignin thermal degradability occurred in the temperature range of 300–450 °C, while the hemicellulose degradation occurred in the temperature range (200–300 °C). Over 60% of the thermal degradation of natural fibres occurs between 230 and 350 °C due to hemicellulose and lignin [65]. The components' first breakdown explains the decreased thermal stability [66]; however, Abrupt weight loss at elevated temperatures is more in line with thermal stability, which is a feature of oxidative breakdown [67].

7 Economic Viability of Functionalized Cellulose

Functionalized cellulose and specifically extraction and modification products through enzymes are broadly discussed regarding their economic viability. Many pulp and paper industries, research organizations, and people adhere to the main concept of a cellulose economy, which is to run a cycle for all cellulose-based products/metals. It is a scientifically and logically based argument as well as a sustainable approach to environmental protection and waste minimization. The advantages of functionalized cellulose and modification with enzymes are its environmental sustainability, customization, costs, and market opportunities [18]. The enzymatic process for isolating and functionalizing cellulose is environmentally friendly as the process involves biodegradable catalysts and mild operating conditions. Customization can help production as needed, which can be crucial for a range of industrial needs. It can help stimulate applications and innovations within different fields as both active and inactive materials [68]. Costs might be efficient if lower amounts of capital are used, and operating expenditures are minimized.

The market for sustainable materials is currently booming and various markets such as packaging, automotive, and biomedical applications will be becoming more popular. Nanocellulose products deliver superior toughness, biodegradability, and protective qualities, offering plenty of high-value applications. The industry features a variety of distinct products constructed from nanocellulose. This product development has been supported by advances in chemical, mechanical, and microbiological science. The demand for environmentally compatible products has been increasingly enforced in recent years, and the number of consumers who appreciate green products is increasing. The evolving e-commerce segment is fostering transparency and new business models in the R&D and pilot biotechnology manufacturing market. However, the industrial process of enzyme medium functionalized cellulose, the cost, and its use determines the economic viability of this method.

Zeolitic Imidazolate Framework-67@Cellulose aerogel (ZIF-67@CA) is a green, reusable and non-toxic material, and it can be applied for bioremediation of textile-based dyes, leading to reduced operational costs and liabilities [69]. It can be recovered and reused, making it economically beneficial. On the other hand, green plastics production, such as cellulose and functionalized graphene composite plastics give environment-friendly technology with the use of traditional, synthetic plastics. This technology utilizes cellulose and functionalized graphene, reserving them for high-performance applications. The total process takes place at low temperatures, reducing energy consumption. cellulose/functionalized graphene composite plastics (CGP) are sustainable due to the fact they biodegrade, while cellulose is renewable [70]. On the other hand, traditional plastics receive exceptional downsides due to pollution and difficulty in recycling, leading to environmentally and ecologically damaging consequences in the long run.

Microwave-assisted production of functionalized cellulose from rice straw is a practical and ecologically conscious process for removing heavy metals like copper, lead, and nickel from water.. That must use extreme quantities of rice straw as a very prospective, capacitive adsorbent, effectively prepared with rapid, as well as energy-effective method. The final product of functionalized cellulose used was truly superior to all other adsorbents and had a very high adsorption capacities for almost all heavy metals. It was also extremely rapid, consumed fewer toxic chemicals, and save energy [71]. The conventional adsorbent preparation techniques either use costly or hard-to-find materials, raising the initial investment costs.

8 Future Scope

Cellulose derivatives have versatile applications in various industries, providing sustainable solutions for pharmaceuticals, food, packaging, textiles, and more, due to their eco-friendly properties. Since it is the most promising candidate feedstock for producing bio-based polymeric materials, cellulose has garnered much attention. Significant advancements in synthesizing biopolymers based on various cellulosic forms have been made throughout the last ten years [72].

Along with lignin and hemicellulose, cellulose comprises the majority of lignocellulosic biomass, comprising 35–50% of the total biomass [73]. Due to its renewability, biocompatibility, and biodegradability, it is considered the most powerful candidate to replace petroleum-based polymers. In 2013, the manufacturing capacity of sustainable biopolymers based on cellulose constituted 61.8% of the total output of bio-based structural polymers. The growth in annual production is likewise rising [74]. Unmodified cellulose fibers chemically altered the cellulose (such as cellulose derivative), and nanocellulose can all be processed in different ways to create biodegradable polymeric materials, including films, composites, and nanocomposite systems.

Cellulose, which includes C6 glucose monosaccharide and its derivatives, is largely composed of glycopolymers and polymer systems. By conjugating sugar moieties, glucose and its derivatives can be used as pendant groups to synthesize glycopolymers. Alternatively, it can be employed as a flexible building block to create sugar-linked polymers by integrating them into a polymer backbone [75]. LA is currently produced commercially by fermenting sugars, but it has high potential as a versatile cellulosic biomass-derived platform chemical [76] when used in large amounts, a new type of bacteria called homofermentative, facultative anaerobe *Enterococcus faecalis* CBRD01 can be an effective.

8.1 Obtaining Favourable Properties with Sustainable Modifications

Natural cellulose-based material is used in the engineering sectors, forest products, textiles, paper, and packing materials [77]. Natural materials have excellent mechanical strength and flexibility because of their hierarchical structural design, which spans nanoscale to macroscopic scales [78, 79]. Natural cellulose, which is mostly derived from plant and bacterial sources, is gaining increasing interest from researchers as a very promising scaffolding material for a range of restorative objectives. When it comes to mimicking native tissues, cost-effectiveness, availability, biocompatibility, moderate immunological reaction, low toxicity, and low toxicity, natural cellulose offers various advantages over synthetic cellulose. Because of its extraordinary utility, cellulose is an environmentally occurring polymer that is both renewable and biodegradable. Moreover, it has a large specific surface area, high porosity and low density [80]. As a result, it has numerous uses in a variety of industries, including adsorption, thermal insulation, oil/water separation, and biological applications [81].

CMC, a powder that is white, odourless, and tasteless, is a cellulose derivative that is extremely soluble in alkaline solutions or water. Due to its many uses, CMC has emerged as a synthetic plastic substitute. qualities such as biodegradability, biocompatibility, non-toxicity, and high-water retention qualities can be linked to the usage of CMC in these kinds of uses There are several drawbacks to pure carboxymethyl cellulose, including limited flexibility, gel fraction, thermal stability, and mechanical strength. To get beyond these restrictions, CMC can be mixed with other biomaterials, nanoparticles, enzymes, peptides, amino acids, and other polymers [82]. Applications for CMC-based composites include 3D bioprinting, organic electrical devices, food packaging, agriculture, medicinal delivery, textiles, sensors and tissue engineering [83].

8.2 Enhancement in Smart Electronics

Compelled by the need to integrate these kinds of technologies into a wide range of cheap, reusable goods that are part of our daily lives and as a means of connecting the digital and physical realms, a unique way of flexible photonic smart applications is currently evolving. These applications make use of sensors and photovoltaic systems. Nowadays, a variety of flexible polymeric materials are being researched for use as mechanical supports in a range of applications. Since cellulose is the most common organic polymer on the Earth and is used to produce paper, researchers have focused a lot of attention on it. This is because of its many benefits, which include being quite affordable in comparison to other materials, recyclable, flexible, lightweight, and biocompatible [84]. In 2022, CMC, which held approximately one-third of the market, and hydroxypropyl methylcellulose (HPMC), which had a market share as high as 40%, were the two most successful cellulose derivatives [85]. To store hydrogen energy, scientists created plasticized solid biopolymer electrolytes (SBEs) based on CMC. Diffusion and the movement of ionic species were the main factors influencing the conductivity of SBEs. The results of the characterizations showed that an SBE sample plated on the surface of a solidified protonic cell can be used to demonstrate effective rechargeability for ten cycles and an open-loop voltage of 1.1 V at ambient temperature [86]. Organic memory devices with extraordinarily large storage capacity are in great demand for future technologies due to the rapid growth of information technology. Among the many different types of storage media, resistance random access memory (RRAM), a memory that is not volatile type with clear systems, simple operation, low cost, and mechanical flexibility, has garnered a lot of interest [87]. Based on CMCs, ion-conductive hydrogels were created for use in sensing apparatuses. CMC and phytic acid were cross-linked in a single pot as part of the synthesis. There was little residual strain and an elevated level of mechanical strength in the matrix. Furthermore, sensors built using these modified hydrogels demonstrated outstanding recovery rates, tensile strengths, and sensitivity. These characteristics therefore supported its application in sensing devices [88]. CMC, also known as sodium salt, CMC-Na, is a typical thickening and bulking ingredient in the cosmetic industry. Its ability to act as an emulsifier and stabilizer is explained by its anionic nature [89]. The market size of cationic cellulosic derivatives is currently significantly smaller than that of CMC. Among other factors, the chemicals required for manufacturing carboxylated (anionic) derivatives are typically more affordable compared to those needed for producing cationic cellulose. For instance, in March 2023, Free-on-board (Hamburg) prices stood at USD 1220 per ton for chloroacetic acid in Europe, a chemical directly utilized in CMC production, and at USD 2970 per ton for epichlorohydrin, necessary for synthesizing tertiary amines and cationizing reagents [90]. Contrary to many assumptions about the procedure, a life cycle study carried out by our research indicates that this method of producing CNFs can be more ecologically friendly compared to the enzymatic way [91].

8.3 Research and Development to Broaden the Viability of Cellulose Material Production

Fibrillated cellulose is biocompatible and possesses adjustable, appealing qualities that point to its potential for commercialization and real-world use. Furthermore, the cost of fibrillated cellulose is significantly lower than that of petroleum-based nanomaterials (about US$20 per dry kg for nanoscale fibrillated cellulose & roughly

US$0.60 per dry kg to paper make-grade fibrillated cellulose in 2020 [92] it has the potential to be produced on an ind+ustrial scale, offering still another financial benefit.

In recent times, there has been a preference for intelligent or smart food packaging, which mostly uses halochromic sensors or colourimetric indicators. These packaging films are capable of accurately detecting and tracking the quality, freshness, and spoilage of packed food items across the whole supply chain. Because of their superior biodegradability, appropriate miscibility qualities, compatibility, non-toxicity, eco-friendliness, and environmental suitability, biopolymers are widely acknowledged as a possible replacement for synthetic plastics. For use in smart/intelligent packaging films, cellulose and its derivatives present a suitable packaging material [93].

The creation of innovative biodegradable plastic compounds from renewable resources is one of the main challenges the plastics industry has in its attempt to provide sustainable alternatives for plastic derived from petroleum. Bio-based polymers are usually derived from plant biomass, which includes higher crops, plants and recently, algae [94]. The usage of algal cellulose and nanocellulose has been introduced for the application of bioplastics. Fast-growing microorganisms, terrestrial crop biomass, and agricultural waste are examples of organic renewable sources used to make bioplastics. Sadly, tensile strength, water and oxygen permeability, and other mechanical characteristics that are important to assess an application's viability are typically lacking in biodegradable bioplastics [95].

A key factor in improving the viability of cellulose material production is research and development. Employing constant innovation and cooperation, we may unleash novel approaches, enhance efficacy, and alleviate the effects on the environment. By ensuring sustainable growth, this commitment opens up new possibilities for cellulose materials in a variety of industries, contributing to a more environmentally sensitive and optimistic future.

9 Conclusion

Green chemistry is in line with the use of cellulose, a naturally occurring polymer that is abundant and renewable, as a precursor towards biocompatible derivatives. Over the past few decades, several research studies have produced data regarding the modification and application of cellulosic materials. Cellulose has become utilized for many different biomedical applications for its outstanding characteristics and broad expansion potential. Recent developments demonstrate that cellulosic compounds offer inherent properties that can be modified and customized for tissue engineering, medication delivery, and wound dressing systems, among other biomedical uses. The many chemical transformations as well as derivatizations described below have been used to change particular cellulose characteristics to increase the number of possible uses, particularly in biomedical applications. Specifically, CMC and hydroxyethyl cellulose are utilized because they make excellent, adaptable substances for tissue engineering applications. Green chemistry techniques reduce the amount of chemical waste produced, energy used, and pollution to the environment. Examples of these techniques include enzyme modification, solvent-free responses and aqueous-based procedures. Through the emphasis on non-toxic compounds and feedstocks that are renewable, these techniques help lower the ecological load and carbon footprint associated with the manufacture of cellulose derivatives. Biocompatible cellulose derivatives find sustainable use not just in biomedical sciences but also in a variety of industrial sectors. For example, cellulose-based polymers include barrier qualities, biodegradability, and composability that make them environmentally preferable to plastics made from petroleum. Furthermore, cellulose derivatives are useful in textiles, coverings, and adhesives because of their biodegradability and renewable nature, which meets the increasing need for sustainable products. The development of improved cellulosic materials remains a problem from both an academic and commercial standpoint. Overcoming this challenge requires a versatile approach which includes an increasing variety of innovative procedures that have been created to prepare various chemical molecules. All things considered, the incorporation of green chemistry concepts into the production and use of biocompatible derivatives of cellulose highlights its capacity to promote favorable environmental and social consequences. These sustainable materials are the key to solving global issues and fostering a more resilient and sustainable future for future generations through further study, technical developments, and collaboration.

References

1. Hon, D.N.-S.: Cellulose: a random walk along its historical path. Cellulose **1**, 1–25 (1994)
2. Mais, U., Knaus, S., Binder, W.H., Gruber, H.: Functionalization of cellulose. Lenzinger Berichte **79**, 71–76 (2000)
3. Zhbankov, R.G., Firsov, S.P., Buslov, D.K., Nikonenko, N.A., Marchewka, M.K., Ratajczak, H.: Structural physico-chemistry of cellulose macromolecules. Vibrational spectra and structure of cellulose. J. Mol. Struct. **614**(1–3), 117–125 (2002)
4. Voicu, S.I., Thakur, V.K.: Aminopropyltriethoxysilane as a linker for cellulose-based functional materials: New horizons and future challenges. Curr. Opin. Green Sustain. Chem. **30**, 100480 (2021)
5. Khanjanzadeh, H., Behrooz, R., Bahramifar, N., Gindl-Altmutter, W., Bacher, M., Edler, M., Griesser, T.: Surface chemical functionalization of cellulose nanocrystals by 3-aminopropyltriethoxysilane. Int. J. Biol. Macromol. **106**, 1288–1296 (2018)
6. Seddiqi, H., Oliaei, E., Honarkar, H., Jin, J., Geonzon, L.C., Bacabac, R.G., Klein-Nulend, J.: Cellulose and its derivatives: towards biomedical applications. Cellulose **28**, 1893–1931 (2021). https://doi.org/10.1007/s10570-020-03674-w
7. Tang, J., Sisler, J., Grishkewich, N., Tam, K.C.: Functionalization of cellulose nanocrystals for advanced applications. J. Colloid Interface Sci. **494**, 397–409 (2017)
8. Bethke, K., Palantöken, S., Andrei, V., Roß, M., Raghuwanshi, V.S., Kettemann, F., Greis, K., et al.: Functionalized cellulose for water purification, antimicrobial applications, and sensors. Adv. Funct. Mater. **28**(23), 1800409 (2018)
9. Venkataramanan, S.N., Matsui, K., Kawanami, H., Ikushima, Y.: Green Chem. **7**, 18 (2007)
10. Wittmar, Vorat, D., Ulbricht, M.: RSC Adv. **5**, 88070 (2015)
11. Carapeto, P., Ferraria, A.M., Botelho do Rego, A.M.: Polym. Test. **58**, 236 (2017)

12. Zabetakis, D., Dressick, W.J., Appl, A.C.S.: Mater. Interfaces **4**, 2358 (2012)
13. Shchipunov, Y., Postnova, I.: Cellulose mineralization as a route for novel functional materials. Adv. Func. Mater. **28**(27), 1705042 (2018)
14. Heinze, Th., Liebert, T.: Unconventional methods in cellulose functionalization. Prog. Polym. Sci. **26**(9), 1689–1762 (2001)
15. Moon, D., Sagisaka, M., Tahara, K., Tsukahara, K.: Progress towards sustainable production: environmental, economic, and social assessments of the cellulose nanofiber production process. Sustainability **9**(12), 2368 (2017)
16. Uusi-Tarkka, E.-K., Skrifvars, M., Haapala, A.: Fabricating sustainable all-cellulose composites. Appl. Sci. **11**(21), 10069 (2021)
17. Kedzior, S.A., Zoppe, J.O., Berry, R.M., Cranston, E.D.: Recent advances and an industrial perspective of cellulose nanocrystal functionalization through polymer grafting. Curr. Opin. Solid State Mater. Sci. **23**(2), 74–91 (2019)
18. Arantes, V., Dias, I.K.R., Berto, G.L., Pereira, B., Marotti, B.S., Nogueira, C.F.O.: The current status of the enzyme-mediated isolation and functionalization of nanocelluloses: production, properties, techno-economics, and opportunities. Cellulose **27**, 10571–10630 (2020)
19. Pandey, A.: Pharmaceutical and biomedical applications of cellulose nanofibers: a review. Environ. Chem. Lett. **19**(3), 2043–2055 (2021)
20. Seddiqi, H., Oliaei, E., Honarkar, H., Jin, J., Geonzon, L.C., Bacabac, R.G., Klein-Nulend, J.: Cellulose and its derivatives: towards biomedical applications. Cellulose **28**(4), 1893–1931 (2021)
21. de Amorim, J.D.P., de Souza, K.C., Duarte, C.R., da Silva Duarte, I., de Assis Sales Ribeiro, F., Silva, G.S., de Farias, P.M.A., et al.: Plant and bacterial nanocellulose: production, properties and applications in medicine, food, cosmetics, electronics and engineering. A review. Environ. Chem. Lett. **18**, 851–869 (2020)
22. Zennifer, A., Senthilvelan, P., Sethuraman, S., Sundaramurthi, D.: Key advances of carboxymethyl cellulose in tissue engineering & 3D bioprinting applications. Carbohyd. Polym. **256**, 117561 (2021)
23. Bäckdahl, H., Esguerra, M., Delbro, D., Risberg, B., Gatenholm, P.: Engineering microporosity in bacterial cellulose scaffolds. J. Tissue Eng. Regen. Med. **2**(6), 320–330 (2008)
24. Mabrouk, M., Beherei, H.H., Das, D.B.: Recent progress in the fabrication techniques of 3D scaffolds for tissue engineering. Mater. Sci. Eng. C **110**, 110716 (2020)
25. Sen, C.K.: Human wounds and its burden: an updated compendium of estimates. Adv. Wound Care **8**(2), 39–48 (2019)
26. Wang, X., Cheng, F., Liu, J., Smått, J.-H., Gepperth, D., Lastusaari, M., Xu, C., Hupa, L.: Biocomposites of copper-containing mesoporous bioactive glass and nanofibrillated cellulose: biocompatibility and angiogenic promotion in chronic wound healing application. Acta Biomater. **46**, 286–298 (2016)
27. Reckhenrich, A.K., Kirsch, B.M., Wahl, E.A., Schenck, T.L., Rezaeian, F., Harder, Y., Foehr, P., Machens, H.-G., Egaña, J.T.: Surgical sutures filled with adipose-derived stem cells promote wound healing. PloS One **9**(3), e91169 (2014)
28. Li, B., Moriarty, T.F., Webster, T., Xing, M.: Racing for the surface. Springer International Publishing (2020)
29. Pang, M., Huang, Y., Meng, F., Zhuang, Y., Liu, H., Du, M., Ma, Q., et al.: Application of bacterial cellulose in skin and bone tissue engineering. Eur. Polym. J. **122**, 109365 (2020)
30. Gorgieva, S., Trček, J.: Bacterial cellulose: Production, modification and perspectives in biomedical applications. Nanomaterials **9**(10), 1352 (2019)
31. Mishra, K., Siwal, S.S., Sithole, T., Singh, N., Hart, P., Thakur, V.K.: Biorenewable materials for water remediation: The central role of cellulose in achieving sustainability. J. Bioresour. Bioprod. (2023)
32. Masoom, W., Khan, A., Sarwar, A., Musaddiq, S., Farooqi, Z.H., Iqbal, S.: Removal of arsenic (III) and (V) from water bodies by using biopolymers via adsorption: a review. Express Polym. Lett. **18**(3) (2024)

33. Motloung, M.T., Magagula, S.I., Kaleni, A., Sikhosana, T.S., Lebelo, K., Mochane, M.J.: Recent advances on chemically functionalized cellulose-based materials for arsenic removal in wastewater: a review. Water **15**(4), 793 (2023)
34. Etale, A., Nhlane, D.S., Mosai, A.K., Mhlongo, J., Khan, A., Rumbold, K., Nuapia, Y.B.: Synthesis and application of cationised cellulose for removal of Cr (VI) from acid mine-drainage contaminated water. AAS Open Res. **4** (2021)
35. Lee, K.-Y., Blaker, J.J., Bismarck, A.: Surface functionalisation of bacterial cellulose as the route to produce green polylactide nanocomposites with improved properties. Compos. Sci. Technol. **69**(15–16), 2724–2733 (2009)
36. Holt, B.L., Stoyanov, S.D., Pelan, E., Paunov, V.N.: Novel anisotropic materials from functionalised colloidal cellulose and cellulose derivatives. J. Mater. Chem. **20**(45), 10058–10070 (2010)
37. Sugiarto, S., Pong, R.R., Tan, Y.C., Leow, Y., Sathasivam, T., Zhu, Q., Loh, X.J., Kai, D.: Advances in sustainable polymeric materials from lignocellulosic biomass. Mater. Today Chem. **26**, 101022 (2022)
38. Hashemzehi, M., Mesic, B., Sjöstrand, B., Naqvi, M.: A comprehensive review of nanocellulose modification and applications in papermaking and packaging: challenges, technical solutions, and perspectives. BioResources **17**(2), 3718–3780 (2022)
39. Rull-Barrull, J., d'Halluin, M., Le Grognec, E., Felpin, F.-X.: Chem. Commun. **52**, 6569 (2016)
40. Foroughi, F., Ghomi, E.R., Dehaghi, F.M., Borayek, R., Ramakrishna, S.: A review on the life cycle assessment of cellulose: from properties to the potential of making it a low carbon material. Materials **14**(4), 714 (2021)
41. Kamalesh, T., Senthil Kumar, P., Hemavathy, R.V., Rangasamy, G.: A critical review on sustainable cellulose materials and its multifaceted applications. Ind. Crops Prod. **203**, 117221 (2023)
42. Araujo, D., Castro, M.C.R., Figueiredo, A., Vilarinho, M., Machado, A.: Green synthesis of cellulose acetate from corncob: physicochemical properties and assessment of environmental impacts. J. Clean. Prod. **260**, 120865 (2020)
43. Paunonen, S., Kamppuri, T., Katajainen, L., Hohenthal, C., Heikkilä, P., Harlin, A.: Environmental impact of cellulose carbamate fibers from chemically recycled cotton. J. Clean. Prod. **222**, 871–881 (2019)
44. Hammar, T., Peñaloza, D., Hanning, A.-C., Haatanen, N., Pakkasmaa, J.: Life cycle assessment of textile fibre-to-fibre recycling by cellulose carbamate technology. J. Clean. Prod. **426**, 139189 (2023)
45. Collinson, S.R., Thielemans, W.: The catalytic oxidation of biomass to new materials focusing on starch, cellulose and lignin. Coord. Chem. Rev. **254**(15–16), 1854–1870 (2010)
46. Nasrollahzadeh, M., Shafiei, N., Nezafat, Z., Bidgoli, N.S.S., Soleimani, F.: Recent progresses in the application of cellulose, starch, alginate, gum, pectin, chitin and chitosan based (nano) catalysts in sustainable and selective oxidation reactions: a review. Carbohyd. Polym. **241**, 116353 (2020)
47. Duceac, I.A., Tanasa, F., Coseri, S.: Selective oxidation of cellulose—a multitask platform with significant environmental impact. Materials **15**(14), 5076 (2022)
48. Guidi, P., Bernardeschi, M., Palumbo, M., Genovese, M., Scarcelli, V., Fiorati, A., Riva, L., Punta, C., Corsi, I., Frenzilli, G.: Suitability of a cellulose-based nanomaterial for the remediation of heavy metal contaminated freshwaters: A case-study showing the recovery of cadmium induced dna integrity loss, cell proliferation increase, nuclear morphology and chromosomal alterations on Dreissena polymorpha. Nanomaterials **10**(9), 1837 (2020)
49. Punta, C., Fiorati, A., Riva, L., Grassi, G., Liberatori, G., Corsi, I.: Sustainable and eco-safe nanocellulose-based materials for water nano-treatment. Sustain. Nanotechnol.: Strat. Prod. Appl. 143–158 (2022)
50. Tanasă, F., Teacă, C.A., Nechifor, M., Zănoagă, M.: Multicomponent polymer systems based on agro-industrial waste. In: Bioplastics for Sustainable Development, pp. 467–513 (2021)
51. Tanasă, F., Teacă, C.-A., Nechifor, M.: Lignocellulosic waste materials for industrial water purification. In: Sustainable Green Chemical Processes and their Allied Applications, pp. 381–407 (2020)

52. Ngo, H.H., Guo, W., Zhang, J., Liang, S., Ton-That, C., Zhang, X.: Typical low cost biosorbents for adsorptive removal of specific organic pollutants from water. Bioresour. Technol. **182**, 353–363 (2015)
53. Dhabhai, R., Niu, C.H., Dalai, A.K.: Agricultural byproducts-based biosorbents for purification of bioalcohols: a review. Bioresour. Bioprocess. **5**(1), 1–14 (2018)
54. Khadir, A., Motamedi, M., Pakzad, E., Sillanpää, M., Mahajan, S.: The prospective utilization of Luffa fibres as a lignocellulosic biomaterial for environmental remediation of aqueous media: a review. J. Environ. Chem. Eng. **9**(1), 104691 (2021)
55. Isogai, A., Bergström, L.: Preparation of cellulose nanofibers using green and sustainable chemistry. Curr. Opin. Green Sustain. Chem. **12**, 15–21 (2018)
56. Zhang, S., Li, J., Chen, S., Zhang, X., Ma, J., He, J.: Oxidized cellulose-based hemostatic materials. Carbohyd. Polym. **230**, 115585 (2020)
57. Caschera, D., Toro, R.G., Federici, F., Montanari, R., de Caro, T., Al-Shemy, M.T., Adel, A.M.: Green approach for the fabrication of silver-oxidized cellulose nanocomposite with antibacterial properties. Cellulose **27**, 8059–8073 (2020)
58. Kwon, G., Lee, K., Kim, D., Jeon, Y., Kim, U.-J., You, J.: Cellulose nanocrystal-coated TEMPO-oxidized cellulose nanofiber films for high performance all-cellulose nanocomposites. J. Hazard. Mater. **398**, 123100 (2020)
59. Gao, P., Cha, R., Luo, H., Xu, Y., Zhang, P., Han, L., Wang, X., Zhang, Z., Jiang, X.: Development of antimicrobial oxidized cellulose film for active food packaging. Carbohyd. Polym. **278**, 118922 (2022)
60. Torgbo, S., Sukyai, P.: Biodegradation and thermal stability of bacterial cellulose as biomaterial: the relevance in biomedical applications. Polym. Degrad. Stab. **179**, 109232 (2020)
61. Akampumuza, O., Wambua, P.M., Ahmed, A., Li, W., Qin, X.-H.: Review of the applications of biocomposites in the automotive industry. Polym. Compos. **38**(11), 2553–2569 (2017)
62. Sarasini, F., Tirillò, J., Sergi, C., Seghini, M.C., Cozzarini, L., Graupner, N.: Effect of basalt fibre hybridisation and sizing removal on mechanical and thermal properties of hemp fibre reinforced HDPE composites. Compos. Struct. **188**, 394–406 (2018)
63. Asim, M., Paridah, M.T., Chandrasekar, M., Shahroze, R.M., Jawaid, M., Nasir, M., Siakeng, R.: Thermal stability of natural fibers and their polymer composites. Iran. Polym. J. **29**, 625–648 (2020)
64. Methacanon, P., Weerawatsophon, U., Sumransin, N., Prahsarn, C., Bergado, D.T.: Properties and potential application of the selected natural fibers as limited life geotextiles. Carbohyd. Polym. **82**(4), 1090–1096 (2010)
65. Yao, F., Wu, Q., Lei, Y., Guo, W., Xu, Y.: Thermal decomposition kinetics of natural fibers: activation energy with dynamic thermogravimetric analysis. Polym. Degrad. Stab. **93**(1), 90–98 (2008)
66. Beg, M.D.H., Pickering, K.L.: Accelerated weathering of unbleached and bleached Kraft wood fibre reinforced polypropylene composites. Polym. Degrad. Stab. **93**(10), 1939–1946 (2008)
67. Badiei, M., Asim, N., Jahim, J.M., Sopian, K.: Comparison of chemical pretreatment methods for cellulosic biomass. APCBEE Proc. **9**, 170–174 (2014)
68. Shi, S.-C., Chen, X.-A.: Cellulose circular economy: amino-functionalized graphene quantum dots as highly sensitive vaccine indicators. Ind. Crops Prod. **206**, 117694 (2023)
69. Song, W., Zhu, M., Zhu, Y., Zhao, Y., Yang, M., Miao, Z., Ren, H., Ma, Q., Qian, L.: Zeolitic imidazolate framework-67 functionalized cellulose hybrid aerogel: an environmentally friendly candidate for dye removal. Cellulose **27**, 2161–2172 (2020)
70. Song, N., Hou, X., Chen, L., Cui, S., Shi, L., Ding, P.: A green plastic constructed from cellulose and functionalized graphene with high thermal conductivity. ACS Appl. Mater. Interfaces **9**(21), 17914–17922 (2017)
71. Du, Z., Chen, H., Guo, X., Qin, L., Lin, D., Huo, L., Yao, Y., Zhang, Z.: Mechanism and industrial application feasibility analysis on microwave-assisted rapid synthesis of amino-carboxyl functionalized cellulose for enhanced heavy metal removal. Chemosphere **268**, 128833 (2021)
72. Rinaldi, R., Schüth, F.: Design of solid catalysts for the conversion of biomass. Energy Environ. Sci. **2**(6), 610–626 (2009)

73. Aeschelmann, F., Carus, M.: Biobased building blocks and polymers in the world: capacities, production, and applications–status quo and trends towards 2020. Ind. Biotechnol. **11**(3), 154–159 (2015)
74. Zhang, Y., Chan, J.W., Moretti, A., Uhrich, K.E.: Designing polymers with sugar-based advantages for bioactive delivery applications. J. Control. Release **219**, 355–368 (2015)
75. Subramanian, M.R., Talluri, S., Christopher, L.P.: Homofermentative lactic acid production. Microb. Biotechnol. **8**(2), 221–229 (2015)
76. Garcia, A., Gandini, A., Belgacem, N., Bras, J.: Modification of nanocellulose with natural molecules: a green perspective for cellulose based materials with active properties. In: Abstracts of papers of the American chemical society, vol. 249. 1155 16TH ST, NW, WASHINGTON, DC 20036 USA: AMER CHEMICAL SOC (2015)
77. Aziz, T., Fan, H., Haq, F., Khan, F.U., Numan, A., Ullah, A., Wazir, N.: Facile modification and application of cellulose nanocrystals. Iran. Polym. J. **28**, 707–724 (2019)
78. Aziz, T., Fan, H., Zhang, X., Khan, F.U., Fahad, S., Ullah, A.: Adhesive properties of bio-based epoxy resin reinforced by cellulose nanocrystal additives. J. Polym. Eng. **40**(4), 314–320 (2020)
79. Aziz, T., Farid, A., Haq, F., Kiran, M., Ullah, A., Zhang, K., Li, C., et al.: A review on the modification of cellulose and its applications. Polymers **14**(15), 3206 (2022)
80. Zheng, J., Aziz, T., Fan, H., Haq, F., Khan, F.U., Ullah, R., Ullah, B., Khattak, N.S., Wei, J.: Synergistic impact of cellulose nanocrystals with multiple resins on thermal and mechanical behavior. Z. für Physikalische Chem. **235**(10), 1247–1262 (2021)
81. Tyagi, V., Thakur, A.: Applications of biodegradable carboxymethyl cellulose based composites. Results Mater. 100481 (2023)
82. Zhang, K., Wang, Y., Wei, Q., Li, X., Guo, Y., Zhang, S.: Design and fabrication of sodium alginate/carboxymethyl cellulose sodium blend hydrogel for artificial skin. Gels **7**(3), 115 (2021)
83. Vicente, A.T., Araújo, A., Mendes, M.J., Nunes, D., Oliveira, M.J., Sanchez-Sobrado, O., Ferreira, M.P., Águas, H., Fortunato, E., Martins, R.: Multifunctional cellulose-paper for light harvesting and smart sensing applications. J. Mater. Chem. C **6**(13), 3143–3181 (2018)
84. Pathak, A., Rao, N.G.R., Grover, P., Sharma, V., Malik, A., Rawat, A.P., Singh, S., Maurya, A.: Ecofriendly pharmaceutical packaging material: a review. Mater. Today: Proc. (2023).
85. Isa, M.I.N., Sohaimy, M.I.H., Ahmad, N.H.: Carboxymethyl cellulose plasticized polymer application as bio-material in solid-state hydrogen ionic cell. Int. J. Hydrog. Energy **46**(11), 8030–8039 (2021)
86. Moxon, S.R., Cooke, M.E., Cox, S.C., Smow, M., Jeys, L., Jones, S.W., Grover, L.M., Smith, A.M.: Suspended manufacture of biological structures. Adv. Mater. **29**(13) (2017)
87. Lu, C., Qiu, J., Sun, M., Liu, Q., Sakai, E., Zhang, G.: Simple preparation of carboxymethyl cellulose-based ionic conductive hydrogels for highly sensitive, stable and durable sensors. Cellulose **28**, 4253–4265 (2021)
88. Wennman, M., Hellberg, M., Svagan, A.J., Hedenqvist, M.S.: A biobased binder of emulsion type that provides unique and durable wet strength and hydrophobicity to paper and nonwoven. Ind. Crops Prod. **193**, 116126 (2023)
89. ChemAnalyst (2023) Monochloroacetic acid price trend and forecast [WWW Document]. https://www.chemanalyst.com/Pricing-data/monochloroacetic-acid-1156
90. Arfelis, S., Aguado, R.J., Civancik, D., Fullana-i-Palmer, P., Pèlach, À., Tarrés, Q., Delgado-Aguilar, M.: Sustainability of cellulose micro-/nanofibers: A comparative life cycle assessment of pathway technologies. Sci. Total Environ. **874**, 162482 (2023)
91. Isogai, A.: Present situation and future prospects of nanocellulsoe R&D in Japan. In: 2018 TAPPI international conference on nanotechnology for renewable materials, TAPPI (2018)
92. Yekta, R., Abedi-Firoozjah, R., Salim, S.A., Khezerlou, A., Abdolmaleki, K.: Application of cellulose and cellulose derivatives in smart/intelligent bio-based food packaging. Cellulose **30**(16), 9925–9953
93. Zanchetta, E., Damergi, E., Patel, B., Borgmeyer, T., Pick, H., Pulgarin, A., Ludwig, C.: Algal cellulose, production and potential use in plastics: challenges and opportunities. Algal Res. **56**, 102288 (2021)

94. Cava, D., Giménez, E., Gavara, R., Lagarón, J.M.: Comparative performance and barrier properties of biodegradable thermoplastics and nanobiocomposites versus PET for food packaging applications. J. Plast. Film & Sheeting **22**(4), 265–274 (2006)
95. Islam, S., Rahman Bhuiyan, M.A., Islam, M.N.: Chitin and chitosan: structure, properties and applications in biomedical engineering. J. Polym. Environ. **25**, 854–866 (2017)

Cellulose and Chitosan-Based Bionanocomposites in Biomedical Applications

Niranjan Patra, Vinith Supriya Gopi, and Pradeep Kumar Brahman

Abstract Bionanocomposites are a promising class of hybrid materials made from natural and manmade biodegradable polymers with organic and inorganic fillers that have emerged with varied scope of promising bioapplications in fields like tissue engineering, drug delivery, gene delivery, wound healing, anti fouling and other biotechnological fields. In order to design and develop them, a thorough understanding of their structure, properties, polymer-matrix interaction, and additional modifications based on fundamental requirements is required. As a corollary, the assessment of biological interactions and the management of physical and chemical properties are crucial for their potential uses. This chapter emphasizes on the use biopolymer based composites in the medical and pharmaceutical field.

Keywords Bionanocomposites · Cellulose · Chitosan · Pharmaceuticals · Nanomedicine

1 Introduction

As an exciting new field at the intersection of biology, materials science, and nanotechnology, bionanocomposites are a fascinating example of the interdisciplinary nature of today's scientific research. The advances in biomedical science are just one field that is benefiting from the introduction of novel bionanocomposites. Biopolymers are gaining popularity in the food and pharmaceutical industries due to their easy accessibility and abundance. For the reasons that they have excellent properties such as biocompatibility, nontoxicity, biodegradability, flexibility, and enhanced structural and multifunctional features, bionanocomposites have attracted a lot of study [1–6]. Some common biopolymers include chitosan, pectin, alginate, agarose, cellulose, guar gum, agar, carrageenan, and gelatin. For the pharmaceutical industry, biopolymers have a lot of promise because they can be produced

N. Patra (✉) · V. S. Gopi · P. K. Brahman
Department of Chemistry, Koneru Lakshmaiah Education Foundation, Greenfield, Vaddeswaram 522502, Andhra Pradesh, India
e-mail: patraji@gmail.com

K. K. Kesari et al. (eds.), *Functionalized Cellulose Materials*, Engineering Materials,
https://doi.org/10.1007/978-3-031-76953-5_8

cheaply, have several uses, and are a renewable resource [7–12]. Due to their pervasiveness and quantity, biopolymers can be broadly categorised into three groups such as polynucleotides, polypeptides/poly amino acids, and polysaccharides [13–18] while fillers like carbon nanostructures, iron oxide, hydroxyapatite, clay, and metal nanoparticles are examples of existing fillers [19–23]. Filler components in nanocomposites can function as molecular bridges in the polymer matrix thanks to their nanoscale interactions. This is the reason why the nanocomposite has better mechanical properties than regular microcomposites [16]. To this list of advantageous qualities, bionanocomposites add the fact that they are biocompatible and/or biodegradable.

As a result of the wide variety of biomedical applications and the widely varying functional needs of these applications, polymer nanocomposites are well positioned to take advantage of a wealth of market opportunities. In the case of internal bone fixation, for instance, the screws and rods utilised to bring the bone close-by surfaces to aid in recovery. This fix must be firm for several weeks to months without becoming brittle or loose. When transferring loads to the implant, it's best if its modulus is comparable to that of the bone. The rods and screws should be simple to remove and not harmful in any way. Therefore, a nanocomposites biocompatibility, biodegradability, mechanical qualities, and even aesthetic requirements are only some of the design and functionality criteria that an implant must meet [24, 25]. Bionanocomposites have a wide variety of potential uses, but the optimal considerations for the matrix, polymer, filler, and matrix-filler interaction, is not straight forward.

This chapter offers a comprehensive view as well recent development of biopolymer composites in pharmaceuticals and nanomedicine, focusing on two specific biopolymers (cellulose and chitosan). The potential of composites made of cellulose and chitosan is examined, along with its applications in various field such as in tissue engineering, drug delivery, antifouling, wound healing, and antibacterial agents.

2 Cellulose

Anhydroglucose monomers ($(C_6H_{10}O_5)_n$; n is the degree of polymerization; n = 10,000 to 5,000 depending on the source used for cellulose extraction) are the monomers that make up cellulose [26–28]. Cellulose is mostly derived by bacteria, marine algae, sugarcane bagasse, plants, tunicates [29–31]. More than a few hundred billion tonnes of cellulose are produced each year. Demand in the market has risen steadily in recent years. Cellulose has excellent physicomechanical, and chemical properties, including stability in strong acidity, strength, chirality and modulus, lightweight, numerous surface hydroxyl functional groups allow for simple chemical modification and excellent biodegradability. Therefore, cellulose offers cutting-edge applications in the fields of energy, the environment, and health [4, 26, 28, 30–36].

2.1 Cellulose Based Biocomposites

Composites made from cellulose have been found to have sophisticated clinical relevance. Antimicrobials, drug delivery, wound healing materials, tissue engineering, artificial blood vessels, and UV radiation protection are all uses that have been documented for these materials [4, 34, 37–47]. Hydrogels, aerogels, membranes, and three-dimensional (3D) scaffolds are only some of the types that cellulose can be processed [34, 48]. Several functional groups and chemicals can alter the surface properties of cellulose materials. The applications of cellulose-based materials in medicine are promising and have a lot of future potentials [14, 26, 27, 29, 30, 32, 37–39, 41–44, 49, 50].

This section focuses on various ways in which cellulose-based materials have been used in the field of nanomedicine and pharmaceuticals. Discussion of cellulose and cellulose composites as possible drug delivery vehicles is included. Antimicrobials, wound dressings, antifouling coatings, and tissue engineering (Fig. 1).

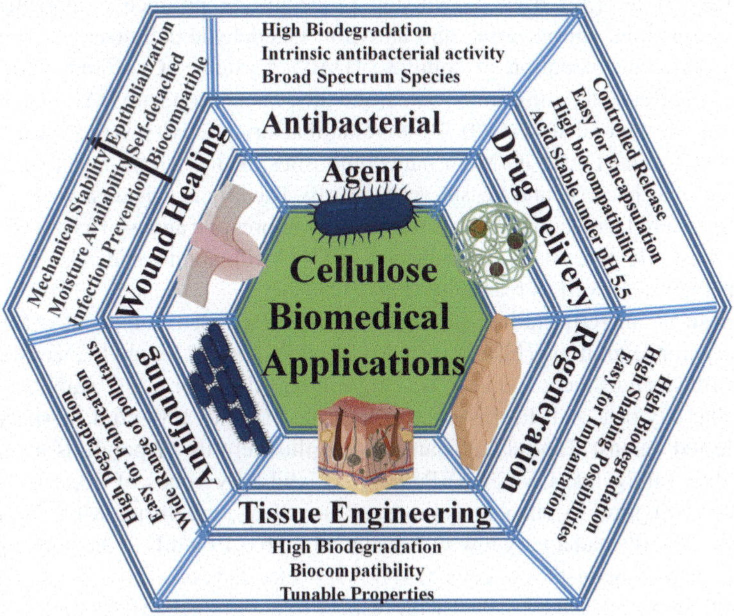

Fig. 1 Biomedical applications of cellulose materials in different ways (Adapted from [51] under creative commons attribution CC BY.)

2.2 *Cellulose Based Composites for Antibacterial Applications*

Antimicrobial efficacy of cellulose can be improved by polymer grafting. Both simple techniques of mixing and in-situ polymerizations allow for the direct blending (anchoring/grafting) of polymers with cellulose (in situ procedure). To alter its properties, cellulose was treated with various polymers [52–58], including polypyrrole, PP, polyethyleneimine, polyhexamethylene guanidine hydrochloride (PHMG-Cl), 2-aminoethylmethacrylate, and poly(3-hydroxyacetylthioalkanoate-co-3-hydroxy alkanoates [52–58].

In situ polymerization of 2-aminoethyl methacrylate with varying concentrations of N,N-methylenebis(acrylamide) (MBA) as a cross-linker was used to create antimicrobial cellulose/poly(2-aminoethyl methacrylate) (BC/PAEM) nanocomposites [58]. As a result, the composite membranes produced were far more see-through than those made from pure bacterial cellulose, in addition to having superior thermal and mechanical qualities.

Hashem et al. [59] used a two-step procedure to produce composites that are biocompatible, antibacterial, and antiviral. Dialdehyde cellulose was obtained through periodate oxidation of cellulose (DAC). Cysteine (Cys) and methionine (Meth) are sulfur-containing amino acids, and their interaction with DAC in graphene oxide was the second step (GO). Positive antibacterial activity was shown by the DAC/GO/Cys and DAC/GO/Meth nanocomposites against Gram-negative, E.Coli and unicellular bacteria and yeasts, respectively. The efficacy against viruses of all of the engineered nanocomposites is particularly impressive against Herpes (HSV-1) at the lowest effective dose. The inclusion of DAC with amino acids also improved their interaction as well as reactivity (Fig. 2).

Doghish et al. [60] studied a cellulose/gold NPs based composites. Minimal inhibitory concentrations (MICs) for CMC-AuNPs against Staphylococcus aureus and Bacillus cereus were found to be 25 g/ml, but MICs for CMC-AuNPs against Klebsiella oxytoca and Escherichia coli were 50 and 100 g/ml, respectively. When tested against Candida albicans, Aspergillus terreus, Aspergillus niger, and Aspergillus fumigatus, CMC-AuNPs showed inhibitory zones of 20, 13, 23, and 26 mm at 500 g/ml. In vitro analysis using breast cancer cell line MCF-7 showed that CMC-AuNPs had a very low IC50 value of 2.56 0.19 g/mL, indicating a potent cytotoxic action against these cells. CMC-AuNPs treated cells showed a higher rate of overall apoptosis (38.81 1.9%) with respect to control cells (0.61 0.022%). The result reveals, that the anticancer behaviour of these composites is due to the stimulation of necrosis and apoptosis in breast cancer cells through increased caspase-8 and 9 activity and decreased VEGFR-2 levels (Fig. 3).

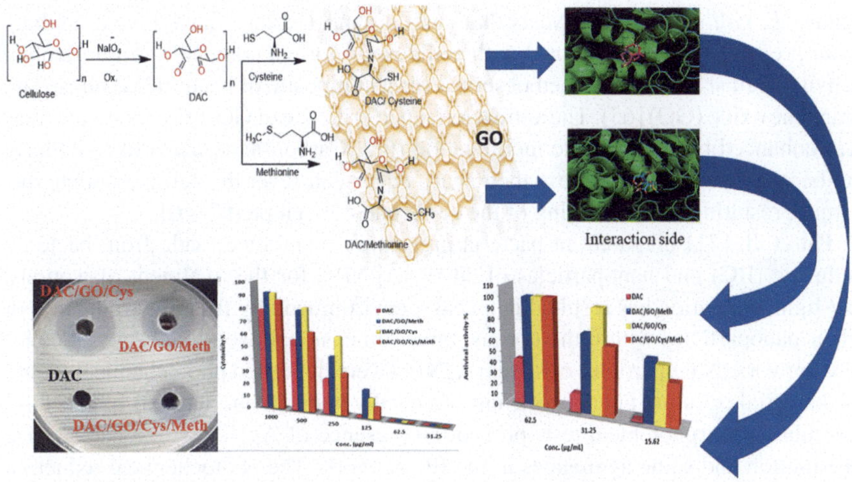

Fig. 2 Schematic process to produce the bionanocomposites with antibacterial results (Adapted with permission from Elsevier [59])

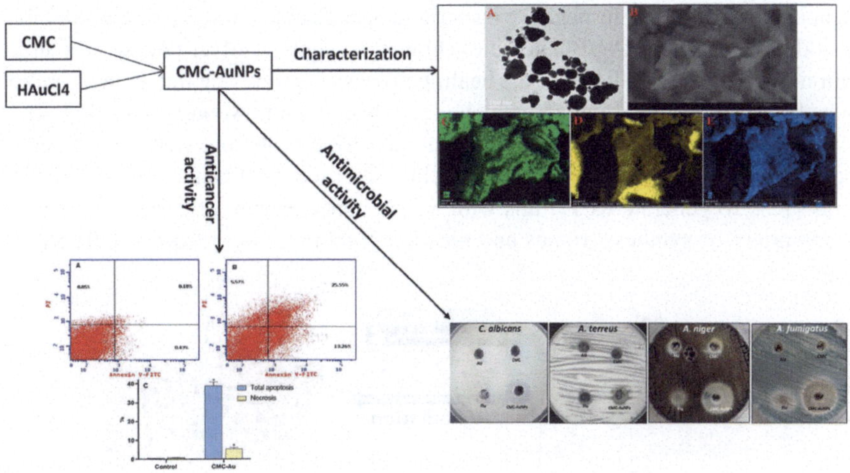

Fig. 3 Process with morphological and antibacterial results (Adapted with permission from Elsevier [60])

2.3 *Cellulose–Inorganic Nanoparticles Bionanocomposites for Antibacterial Applications*

Carbon nanostructures and other metallic nanoparticles are just some of the inorganic nanoparticles that have been used to alter cellulose [61–64]. Many of these materials have effective antibacterial properties on their own, making them useful in the fight against *E. coli*. The study reveals that carbon nanostructures have potent antibacterial properties. They combined with cellulose to form a conjugate. Antimicrobial activity against E. coli has been observed for a composite of bacterial cellulose and graphene oxide (GO) [65]. The antibacterial performance of GO/BC nanocomposites was enhanced by electrostatic modification [66]. Carbon nanosheets like GO have antibacterial properties because their sharp edges can sever the cell membrane like a knife, resulting in the bursting of the cell's outer envelope [67–70].

Pal et al. [71] prepared an bacteria-preventing membrane made from bacterial cellulose (BC) and nanoparticles of silver (AgNPs) for the treatment of wounds. UV light irradiation has resulted in an easy, environmentally friendly production of silver nanoparticles within the porous, three-dimensional, web-like bacterial cellulose network. Nanoparticles of silver (AgNPs) were deposited photochemically onto the bacterial cellulose gel network and chemically bound to the surfaces of the cellulose fibres. Morphological tests proved the presence of AgNPs with a narrow size distribution and some aggregates in the BC network. The photochemical reduction of Ag + Ag0 in Ag/ bacterial cellulose composites was verified by the presence of metallic silver, and X-ray diffraction investigations revealed the bacterial cellulose membranes to be of a highly crystalline form. As demonstrated by disc diffusion and growth dynamics techniques, hybrid composites like pellicles are extremely efficient in eliminating Gram-negative bacteria such as E. coli. Even after a considerable period of soaking, the Ag/BC pellicles showed no signs of silver release. The utilisation of composite pellicles in the healing process is promising due to their ability to be kept in a wet environment, which is also beneficial to wound repair (Fig. 4).

In a single process, cellulose carbamate-NaOH/ZnO solutions were coagulated in Na2SO4 aqueous solutions to produce cellulose based ZnO nanocomposite (RCZ) films [72]. To generate RCZ films with a ZnO concentration of 2.7%-15.1 wt%, a wide variety of synthesis routes and reaction conditions were explored. The ZnO

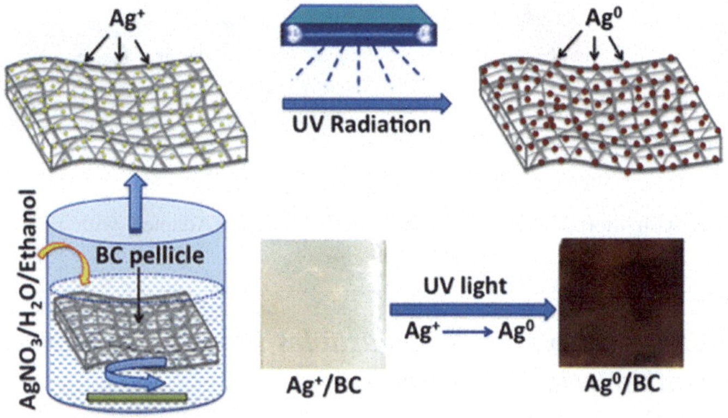

Fig. 4 Illustration of the formation of silver nanoparticles on a nanofibrillated cellulose matrix including before and after pictures of pellicles exposed to ultraviolet light (Adapted with permission from American Chemical Society [71] under creative commons attribution CC BY)

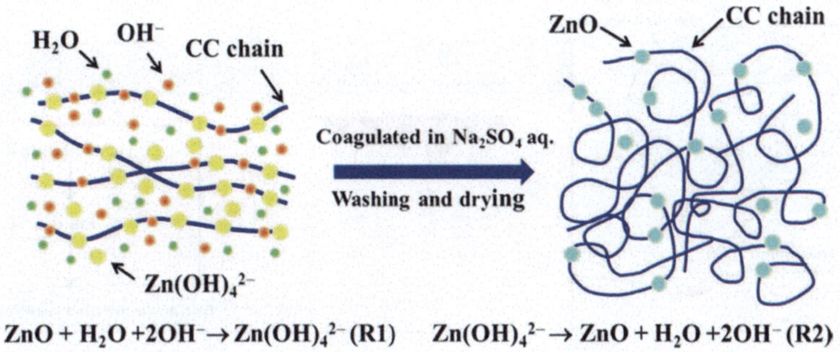

Fig. 5 Schematic representation of one-step coagulation formation of composites films (Adapted with permission from American Chemical Society [72])

nanoparticles, which have a hexagonal wurtzite structure, aggregated into big particles and became securely enmeshed in the cellulose matrix. Excellent mechanical characteristics and excellent thermal stability were demonstrated using RCZ films. Further, the coatings were highly effective at blocking UV light and preventing the growth of microorganisms like Staphylococcus aureus and Escherichia coli. By only 3 h of exposure, the number of live bacteria had dropped significantly, and after 6 h, it had been eliminated entirely. This study established a unique and straightforward method for the fabrication of regenerated cellulose films doped with ZnO nanoparticles for use as biofunctional materials (Fig. 5).

2.4 *Cellulose Based Bionanocomposites for Wound Dressing*

Wound dressings made of cellulose composites have been frequently documented. Lin [73] prepared bacterial cellulose and chitosan composite membrane where bacterial cellulose was submerged in a CTS solution and then freeze-dried. In other words, the method can be used for mass production. CTS and other antibacterial drugs dramatically reduced the proliferation of germs like E. coli and Staph aureus. When compared to a pure bacterial cellulose membrane, the membrane showed considerably more efficacy against bacteria. The efficacy of antibacterial activity of bacterial cellulose and bacterial cellulose-chitosan composite membranes was ~49 and ~100% against E. coli and ~30 and ~100% against S. aureus, respectively. In comparison to wounds dressed with bacterial cellulose alone or a commercial materials like TegadermTM, those treated with the bacterial cellulose-chitosan membrane exhibited significantly higher epithelialization and regeneration rates during the healing process.

An important contributing factor that slows wound treatment in diabetics is the abnormally high production of matrix metalloproteinase-9. Therefore, it may be possible to potential to quicken the repairing of a diabetic wound by suppressing MMP-9 expression. Li et al. [74] studied the feasibility of own-treatment using

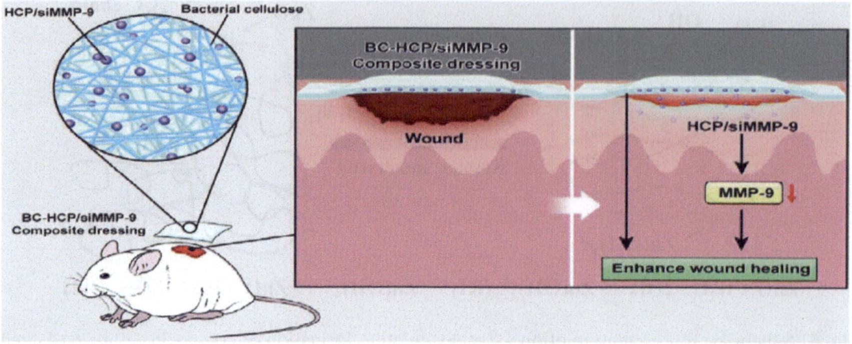

Fig. 6 Schematic representation of the work in brief (Adapted with permission from Elsevier [74])

wound treatment containing cationic hyperbranched polysaccharide derived from bacterial cellulose derivatives that enclose siRNA (BC-HCP/siMMP-9). The study employed four different HCPs (Gly-DMAPA, Gly-D4, Amyp-DMAPA, and Amyp-D4) as vectors for genetic material. Bacterial cellulose (BC) was found to slowly release HCP/siMMP-9 when exposed to bacterial cellulose. A living immortalised epithelial cell line (HaCAT) and diabetic rat wounds showed significant reductions in MMP-9 gene expression and protein levels after treatment with the released siMMP-9. The BC-HCP/siMMP-9 composite dressing significantly improved wound repair in diabetic rat models, showing its potential as a feasible and efficacious dressing for promoting wound repair in diabetic animals (Fig. 6).

Though bacterial cellulose (BC) is an extremely versatile biopolymer but it can't be used as a wound dressing because it has no built-in defences against bacteria. Copper sulphate ($CuSO_4$), chitosan, and sodium alginate were used in bacterial cellulose with the intention of giving it microstatic or antibacterial activity (Shown in Fig. 7), hence expanding its use as an active wound dressing [75]. Weight distribution, tensile strength, swelling ratio, and releasing properties are all associated with Cu particles. Staphylococcus aureus (MRSA) and E. coli that have developed resistance to methicillin were both killed by the BC/AG/CS-Cu composites that were produced (E.coli). The produced composites showed great promise as antimicrobial wound dressings.

2.5 Cellulose Based Bionanocomposites in Drug and Gene Delivery

Nanosize, rich in active functional groups, biodegradable, inexpensive, and biocompatible, nanocrystals of cellulose are one of the most potential naturally available nanomaterials. In light of these features, CNCs and their composites have attracted a lot of study for a wide range of purposes. Modifying the nanomaterial's surface

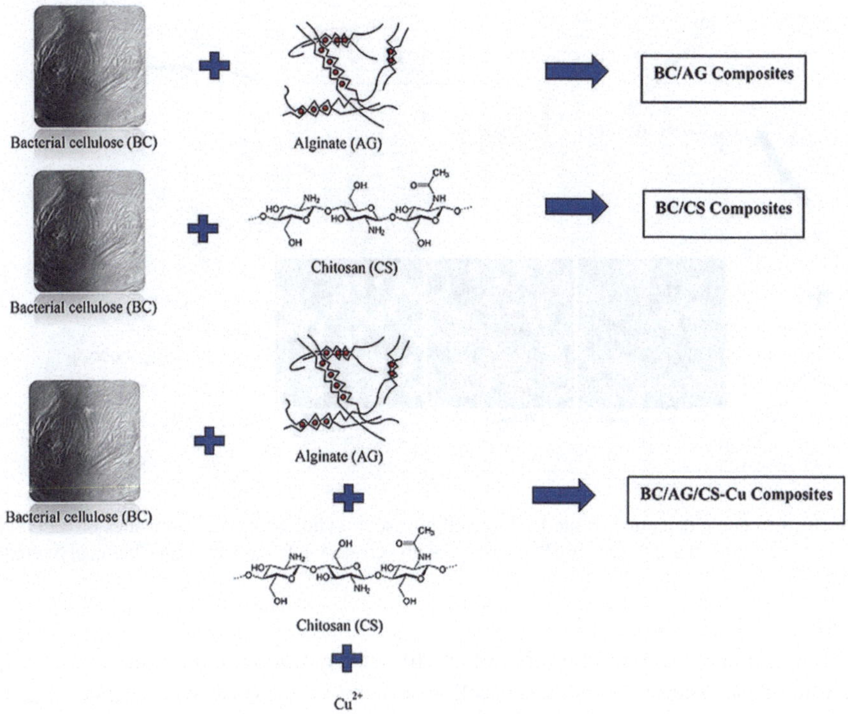

Fig. 7 Scheme of different composites made by using chitosan, sodium alginate and copper sulphate (Adapted with permission from Elesvier [75])

is essential for tailoring its surface properties and bestowing innovative capabilities for targeted uses. Cellulose is an excellent vehicle for transporting hydrophobic medicines such paclitaxel (PTX), etoposide, and docetaxel [76–81].

Wan et al. [82] used an unique approach to prepare CNCs with hyperbranched polymer functionalization via anionic polymerization combined with glycidol as the monomer and the hydroxyl groups on the CNCs' surfaces as the initiators. As a result of forming pH-responsive hydrazone bonds at the functional groups at the ends of these functionalized CNCs, they might be used for loading anticancer medicines like epirubicin (EPI). The study shows that CNCs could have hyperbranched polymers attached to their surfaces by analysing their 1H NMR spectra, The results of the drug release assay, cell survival assay, and cell imaging assay all revealed that CNCs-based carriers exhibit pH-responsive behaviour, which may allow for the release of EPI and that the resulting complexes containing drugs retain their anticancer power. Finally, a new method has been developed for the surface functionalization of CNCs using hyperbranched polymers that are pH-responsive., and it is based on anionic polymerization. The generated functionalized CNCs may be useful as potential candidates for intracellular drug delivery that can be precisely regulated (Fig. 8).

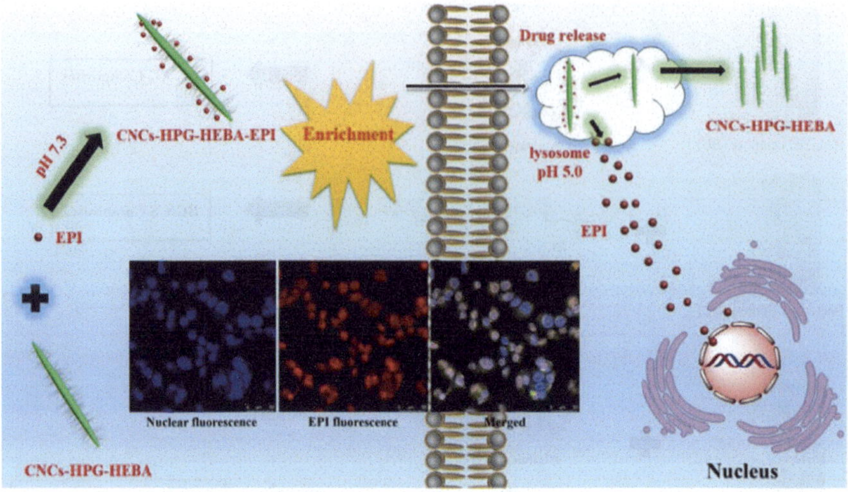

Fig. 8 Schematic depiction of the loading and release of cellulose based composites on EPI and its use in intracellular delivery of EPI (Adapted with permission from American Chemical Society [82])

Singhsa et al. [83] studied the effects of acid hydrolysis conditions and bacterial cellulose sources on bacterial cellulose nanocrystal quality of output, shape, and chemical composition were analysed. The unique properties of these rodlike nanocrystals were made possible by BCNC manufacture using these tunable parameters. The biomedical utility of bacterial cellulose bionanocompositess as nucleic acid delivery systems was demonstrated by testing their ability to complex siRNA and their cytotoxicity in HeLa cells after developing cationic bacterial cellulose nanocrystals via adsorption of amines and amine-containing polymers onto the surface of the resulting sulfuric-hydrolyzed BCNC creates a cationic surface. At a concentration of 0.1 mg/mL, both unmodified and cationic-modified bacterial cellulose nanocrystals samples showed low toxicity, demonstrating their good biocompatibility; in addition, siRNA was completely complexed with polymers of methacrylamide that have been treated with cationic bacterial cellulose nanocrystals. Thus, the results of this study imply that by adjusting the acid and bacterial cellulose source, a BCNC with the necessary characteristics can be created; furthermore, using a simple cationic surface modification, cationic functionalized BCNCs were produced and demonstrated use as nucleic acid nanocarriers (Fig. 9).

To accomplish sustained, stimuli-driven, and slow-release localised drug delivery, biocompatible, adjustable, and injectable hydrogels based on cellulose were created by Chen et al. [84]. Two different kinds of oxidised carboxymethyl cellulose (CMC), one of which was hydrazide-modified carboxymethyl cellulose (CMC-NH_2), and the other of which was oxidised carboxymethyl cellulose with a range of oxidation levels. After that, atom transfer radical polymerization was used to create pH-sensitive

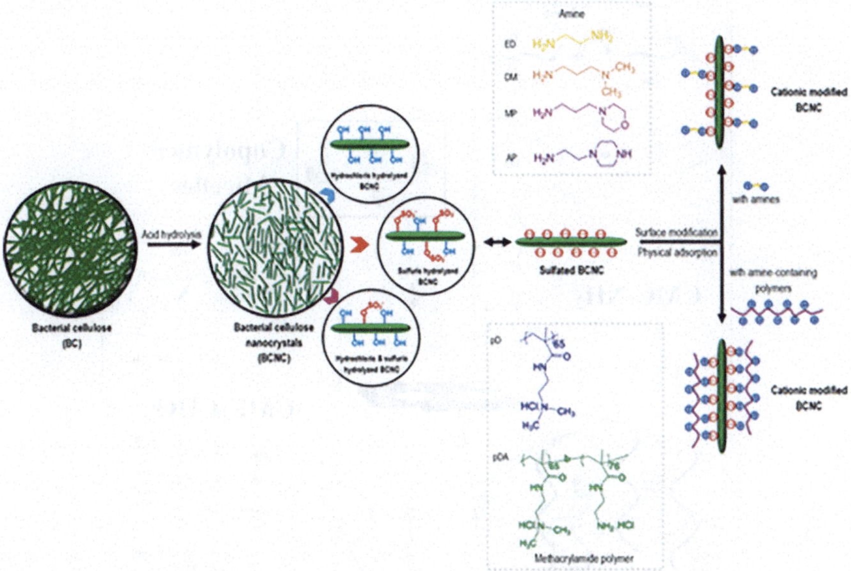

Fig. 9 Surface cationization of a sulfated bacterial cellulose nanocrystals via the use of amines and polymers containing amines in physical adsorption, and production of bacterial cellulose nanocrystals from bacterial cellulose via acid hydrolysis using various acid types (Methacrylamide Polymers) are depicted schematically (Adapted with permission from American Chemical Society [83])

poly (ethylene oxide)-block-poly (2-(diisopropylamino) ethyl methacrylate) (PEO-b-PDPA) copolymers as micelle cores to transport hydrophobic compounds (ATRP). After combining polymer suspensions of CMC-NH_2 and CMC-CHO containing PEO-b-PDPA copolymer micelles in a Schiff base reaction, an injectable hydrogel composite system was developed. The release test with Nile Red dye and doxorubicin demonstrates that this newly synthesised, adaptable, a pH-activated double-barrier cellulose system, extended, and gradual onset of action. Similar storing moduli and controllable degradation were also seen in the hydrogel system (Fig. 10).

2.6 *Cellulose Based Bionanocomposites in Scaffolds for Bone, Skin, and Tissue Engineering*

To aid in tissue regeneration, scaffolds made from nanocellulose have many useful properties. They have excellent mechanical qualities, transparency to light, water absorption/retention, and biocompatibility. Processes include as electrospinning, lipholizing and casting, and even 3D printing are just some of the technologies that can be used to create them. When it comes to tissue engineering, repairing damaged tissues or organs, cellulose materials provide simple, tailor-made procedures. Structure, pore size, surface functional groups, and mechanical properties can

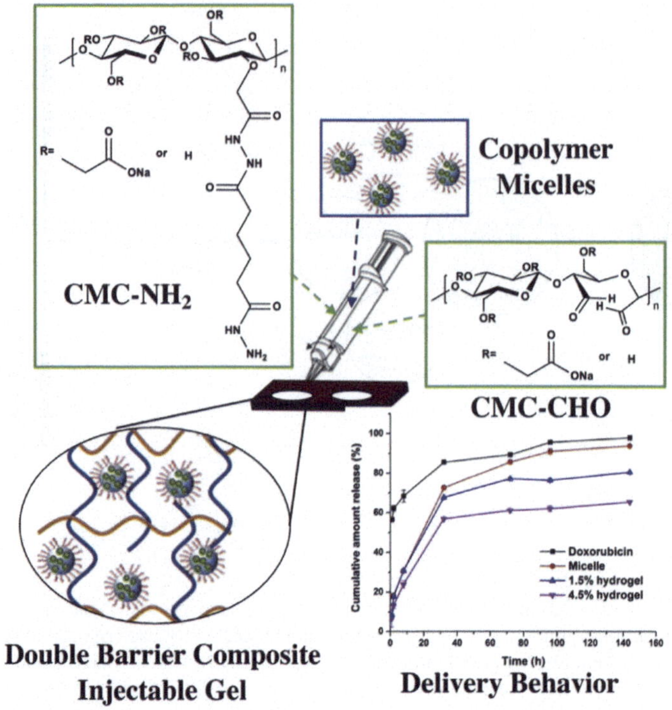

Fig. 10 Hydrogel formation and its delivery behaviour (Adapted with permission from American Chemical Society [84])

all be tailored to meet specifications. Therefore, cellulose-based materials hold great potential for use in the engineered construction of blood arteries, skin, and organs. A high level of biocompatibility is necessary for scaffolds to successfully replicate the native extracellular matrix (ECM). Further, they need to encourage cell division, proliferation, and differentiation [35, 76, 85].

Several criteria relating to mechanical and biological qualities must be met during the fabrication of a suitable skin scaffold. Natural or eco-friendly scaffolds that are biodegradable, biocompatible, and physiologically sound but lack adequate mechanical qualities are rare. Hence, for effective skin tissue engineering and to simulate the actual functions, we need to incorporate synthetic polymers and/or additives in addition to green polymers. Green nanocomposites (nanoscale natural macromolecules or biopolymers incorporating nanoparticles) are a type of scaffolds that have an appropriate range of biomedical properties and in order to reach the stage of skin regeneration and wound repair (Therapeutics administration and tissue engineering for the heart, nervous system, bones, cartilage, and skin).

3 Chitosan

Chitosan is a linear copolymer polysaccharide that has a cationic charge consisting of a non-random arrangement of β(1–4) linked 2-amino-2-deoxy-D-glucosamine (D-glucosamine) and 2-acetamido-2-deoxy-D-glucosamine (N-acetyl-D-glucosamine) units. Figure 1 depicts the structure, which is very similar to cellulose but with acetamido residues replacing the C-2 hydroxyl groups. However, chitosan exhibits far more commercial potential than synthetically substituted cellulose (1.2%) because of the existence of a large concentration of nitrogen (6.69%). Because of this, chitosan gains chelating abilities. Extensive deacetylation of chitin, a plentiful polymer found in crab shells, is the usual method for obtaining chitosan (Fig. 11).

As a bioactive compound, it's shown that it can do a lot of different things such as biological effects, including antitumor, immunoenhancing, antifungal, antimicrobial, antioxidant, and wound healing. These characteristics, in addition to others features, including non-toxicity, biodegradability, biocompatibility, non-antigenicity, and low cost, have led to its widespread use in the pharmaceutical industry, which include biomedicine with the potential for prescription drugs, medication delivery methods applications in the fields of biomedical, food science, bioimaging, implants, contact lenses, gene transfer, protein binding, wound healing, and textiles [2, 6, 13, 23, 87–92].

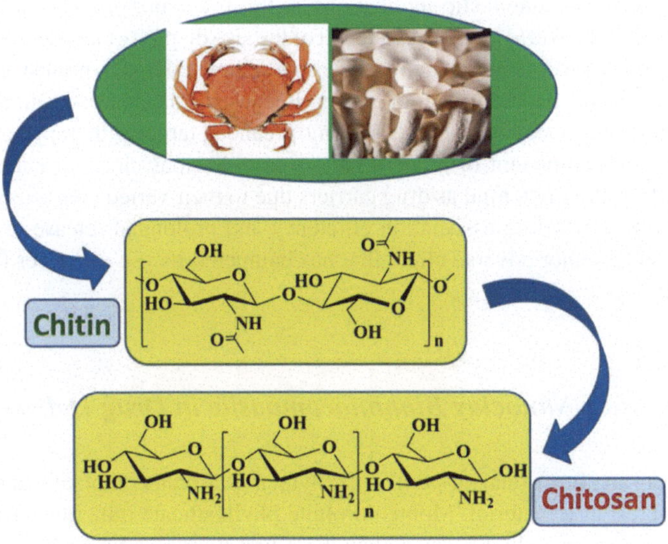

Fig. 11 Chitin and chitosan are generated from the N-deacetylation of chitin, which occurs naturally. Adapted with permission from springer verlag [86]

3.1 Chitosan Based Bionanocomposites in Biomedical Applications

Nanofillers (with dimensions within 100 nm) are added to natural polymers like chitosan, the resulting materials show great promise due to their enhanced biological capabilities, primarily used in the biomedical sector. Fillers in nanometer scale integrated into the chitosan polymer have proven as a potent method for addressing the common issues often associated with biopolymer. Numerous sectors may benefit from the use of chitosan-based nanocomposites [3, 6, 23, 87, 88, 92]. These sectors include the medicine, agriculture, biotechnology, cosmetics, textiles, environmental protection, and many more. Chitosan-based nanocomposites indicate good enhancement in features even at very low level of nanofillers (5–10 wt.%), when compared to conventional composites with high filler content (50%). Products with high bioactivity and biocompatibility can be achieved by maintaining a high chitosan proportion in the composites.

The goal of new medication delivery methods is to reduce the systemic distribution of chemotherapeutic drugs while increasing their local concentration at the site of disease progression (targeted delivery). A safe, effective, long-lasting, precisely targeted, and biodegradable delivery system, chitosan based bionanocomposites played a crucial role in these areas. Nanocomposites, which can be easily produced by introducing therapeutic drugs and imaging agents, are getting a lot of attention for use in target-specific diagnostic applications, as this is a major public health concern in many parts of the world [6, 20]. As a result of the size disparity electrical, magnetic, and optical properties, specific nanoparticles have attracted much interest, especially imaging, diagnosis, and treatment of cancer. Because of its usefulness in bolstering imaging and therapy-centered efforts to combat cancer, targeted therapy has quickly become a vital component of modern cancer care. Various chitosan nanocomposites studied for their potential as drug carriers due to their varied characteristics and potential uses. High drug association efficiency and prolonged release profiles are achieved, and a major adverse effect that has hampered its use in cancer therapy is eliminated.

3.2 Chitosan-Nanoclay Bionanocomposite in Drug Delivery

The structure and functional properties of polymer–clay nanocomposites is attracting a lot of interest as a material. Montemorolnite phyllosilicate minerals having a 2:1 layered structure; so-called because they were first discovered near Montmorillon, France. There are layers of silicate tetrahedra with sheets of alumina octahedra in between [19]. Adsorption of electropositive alkaline earth metal ions in the interlayer space counteracts the negative charge distribution inside the plane caused by dislocations and isomorphous substitution in the crystal lattice [19]. These flaws ultimately ends up being what triggers activity and exchange interactions with organic molecules. Because of its high surface area, Mnt is effective at cation exchange, adsorption, adhesion, and drug delivery. Chitosan can be altered at its chemically active groups, such as its NH_2 and OH. A variety of nano-clays were used for the drug release investigation, each with its own unique properties of counter ions and their ability to bind to cations. The surface roughness and tensile strength of the chitosan/Cloisite 30B nanocomposite were improved, whereas the elongation at break was decreased. Further, the inclusion of clay does not significantly alter thermal stability or melt behaviour [93].

Liu et al. [23] produced a nanocomposites hydrogel made of chitosan and montmorillonite to investigate the behaviour of drug release after electrostimulation. Major constraints for clinical applications include CS's structural instability, which makes it difficulty in regulating the release of medication during cyclic electrostimulation, and its sensitivity and reversibility to repeated on–off switching operations. To get around these constraints, the anti-fatigue effect and accompanying long-term steady release kinetics of CS were improved by including an inorganic phase, MMT. The MMT integrated into the nanohydrogel was found to have an exfoliated nanostructure, as evidenced by XRD studies and time-varying optical absorbance. Vitamin B12 release in response to electrical stimulation is highly dependent on the cross-linking density established by the exfoliated silica nanosheets, which serve as cross-linkers to construct a network structure between the CS and MMT. Under electrostimulation, the lower the MMT concentration (1 wt.%), the more mechanism of release shifts switching from a mode controlled by diffusion to one controlled by swelling, creating a situation of "pseudo-zero-order release" kinetics for vitamin B12. Both the diffusion exponent n and the nanohydrogel's response to electrostimulation were diminished as the MMT concentration was raised. Furthermore, the nanohydrogel sensitivity to electrical currents with greater MMT concentrations was diminished, but its anti-fatigue behaviour was significantly enhanced, as shown by a series of "on" and "off" operations performed in rapid succession. The nanohydrogel containing 2 wt.% MMT outperformed pure CS in terms of mechanical reliability, practical desirability, and good anti-fatigue performance.

3.3 Chitosan-Magnetic Nanoparticle Bionanocomposites in Drug Delivery

As a novel and potentially useful resource for the medical field, organic–inorganic hybrid nanocomposites have recently come into the spotlight. Compared to their base materials, hybrids' characteritatstics are vastly superior. Cancer imaging and therapy are two areas where magnetic nanocomposites are garnering a lot of attention in the biomedical field. Using state-of-the-art nanoparticle methods to facilitate their speedy arrival at the intended destination via stimuli or by a specialised recognition mechanism, anticancer medicines can be precisely targeted to the malignant tissue [22, 94]. The physiological properties of drug nanocarriers can be altered in response to an external stimuli (such as magnetic gradient, pH, temperature, ultrasound, etc.), allowing for the sustained, high-concentration release of bioactive chemical at the site of damage or disease [95]. This procedure has the potential to improve the drug's efficacy, specificity, and systemic toxicity. Biocompatibility, minimal immunogenicity and toxicity, a good drug transport capability, and responsiveness to intensity of a magnetic field are all desirable qualities in magnetic nanocarriers. Nanoparticles also have the ability to cause the death of malignant tissues by means of magnetic fluid hyperthermia.

Lin et al. [6] prepared by integration of medico-diagnostic procedures to a unified nanostructures is a promising use of multifunctional nanocomposites. Generation of a nanocomposites by nanoparticles of iron oxide coated with chitosan with polyethylene glycolated methotraxate and near-infrared fluorescent cyanin dye (Cy5.5). Also, MTX-PEG, the anti-cancer prodrug that has been PEGylated, was developed as a ligand that selectively targets cancer cells for treatment that works from the inside out. An orthogonally functionalized, self-directed bionanocomposites exhibited synergistic tumor-selective targeting efficacy in early and late phases, as demonstrated by magnetic resonance and fluorescence imaging. In addition, the bionanocomposites was able to deliver MTX to FA receptors-over expressed cancer cells, demonstrating enhanced anticancer activity with reduced side effects, through a combination of receptor-mediated endocytosis of folic acid (FA) (which can turn cellular uptake "off" in normal cells and "on" in cancer cells) and pH/intracellular protease-mediated hydrolyzing peptide bonds. The combination of MTX, PEG, CS, IONPs, and Cy5.5 shows significant promise for a wide range of future biomedical applications because to its potential as a unified, flexible, and efficient framework for self-targeted cancer therapy that uses dual-model imaging (Fig. 12).

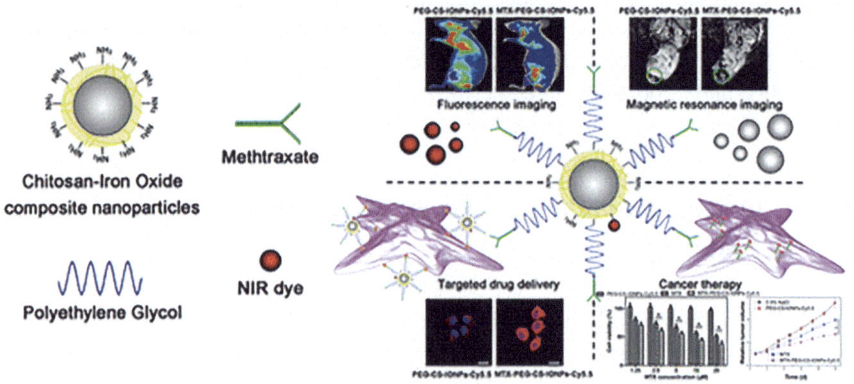

Fig. 12 Schematic representation of chitosan–iron oxide bionanocomposites and some results in brief. Adapted with permission from [6]. 2015, American Chemical Society

3.4 Chitosan-Carbon Nanostructures Bionanocomposite in Drug Delivery

Carbon nanotubes are carbon allotropes with a sp^2 hybridised carbon atom arrangement in a cylinder-shaped nanostructure that was found in 1991 by a Japanese electron microscopist. A high specific area, regular pore structure, and excellent electrical, thermal, and mechanical, properties [96]. There are two ways in which drug delivery can benefit from the use of polymer nanotube composites. Biocompatibility and biodegradability are provided by the polymer, whereas magnetic and electromagnetic properties, cellular uptake, and stability are provided by the CNTs [21]. Although the cytotoxicity issue is a key worry connected with CNTs, they have been thoroughly investigated as a drug delivery system (mostly in cancer therapy) [5].

Graphene oxide (GO) is a new nanovehicle for anticancer drug loading and delivery medicines prepared by Zhang et al. [20]. Sulfonic acid groups are attached to the NGO to make it stable in physiological solution, and then folic acid (FA) molecules are covalently bound to the NGO to make it selectively target MCF-7 cells, human breast cancer cells that express FA receptors. In addition, doxorubicin (DOX) and camptothecin (CPT), two anticancer medicines, are studied in relation to their regulated loading onto the FA-conjugated NGO (FA-NGO) via - stacking and hydrophobic interactions. It is revealed that FA–NGO loaded with the two anticancer medicines demonstrates selective targeting to MCF-7 cells, and surprisingly high cytotoxicity compared to NGO loaded with either DOX or CPT only. Nanocarriers based on graphene may find extensive usage biologically, for controlled dosing and specific delivery of combination anticancer medications, as the use of two or more drugs is a common clinical practise that often shows significantly more effective than a single drug in treatment.

4 Conclusions

Recent developments in the field of new materials range from biocompatible inorganic/ceramic nanocomposites to mechanically enhanced ceramic nanocomposites to nanomaterials based on peptides, wherein the peptides function as both the phases. Bionanocomposites are of interest for a wide variety of reasons, including application-oriented design and the study of complex structure–property relationships. Mechanical strength, biocompatibility, biodegradability, morphology, and a variety of other properties may be necessary functional criteria depending on end use. However, the biopolymer and the filler can be fine-tuned and optimised to meet the requirements. Considering the preparation of innovative bionanocomposites for biomedical purposes, further research into these interactions may prove useful.

References

1. Abdeen, R., Salahuddin, N.: Modified chitosan-clay nanocomposite as a drug delivery system intercalation and in vitro release of ibuprofen. J. Chem. (2013)
2. Zhang, L., Ma, Y., Pan, X., Chen, S., Zhuang, H., Wang, S.: A composite hydrogel of chitosan/heparin/poly (γ-glutamic acid) loaded with superoxide dismutase for wound healing. Carbohydr. Polym. **180**, 168–174 (2018)
3. Shariatinia, Z., Fazli, M.: Mechanical properties and antibacterial activities of novel nanobiocomposite films of chitosan and starch. Food Hydrocoll. **46**, 112–124 (2015)
4. Borkotoky, S.S., Dhar, P., Katiyar, V.: Biodegradable poly (lactic acid)/Cellulose nanocrystals (CNCs) composite microcellular foam: Effect of nanofillers on foam cellular morphology, thermal and wettability behavior. Int. J. Biol. Macromol. **106**, 433–446 (2018)
5. Zhang, W., Zhang, Z., Zhang, Y.: The application of carbon nanotubes in target drug delivery systems for cancer therapies. Nanoscale Res. Lett. **6**, 1–22 (2011)
6. Lin, J., Li, Y., Li, Y., Wu, H., Yu, F., Zhou, S., Xie, L., Luo, F., Lin, C., Hou, Z.: Drug/dye-loaded, multifunctional PEG-chitosan-iron oxide nanocomposites for methotraxate synergistically self-targeted cancer therapy and dual model imaging. ACS Appl. Mater. Interfaces **7**, 11908–11920 (2015)
7. Patra, N., Martinová, L., Stuchlik, M., Černík, M.: Structure-property relationships in Sterculia urens/polyvinyl alcohol electrospun composite nanofibers. Carbohydr. Polym. (2015)
8. Patra, N., Jayaseelan, D.D., Lee, W.E.: Synthesis of ZrB2/SiC composite powders by single-step solution process from organic-inorganic hybrid precursor. Adv. Appl. Ceram. **115**, 36–42 (2016)
9. Patra, N., Cernik, M., Salerno, M.: Advances in electrospun nanofibers, J. Nanomater. (2016)
10. Patra, N., Vojtová, L., Martinová, L.: Deacetylation-induced changes in thermal properties of Sterculia urens gum. J. Therm. Anal. Calorim. (2015)
11. Patra, N., Jayaseelan, D.D., Lee, W.E.: Synthesis of biopolymer-derived zirconium carbide powder by facile one-pot reaction. J. Am. Ceram. Soc. (2015)
12. Patra, N., Al Nasiri, N., Jayaseelan, D.D., Lee, W.E.: Synthesis, characterization and use of synthesized fine zirconium diboride as an additive for densification of commercial zirconium diboride powder. Ceram. Int. (2016)
13. Saikia, C., Gogoi, P.: Chitosan: a promising biopolymer in drug delivery applications. J. Mol. Genet. Med. **s4** 1–10 (2015)
14. Julkapli, N.M., Bagheri, S.: Progress on nanocrystalline cellulose biocomposites. React. Funct. Polym. **112**, 9–21 (2017)
15. Aththanayaka, S., Thiripuranathar, G., Ekanayake, S.: Emerging advances in biomimetic synthesis of nanocomposites and potential applications. Mater. Today Sustain. **20**, 100206 (2022)
16. Chakoli, A.N., Sadeghzadeh, M.: Recent trends in biomedical and pharmaceutical industry due to engineered nanomaterials. Elsevier Inc. (2018)
17. Alexander, A., Ajazuddin, P.R.J., Saraf, S., Saraf, S.: Recent expansion of pharmaceutical nanotechnologies and targeting strategies in the field of phytopharmaceuticals for the delivery of herbal extracts and bioactives. J. Control. Release **241**, 110–124 (2016)
18. Ali, A., Ahmed, S.: A review on chitosan and its nanocomposites in drug delivery. Int. J. Biol. Macromol. **109**, 273–286 (2018)
19. Gutiérrez, T.J., Ponce, A.G., Alvarez, V.A.: Nano-clays from natural and modified montmorillonite with and without added blueberry extract for active and intelligent food nanopackaging materials. Mater. Chem. Phys. **194**, 283–292 (2017)
20. Zhang, L., Xia, J., Zhao, Q., Liu, L., Zhang, Z.: Functional graphene oxide as a nanocarrier for controlled loading and targeted delivery of mixed anticancer drugs. Small **6**, 537–544 (2010)
21. Cirillo, G., Hampel, S., Spizzirri, U.G., Parisi, O.I., Picci, N., Iemma, F.: Carbon nanotubes hybrid hydrogels in drug delivery: a perspective review. Biomed Res. Int. (2014)
22. Arias, J.L., Reddy, L.H., Couvreur, P.: Fe 3O 4/chitosan nanocomposite for magnetic drug targeting to cancer. J. Mater. Chem. **22**, 7622–7632 (2012)

23. Liu, K.H., Liu, T.Y., Chen, S.Y., Liu, D.M.: Drug release behavior of chitosan-montmorillonite nanocomposite hydrogels following electrostimulation. Acta Biomater. **4**, 1038–1045 (2008)
24. Chee, B.S., de Lima, G.G., Devine, D., Nugent, M.J.D.: Electrospun hydrogels composites for bone tissue engineering. Elsevier Inc. (2018)
25. Bishi, D.K., Mathapati, S.S., Padhi, P., Venkataprasanna, K.S., Ezhilarasu, H., Ramalingam, R., Rahim, M.H.A., Venugopal, J.R.: Advances and applications of biofiber polymer composites in regenerative medicine (2021)
26. Moon, R.J., Martini, A., Nairn, J., Simonsen, J., Youngblood, J.: Cellulose nanomaterials review: Structure, properties and nanocomposites, vol. 40 (2011)
27. Lam, E., Hemraz, U.D.: Preparation and surface functionalization of carboxylated cellulose nanocrystals. Nanomaterials **11** (2021)
28. Klemm, D., Kramer, F., Moritz, S., Lindström, T., Ankerfors, M., Gray, D., Dorris, A.: Nanocelluloses: a new family of nature-based materials. Angew. Chemie - Int. Ed. **50**, 5438–5466 (2011)
29. Abdel-Halim, E.S.: Chemical modification of cellulose extracted from sugarcane bagasse: preparation of hydroxyethyl cellulose. Arab. J. Chem. **7**, 362–371 (2014)
30. Hafid, H.S., Omar, F.N., Zhu, J., Wakisaka, M.: Enhanced crystallinity and thermal properties of cellulose from rice husk using acid hydrolysis treatment. Carbohydr. Polym. **260**, 117789 (2021)
31. Baghel, R.S., Reddy, C.R.K., Singh, R.P.: Seaweed-based cellulose: applications, and future perspectives. Carbohydr. Polym. **267**, 118241 (2021)
32. Mo, M., Chen, C., Gao, H., Chen, M., Li, D.: Wet-spinning assembly of cellulose nanofibers reinforced graphene/polypyrrole microfibers for high performance fiber-shaped supercapacitors. Electrochim. Acta **269**, 11–20 (2018)
33. Tammela, P., Wang, Z., Frykstrand, S., Zhang, P., Sintorn, I.M., Nyholm, L., Strømme, M.: Asymmetric supercapacitors based on carbon nanofibre and polypyrrole/nanocellulose composite electrodes. RSC Adv. **5**, 16405–16413 (2015)
34. Huo, Y., Liu, Y., Xia, M., Du, H., Lin, Z., Li, B., Liu, H.: Nanocellulose-based composite materials used in drug delivery systems. Polymers (Basel) **14** (2022)
35. Fernandes, E.M., Pires, R.A., Mano, J.F., Reis, R.L.: Bionanocomposites from lignocellulosic resources: properties, applications and future trends for their use in the biomedical field. Prog. Polym. Sci. **38**, 1415–1441 (2013)
36. Stanisz, M., Klapiszewski, Ł, Jesionowski, T.: Recent advances in the fabrication and application of biopolymer-based micro- and nanostructures: a comprehensive review. Chem. Eng. J. **397**, 125409 (2020)
37. Ul-Islam, M., Khan, S., Khattak, W.A., Ullah, M.W., Park, J.K.: Synthesis, chemistry, and medical application of bacterial cellulose nanocomposites, vol. 74 (2015)
38. Agarwal, C., Csóka, L.: Surface-modified cellulose in biomedical engineering (2019)
39. Long, W., Ouyang, H., Hu, X., Liu, M., Zhang, X., Feng, Y., Wei, Y.: State-of-art review on preparation, surface functionalization and biomedical applications of cellulose nanocrystals-based materials. Int. J. Biol. Macromol. **186**, 591–615 (2021)
40. Kashani Rahimi, S., Aeinehvand, R., Kim, K., Otaigbe, J.U.: Structure and biocompatibility of bioabsorbable nanocomposites of aliphatic-aromatic copolyester and cellulose nanocrystals. Biomacromol **18**, 2179–2194 (2017)
41. Rajwade, J.M., Paknikar, K.M., Kumbhar, J.V.: Applications of bacterial cellulose and its composites in biomedicine. Appl. Microbiol. Biotechnol. **99**, 2491–2511 (2015)
42. Ullah, H., Wahid, F., Santos, H.A., Khan, T.: Advances in biomedical and pharmaceutical applications of functional bacterial cellulose-based nanocomposites, vol. 150. Elsevier Ltd. (2016)
43. Sunasee, R., Hemraz, U.D., Ckless, K.: Cellulose nanocrystals: a versatile nanoplatform for emerging biomedical applications. Exp. Opin. Drug Deliv. **13**, 1243–1256 (2016)
44. Amalin Kavitha, A., Thomas Paul, K., Anilkumar, P.: Cellulose-derived materials for drug delivery applications (INC) (2020)

45. Jorfi, M., Foster, E.J.: Recent advances in nanocellulose for biomedical applications. J. Appl. Polym. Sci. **132**, 1–19 (2015)
46. Nehra, P., Chauhan, R.P.: Eco-friendly nanocellulose and its biomedical applications: current status and future prospect. J. Biomater. Sci. Polym. Ed. **32**, 112–149 (2021)
47. Abitbol, T., Rivkin, A., Cao, Y., Nevo, Y., Abraham, E., Ben-Shalom, T., Lapidot, S., Shoseyov, O.: Nanocellulose, a tiny fiber with huge applications. Curr. Opin. Biotechnol. **39**, 76–88 (2016)
48. De France, K.J., Hoare, T., Cranston, E.D.: Review of hydrogels and aerogels containing nanocellulose. Chem. Mater. **29**, 4609–4631 (2017)
49. Czaja, W.K., Young, D.J., Kawecki, M., Brown, R.M.: The future prospects of microbial cellulose in biomedical applications. Biomacromol **8**, 1–12 (2007)
50. Du, H., Liu, W., Zhang, M., Si, C., Zhang, X., Li, B.: Cellulose nanocrystals and cellulose nanofibrils based hydrogels for biomedical applications. Carbohydr. Polym. **209**, 130–144 (2019)
51. Abdelhamid, H.N., Mathew, A.P.: Cellulose-based nanomaterials advance biomedicine: a review. Int. J. Mol. Sci. **23** (2022)
52. Wahid, F., Bai, H., Wang, F.P., Xie, Y.Y., Zhang, Y.W., Chu, L.Q., Jia, S.R., Zhong, C.: Facile synthesis of bacterial cellulose and polyethyleneimine based hybrid hydrogels for antibacterial applications. Cellulose **27**, 369–383 (2020)
53. Rong, L., Yang, D., Wang, B., Xiao, D., Lu, M., Mao, Z., Xu, H., Gu, Y., Feng, X., Sui, X.: Durable and effective antibacterial cotton fabric collaborated with polypropylene tissue mesh for abdominal wall defect repair. ACS Biomater. Sci. Eng. **6**, 3868–3877 (2020)
54. Rivero-Buceta, V., Aguilar, M.R., Hernández-Arriaga, A.M., Blanco, F.G., Rojas, A., Tortajada, M., Ramírez-Jiménez, R.A., Vázquez-Lasa, B., Prieto, A.: Anti-staphylococcal hydrogels based on bacterial cellulose and the antimicrobial biopolyester poly(3-hydroxy-acetylthioalkanoate-co-3-hydroxyalkanoate). Int. J. Biol. Macromol. **162**, 1869–1879 (2020)
55. Liu, H., Hu, Y., Zhu, Y., Wu, X., Zhou, X., Pan, H., Chen, S., Tian, P.: A simultaneous grafting/vinyl polymerization process generates a polycationic surface for enhanced antibacterial activity of bacterial cellulose. Int. J. Biol. Macromol. **143**, 224–234 (2020)
56. Kukharenko, O., Bardeau, J.F., Zaets, I., Ovcharenko, L., Tarasyuk, O., Porhyn, S., Mischenko, I., Vovk, A., Rogalsky, S., Kozyrovska, N.: Promising low cost antimicrobial composite material based on bacterial cellulose and polyhexamethylene guanidine hydrochloride. Eur. Polym. J. **60**, 247–254 (2014)
57. Hanif, Z., Shin, D., Choi, D., Park, S.J.: Development of a vapor phase polymerization method using a wet-on-wet process to coat polypyrrole on never-dried nanocellulose crystals for fabrication of compression strain sensor. Chem. Eng. J. **381**, 122700 (2020)
58. Figueiredo, A.R.P., Figueiredo, A.G.P.R., Silva, N.H.C.S., Barros-Timmons, A., Almeida, A., Silvestre, A.J.D., Freire, C.S.R.: Antimicrobial bacterial cellulose nanocomposites prepared by in situ polymerization of 2-aminoethyl methacrylate. Carbohydr. Polym. **123**, 443–453 (2015)
59. Hashem, A.H., Hasanin, M., Kamel, S., Dacrory, S.: A new approach for antimicrobial and antiviral activities of biocompatible nanocomposite based on cellulose, amino acid and graphene oxide. Colloids Surf. B Biointerfaces **209**, 1–4 (2022)
60. Doghish, A.S., Hashem, A.H., Shehabeldine, A.M., Sallam, A.A.M., El-Sayyad, G.S., Salem, S.S.: Nanocomposite based on gold nanoparticles and carboxymethyl cellulose: Synthesis, characterization, antimicrobial, and anticancer activities. J. Drug Deliv. Sci. Technol. **77**, 9–11 (2022)
61. Qian, L., Lei, D., Duan, X., Zhang, S., Song, W., Hou, C., Tang, R.: Design and preparation of metal-organic framework papers with enhanced mechanical properties and good antibacterial capacity. Carbohydr. Polym. **192**, 44–51 (2018)
62. Emam, H.E., Darwesh, O.M., Abdelhameed, R.M.: Protective cotton textiles via amalgamation of cross-linked zeolitic imidazole frameworks. Ind. Eng. Chem. Res. **59**, 10931–10944 (2020)
63. Yang, Y., Guo, Z., Huang, W., Zhang, S., Huang, J., Yang, H., Zhou, Y., Xu, W., Gu, S.: Fabrication of multifunctional textiles with durable antibacterial property and efficient oil-water separation via in situ growth of zeolitic imidazolate framework-8 (ZIF-8) on cotton fabric. Appl. Surf. Sci. **503**, 144079 (2020)

64. Abdelhamid, H.N., Mathew, A.P.: Cellulose–metal organic frameworks (CelloMOFs) hybrid materials and their multifaceted applications: a review. Coord. Chem. Rev. **451**, 214263 (2022)
65. Shao, W., Liu, H., Liu, X., Wang, S., Zhang, R.: Anti-bacterial performances and biocompatibility of bacterial cellulose/graphene oxide composites. RSC Adv. **5**, 4795–4803 (2015)
66. Yang, X.N., Xue, D.D., Li, J.Y., Liu, M., Jia, S.R., Chu, L.Q., Wahid, F., Zhang, Y.M., Zhong, C.: Improvement of antimicrobial activity of graphene oxide/bacterial cellulose nanocomposites through the electrostatic modification. Carbohydr. Polym. **136**, 1152–1160 (2016)
67. Shahnawaz Khan, M., Abdelhamid, H.N., Wu, H.F.: Near infrared (NIR) laser mediated surface activation of graphene oxide nanoflakes for efficient antibacterial, antifungal and wound healing treatment. Colloids Surf. B Biointerfaces **127**, 281–291 (2015)
68. Abdelhamid, H.N., Wu, H.F.: Multifunctional graphene magnetic nanosheet decorated with chitosan for highly sensitive detection of pathogenic bacteria. J. Mater. Chem. B **1**, 3950–3961 (2013)
69. Wu, B.S., Abdelhamid, H.N., Wu, H.F.: Synthesis and antibacterial activities of graphene decorated with stannous dioxide. RSC Adv. **4**, 3722–3731 (2014)
70. Hua, P.Y., Manikandan, M., Abdelhamid, H.N., Wu, H.F.: Graphene nanoflakes as an efficient ionizing matrix for MALDI-MS based lipidomics of cancer cells and cancer stem cells. J. Mater. Chem. B **2**, 7334–7343 (2014)
71. Pal, S., Nisi, R., Stoppa, M., Licciulli, A.: Silver-functionalized bacterial cellulose as antibacterial membrane for wound-healing applications. ACS Omega **2**, 3632–3639 (2017)
72. Fu, F., Li, L., Liu, L., Cai, J., Zhang, Y., Zhou, J., Zhang, L.: Construction of cellulose based ZnO nanocomposite films with antibacterial properties through one-step coagulation, 1–39 (2015)
73. Lin, W.C., Lien, C.C., Yeh, H.J., Yu, C.M., Hsu, S.H.: Bacterial cellulose and bacterial cellulose-chitosan membranes for wound dressing applications. Carbohydr. Polym. **94**, 603–611 (2013)
74. Li, N.: Naturally-occurring bacterial cellulose-hyperbranched cationic polysaccharide derivative/MMP-9 siRNA composite dressing for wound healing enhancement in diabetic rats. Acta Biomater. (2019)
75. Wichai, S., Chuysinuan, P., Chaiarwut, S., Ekabutr, P., Supaphol, P.: Development of bacterial cellulose/alginate/chitosan composites incorporating copper (II) sulfate as an antibacterial wound dressing. J. Drug Deliv. Sci. Technol. **51**, 662–671 (2019)
76. Ullah, H., Santos, H.A., Khan, T.: Applications of bacterial cellulose in food, cosmetics and drug delivery. Cellulose **23**, 2291–2314 (2016)
77. Khine, Y.Y., Stenzel, M.H.: Surface modified cellulose nanomaterials: a source of non-spherical nanoparticles for drug delivery. Mater. Horizons **7**, 1727–1758 (2020)
78. Paukkonen, H., Ukkonen, A., Szilvay, G., Yliperttula, M., Laaksonen, T.: Hydrophobin-nanofibrillated cellulose stabilized emulsions for encapsulation and release of BCS class II drugs. Eur. J. Pharm. Sci. **100**, 238–248 (2017)
79. Dong, S., Cho, H.J., Lee, Y.W., Roman, M.: Synthesis and cellular uptake of folic acid-conjugated cellulose nanocrystals for cancer targeting. Biomacromol **15**, 1560–1567 (2014)
80. Jackson, J.K., Letchford, K., Wasserman, B.Z., Ye, L., Hamad, W.Y., Burt, H.M.: The use of nanocrystalline cellulose for the binding and controlled release of drugs. Int. J. Nanomedicine **6**, 321–330 (2011)
81. Pötzinger, Y., Kralisch, D., Fischer, D.: Bacterial nanocellulose: the future of controlled drug delivery? Ther. Deliv. **8**, 753–761 (2017)
82. Wan, W., Ouyang, H., Long, W., Yan, W., He, M., Huang, H., Yang, S., Zhang, X., Feng, Y., Wei, Y.: Direct surface functionalization of cellulose nanocrystals with hyperbranched polymers through the anionic polymerization for pH-responsive intracellular drug delivery. ACS Sustain. Chem. Eng. **7**, 19202–19212 (2019)
83. Singhsa, P., Narain, R., Manuspiya, H.: Bacterial Cellulose Nanocrystals (BCNC) preparation and characterization from three bacterial cellulose sources and development of functionalized BCNCs as nucleic acid delivery systems. ACS Appl. Nano Mater. **1**, 209–221 (2018)
84. Chen, N., Wang, H., Ling, C., Vermerris, W., Wang, B., Tong, Z.: Cellulose-based injectable hydrogel composite for pH-responsive and controllable drug delivery. Carbohydr. Polym. **225**, 115207 (2019)

85. Elsayed, H., Rebesan, P., Crovace, M.C., Zanotto, E.D., Colombo, P., Bernardo, E.: Biosilicate® scaffolds produced by 3D-printing and direct foaming using preceramic polymers. J. Am. Ceram. Soc. **102**, 1010–1020 (2019)
86. Boominathan, T., Sivaramakrishna, A.: Recent advances in the synthesis, properties, and applications of modified chitosan derivatives: challenges and opportunities, vol. 379. Springer International Publishing (2021)
87. Shariatinia, Z., Zahraee, Z.: Controlled release of metformin from chitosan–based nanocomposite films containing mesoporous MCM-41 nanoparticles as novel drug delivery systems. J. Colloid Interface Sci. **501**, 60–76 (2017)
88. Ahmad, M., Manzoor, K., Singh, S., Ikram, S.: Chitosan centered bionanocomposites for medical specialty and curative applications: a review. Int. J. Pharm. **529**, 200–217 (2017)
89. Agrawal, P., Strijkers, G.J., Nicolay, K.: Chitosan-based systems for molecular imaging. Adv. Drug Deliv. Rev. **62**, 42–58 (2010)
90. Ahsan, S.M., Thomas, M., Reddy, K.K., Sooraparaju, S.G., Asthana, A., Bhatnagar, I.: Chitosan as biomaterial in drug delivery and tissue engineering. Int. J. Biol. Macromol. **110**, 97–109 (2018)
91. Nawrotek, K., Tylman, M., Rudnicka, K., Balcerzak, J., Kamiński, K.: Chitosan-based hydrogel implants enriched with calcium ions intended for peripheral nervous tissue regeneration. Carbohydr. Polym. **136**, 764–771 (2016)
92. Shariatinia, Z., Nikfar, Z.: Synthesis and antibacterial activities of novel nanocomposite films of chitosan/phosphoramide/Fe3O4 NPs. Int. J. Biol. Macromol. **60**, 226–234 (2013)
93. Xu, Y., Ren, X., Hanna, M.A.: Chitosan/clay nanocomposite film preparation and characterization. J. Appl. Polym. Sci. **99**, 1684–1691 (2006)
94. Alomari, M., Almohazey, D., Almofty, S., Alhibshi, A., Almansour, I., Kaewsaneha, C., Badri, W., Fessi, H., Elaissari, A.: Magnetic-responsive polysaccharide-inorganic composite materials for cancer therapeutics. Elsevier Ltd. (2019)
95. Li, M., Mondrinos, M.J., Chen, X., Gandhi, M.R., Ko, F.K., Lelkes, P.I.: Synthesis and temperature response analysis of magnetic-hydrogel nanocomposites. J. Biomed. Mater. Res. Part A **79**, 963–973 (2006)
96. Ali, E., Hadis, D., Hamzeh, K., Mohammad, K., Nosratollah, Z., Abolfazl, A., Mozhgan, A., Younes, H., Woo, J.S.: Carbon nanotubes: properties, synthesis, purification, and medical applications. Nanoscale Res. Lett. **9**, 393 (2014)

Turning Cigarette Butts (CBs) into Valuable Resources: Technical Potential and Drawbacks

Hamza El Fadili, Mohammed Ben Ali, Mohammed El Mahi, and El Mostapha Lotfi

Abstract Every year, 5.8 trillion cigarettes are consumed, leading to the generation of a significant amount of hazardous litter commonly known as cigarette butts (CBs), and considered as the most prevalent discarded item on the planet. This paper aims to collect and discuss comprehensively all available studies on the composition, toxicity and alternative solutions proposed by the scientific community to recycle cigarette butts in various applications. Previously published papers have already disclosed that this small pollutant contains a multitude of toxicants and hazardous chemical compounds, and could pose a serious threat to all forms of life, including human beings, if not managed properly. Several methods have been investigated, including recycling cigarette butts in construction materials (ceramics, fired clay bricks, and asphalt concrete), Sound absorption, corrosion inhibitor, biodiesel, source of porous and charred carbon. The advantages, limitations, and recommendations required for the development of each proposed solution were highlighted. These findings provide a practical guide towards a sustainable tobacco industry and assist decision-makers in enforcing their vision for this small toxic litter.

Keywords Cigarette butts · Hazardous waste · Waste management · Circular economy

1 Introduction

The notable and swift industrial advancement has led to a significant increase in the global volume of waste generated. Projections suggest that the amount of waste is expected to grow by 50%, reaching 3.4 billion tons by the year 2025 [1].

The tobacco industry is notably one of the most widespread sectors and significantly contributes to the high income of many countries. Annually, six trillion

H. El Fadili (✉) · M. Ben Ali · M. El Mahi · E. M. Lotfi
Laboratory of Spectroscopy, Molecular Modeling, Materials, Nanomaterials, Water and Environment, Materials for Environment Team, ENSAM, Mohammed V University, Rabat, Morocco
e-mail: hamza_elfadili2@um5.ac.ma

K. K. Kesari et al. (eds.), *Functionalized Cellulose Materials*, Engineering Materials,
https://doi.org/10.1007/978-3-031-76953-5_9

cigarettes are manufactured and 5.8 trillion are consumed by approximately one billion smokers worldwide [2], Contributing to the production of a significant quantity of toxic items known as cigarette butts [3], it is considered the most prevalent litter items on the planet [4]. Additionally, due to population growth and the rising prevalence of smoking among young individuals annually, it is projected that the number of cigarettes consumed will increase by over 50%, reaching nine trillion by 2025 [5].

Based on the estimation published by the Tobacco Atlas illustrated in Fig. 1, China was the largest consumer and producer of cigarettes worldwide in 2016, approximately 2.35 cigarettes were consumed by Chinese people. The outcome which could be concluded from the figure is that cigarette production is expected to skyrocket, and subsequently CBs litter.

The majority of these butts are discarded by smokers into the surrounding environment [6]. As per the Cigarette Butt Pollution Project in the USA, one-third to two-thirds of cigarette butts are littered by smokers into various environments, including oceans, rivers, public roads, and more.

Despite their small size, cigarette butts (CBs) are acknowledged as environmental hazards for aquatic organisms, insects, infants, and animals. This is attributed to the significant presence of hazardous chemicals such as nicotine, cyanide, polycyclic aromatic hydrocarbons (PAHs), and heavy metals in a substantial proportion in these wastes [7, 8]. In each cigarette, it is possible to identify more than 4000 toxic elements that are produced during burning or distilled from tobacco, in which 69 are classified as carcinogenic [9].

Moreover, due to the utilization of non-biodegradable materials in the manufacturing of cigarette filters, CBs persist in the environment for extended periods [5].

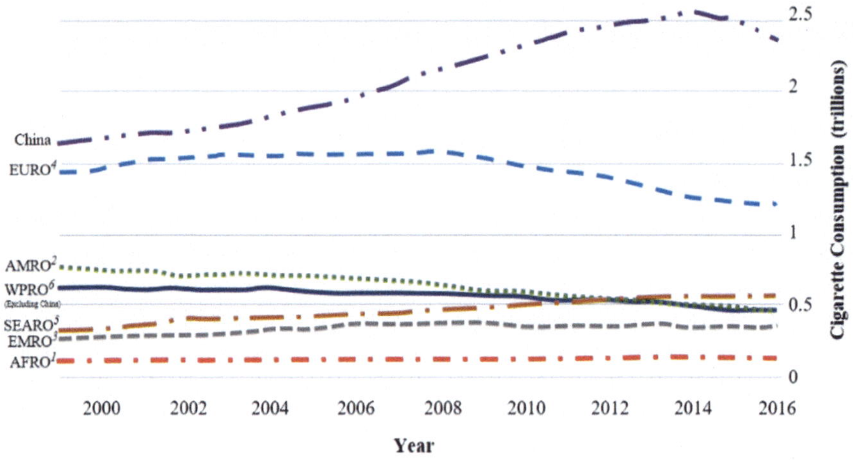

Fig. 1 Global cigarette consumption by WHO region, 1999–2016, in trillions (adapted from The Tobacco Atlas [10]; [1]Africa, [2]Americas, [3]Eastern Mediterranean, [4]European, [5]Southeast Asia, [6]Western Pacific

For example, Cellulose acetate filters exhibit poor biodegradability and can take up to 18 months to decompose under typical environmental conditions [9]. Thus, they have sufficient time to trap pollutants and release them into the environment. To this end, particular emphasis among scientists and decision-makers should be given in this regard.

The huge volume of cigarette butts generated worldwide, the toxic content, and their persistence in the environment after usage, enhance the obligation to establish strategies for efficient disposal to prevent its harmful effect if managed improperly through landfilling and incineration, both techniques are inefficient and unsustainable due to economic and environmental concerns.

The current review involved an examination of relevant articles to identify to main toxic components in cigarette butts, advantages and limitations ahead of CBs recycling reported in the literature. Then, offering some recommendations for future research.

2 Methodology

This study encompasses specific research conducted on CBs using Scopus Database. Specifically, studies related to the toxicity, and proposed recycling and management techniques developed by researchers worldwide. The keywords employed were: "Cigarette Butts" OR "Cigarette Waste" OR "Discarded cigarettes" OR "Cigarette toxicity" OR "recycling cigarette" OR "cigarette management". The identified papers underwent additional filtration, focusing on selecting only those that align with the scope of the study: articles directly investigating the toxicity and potential methods for the recovery and recycling of cigarette butts (CBs).

3 Results and Discussion

3.1 Composition of Cigarette Butts

As illustrated in Fig. 2, CBs consist of five main components: remaining tobacco, wrapping paper, ashes, additives, and a filter made mainly from cellulose acetate [11].

3.1.1 Remaining Tobacco and Additives

The discarded cigarettes discarded into the environment contain unburned tobacco after smoking. In other word, when a cigarette is discarded into the environment,

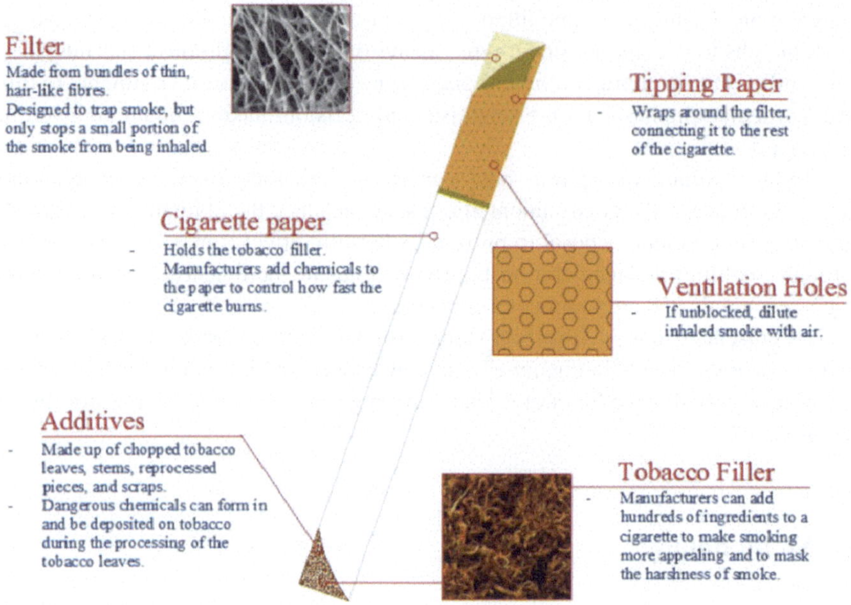

Fig. 2 The main component of discarded cigarette butts [3]

unburned tobacco residue may remain in the cigarette butt and contribute to the release of toxic pollutants (Fig. 3).

Fig. 3 Remaining tobacco from cigarette butts (CBs)

Fig. 4 Tripping paper of cigarette butts

3.1.2 Tripping Paper

The tripping paper is wrapped around the filter of the cigarette and contains small ventilation holes that allow fresh air to mix with the smoke. This tripping paper may take up to 5 months to biodegrade under normal conditions [10] (Fig. 4).

3.1.3 Filters

Cigarette filters are composed mainly of cellulose acetate, which is a naturally polymer of glucose ($C_6H_7O_6$) derived from cotton linters and wood pulp [12], about 97% of the cigarette filters were composed of cellulose acetate, a modified natural polymer [3]. The aspect and molecular structure of cellulose acetate is presented in Fig. 5. Cellulose acetate fibers take several years to degrade when discarded into the environment, because microorganisms do not accept cellulose acetate fibers as a source of food. Additionally, the remaining paper on the CBs inhibits the sunlight UV radiation reaching the fibers.

Filters are composed of a synthetic plastic fiber commonly designed in a way to trap dangerous toxic compounds present in cigarettes during smoking process [13–15]. These filters are well known by its slow biodegradability [3, 16] and can take up to several years to biodegrade under normal conditions. Thus, it has sufficient time to release the entrapped contaminants into the environment [17].

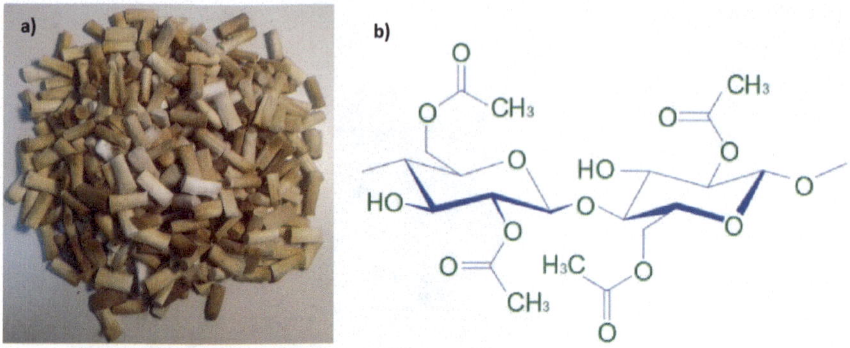

Fig. 5 A cigarette filters, **b)** molecular structure of cellulose acetate

3.2 Pollutants in Cigarette Butts

Several previous studies have been conducted to evaluate the amount of pollutants in discarded cigarettes butts. Through these investigations it has been demonstrated that this small item embodied a myriad of pollutants, including Heavy metals, PAHs, toxic substances, and radioactive elements. The extent of pollution leakage from cigarette butts is contingent on the characteristics of the pollutants, climatic conditions, and physicochemical properties of the cigarette butts [8]. Table 1 presents the primary recognized contaminants in cigarettes butts according to previous studies.

3.2.1 Organic Pollutants Within CBs

Cigarette butts (CBs) have the potential to release polycyclic aromatic hydrocarbons (PAHs), which may have carcinogenic effects on living organisms. Dobaradaran et al. [21] investigated the amount of 16 polycyclic aromatic hydrocarbons (PAHs)

Table 1 Primary recognized contaminants present in cigarette butts

Polluants	Elements	References
Heavy metals	Pb, Ni, Cr, Cd, Cu, Fe, Zn	[6]
PAHs	Acenaphthylene, Fluorene, Chrysene, Pyrene, Anthracene, Chrysene, Dibenz_a_h_anthracene, Benzo_ghi_perylene, Benzo_b_fluoranthene, Benzo_k_fluoranthene	[7, 18]
Toxics substances	Cyanide, Nicotine	[18]
BTEX compounds	Benzene, Ethylbenzene	[19]
Radioactive elements	^{210}Po	[20]

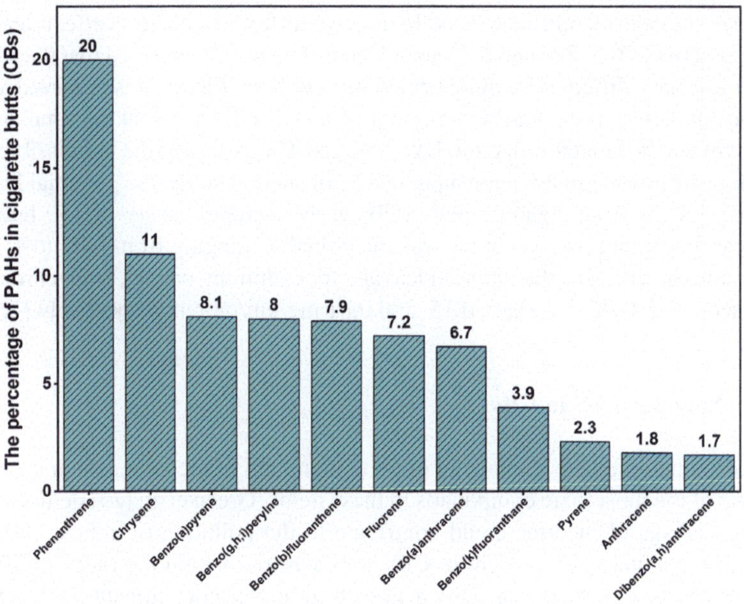

Fig. 6 The percentage of PAHs contained in the cigarette butts [18]

in freshly smoked and discarded cigarettes butts gathered from urban streets and river areas. The total PAHs content was 20 μg/g and out of the studied PAHs, the findings indicated that the average concentration of lower molecular weight PAHs, including acenaphthylene, fluorene, naphthalene, and acenaphthene were the most contributors to the measured level. The Fig. 6 showed the percentage of various PAHs in cigarette butts based on the findings reported by [18]

3.2.2 Heavy Metals in CBs

Heavy metals are the most prevalent pollutants present in discarded cigarette butts, if these pollutants are released into the aquatic environment, they have the potential to contaminate marine ecosystems and enter the food chain, posing a significant carcinogenic and non-carcinogenic risk to all forms of life [21–23]. Chevalier et al. [20] reported that the amount of these pollutants in cigarette butts ranges from several micrograms per gram to several hundred grams per gram.

The investigation of the leachability of heavy metals and metalloids from cigarette butts (CBs), to assess their influence on the adjacent environment was carried out by many authors. Akhbarizadeh et al. [24] assessed the leaching of heavy metal(oid)s, including chromium (Cr), cadmium (Cd), arsenic (As), lead (Pb), iron (Fe), manganese (Mn), copper (Cu), mercury (Hg) and zinc (Zn) from cigarette butts (CBs) into deionized water (DW), tap water (TW), and seawater (SW) in different contact times from 60 min to 60 days. To understand how the pH level affects leaching behavior, a pH of 6.5, 7.5, and 8.3 was selected. The results reflected that the studied metal(oid)s have different leaching behaviors over time. Firstly, it is observed a clear relationship between the leaching amount of metal(oid)s and soaking time and the amount of leached metal(oid)s into DW, SW, and TW exceeded the permissible limit of surface freshwater to maintain aquatic life. In another study, the potential leakage of heavy metals from cigarette butts (CB) at an average density of 150 butts per kilometer per month in road areas was quantified as ranging from 0.02 to 1.7 mg/km/month. Specifically, the reported leakage for cadmium, copper, lead, chromium, and arsenic was 0.02, 1.7, 0.59, 0.15, and 0.81 mg/km/month, respectively [7, 18].

3.2.3 Nanoparticles in CBs

Discarded cigarette butts contain an important amount of nicotine, which is considered one of the most toxic compounds in these items, Green et al. [25] demonstrated that the leakage of nicotine could contribute to the pollution of almost 1000 l of water. The nanoparticles present in CBs are measured within the range of 70–400 ng. These particles have the capacity to absorb additional contaminants and facilitate their transfer to living organisms [26]. In addition, if these nanoparticles leaked to the adjacent environment, their removal would become exceedingly difficult [20].

3.3 Toxicity of Cigarette Butts (CBs)

Previous works have already revealed that this small pollutant contains a myriad of toxicant and hazardous chemical compounds [27], such as nicotine [6], polycyclic aromatic hydrocarbons (PAHs), formaldehyde, BTEX [19], phenol, heavy metals [28], and other toxic compounds [29]. It is claimed that one single cigarette butt can contaminate up to 1000 L of water and they might perhaps bio-accumulate in the wildlife and potentially the food chain. Although, until recently at the regulatory level, the Moroccan catalog of waste (decree 2-07-253) does not propose a specific entry for CBs, and is still considered as a municipal solid waste (MSW). Therefore, disposal of CBs is a very challenging issue because if not managed unscientifically and correctly disposed of, it could pose a serious threat to the quality of soils and aquatic resources as discussed in the above sections [11, 29].

Hence, cigarette smoking is a serious threat to humans, it is estimated that for every cigarette burned, the smoker receives roughly 1.4–2.2 mg of dangerous chemical compounds. A study conducted by Yousefi et al. [30] demonstrated that a standard cigarette contains between 9 and 30 mg of nicotine. The toxicity of nicotine on human health was assessed in a research conducted by Novotny et al. [31], who claimed that the ingestion of 40–60 mg of nicotine is considered fatal in children, and the ingestion of 1–2 mg of nicotine may be toxic, leading to symptoms such as nausea, vomiting, and nerve-related effects for younger. Similarly, for domestic animals, 9.2 mg/kg of nicotine is fatal, but clinical symptoms are observed at 1 mg/kg. In another study, it was demonstrated that tittering CBs in the environment can induce genotoxic, mutagenic, and cytotoxic effects on plants, by taking *Allium cepa* as a case study [32]. Also, Plants exposed to cigarette butt leachate exhibited a low relative growth index, a high inhibition index, a large number of abnormal cells, and a high abnormality frequency during metaphase/anaphase. In the same context, Slaughter et al. [29] highlighted the significant toxicity of this pollutant to fish and other representative marine organisms, including marine bacteria and daphnids.

The toxicity of CBs influences detrimentally nervous, cardiovascular, and reproductive systems. Slaughter et al. [29] demonstrated that one discarded cigarette butt soaked in 1L of clean water for 96 h could reach the lethal concentration LD50 for test fish. Although, the measured toxicity of cigarette butts may vary due to differences in tested organisms, distinct methods of toxicity measurement, variations in cigarette brands, the quantity of cigarette butts, and varying test durations across different studies [8].

Studies on exposures and health outcomes in populations and organisms living showed that there is definitive evidence of negative impacts on their health due to the elevated toxicity of this small pollutant, as widely discussed in a good number of previous studies [33, 34]. However, the development of feasible solutions is still limited [35].

3.4 Research on Valorization of Cigarette Butts

The harmful impact of CBs on all form of life is already evident as demonstrated in several previous studies [4, 11, 19, 36]. Regrettably, in numerous countries worldwide, landfilling and incineration are the prevalent disposal methods for cigarette butts. Nevertheless, they are not efficient and inadequate due to the elevated cost and the possible release of toxic by-products.

In this context, Scientists are urgently seeking eco-friendly and cost-effective ways to recycle the massive amounts of hazardous waste, driven by the growing threat it poses to the environment. Figure 7 shows the various recycling methods proposed in the literature. The developed approaches, include energy recovery through pyrolysis [37], producing corrosion inhibitor [38–40], enhancing sound absorption capacity [41, 42], the synthesis of cellulose pulp [43], porous carbon [44], and construction materials [5, 14, 45, 46].

As discussed above, there are several methods for recycling cigarette butts (CBs), which could be categorized into three main groups: (a) the reuse of chemicals trapped in the cigarette filter following extraction, (b) the recycling of cellulose acetate, and (c) the reuse of CBs without further separation of components. Table 2 presents the characteristics and challenges ahead of CBs recycling reported in the literature.

Cellulose acetate constitutes the main element of filters for cigarettes has been the subject of the majority of research regarding waste recycling. For example, it was converted chemically to produce selective adsorbents and pulp [47–49]. Nevertheless,

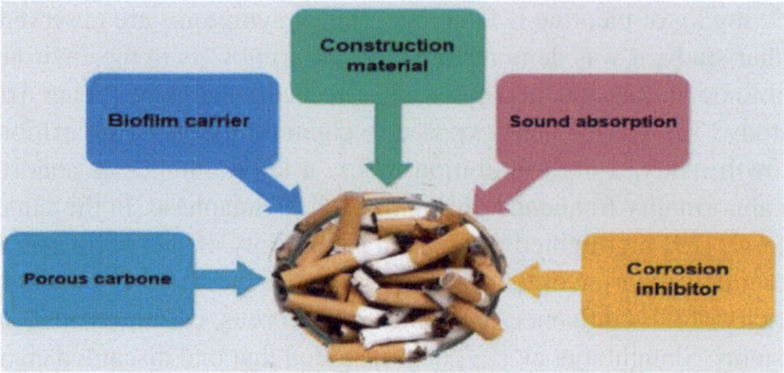

Fig. 7 Recycling methods of discarded cigarette butts (CBs) according to the literature

these methods yield in the subsequent emission of polluted gases and wastewater. Similarly, the recycling of cigarettes in gypsum for building applications, requires the washing of filters, and therefore the generation of a huge amount of polluted water [50]. Also, the limited quantity recycled through this process in comparison to the amount of CBs discarded annually. In the other hand, thermal conversion through pyrolysis was investigated by Youssef et al. [37], this method involves heating cigarette butts in an oxygen-limited environment, converting them into bio-oil or syngas. While promising for energy recovery, concerns remain about potential air pollution and incomplete breakdown of harmful elements.

Despite these efforts, no single method currently offers a complete solution. Challenges like cost, scalability, and ensuring complete breakdown of harmful chemicals persist. Ongoing research and collaboration are crucial to develop efficient and sustainable recycling solutions for this pervasive pollutant.

4 Conclusion and Future Perspectives

The massive quantity, persistence, and toxicity of cigarette butts pose a serious environmental threat to all forms of life. Efficient disposal and recycling are crucial to prevent a global catastrophe of polluted ecosystems. Therefore, this chapter has reviewed the possible pathways to recycle CBs waste and evaluated the viability of these methods.

In conclusion, by turning discarded cigarette butts into valuable resources, we can not only tackle a growing environmental crisis but also unlock a sustainable waste management solution. However, several issues should be addressed such as, conducting LCA to evaluate the environmental feasibility and economic value-added of the proposed solution following a systematic approach.

Table 2 The advantages and limitations ahead of CBs recycling reported in the literature

Recycling method	Process	Advantages	Challenges	References
Gypsum composite for building applications	1. CBs were submerged in water 2. CBs were crushed into small fragments 3. Incorporating crushed CBs in gypsum composite	1. Recycling cigarette filters to produce eco-materials 2. The process does not require chemical addition 3. Improving superficial hardness, mechanical strength, and density of gypsum composite	1. Washing cigarette filters leads to the production of wastewater 2. The limited quantity recycled through this process in comparison to the amount of CBs discarded annually	[50, 51]
Fired clay bricks	1. The CBs were dried at105 °C 2. Crushing of the CBs and turned into the fine powder 3. Produce fired clay bricks containing remaining tobacco	1. Recycling cigarette butts (CBs) to produce fired clay bricks 2. The process does not require chemical addition	1. The limited quantity of recycled process in comparison to the production of CBs annually 2. The release of gaseous pollution	[3, 52]
Production of selective adsorbent	1. Immersing 1.2 g of CBs in a solution containing a special quantity of: Fe $(NO_3)_3{\cdot}9H_2O$ or Zn $(NO_3)_2{\cdot}6H_2O$ or Cu $(NO_3)_2{\cdot}3H_2O$ and 20 mL acetic acid for 0.5 h 2. drying the mixture at 333 K and carbonizing it (873 K; 3 h)	1. The obtained product showed a high adsorption rate 2. Adsorbent separating by the magnet is possible	1. The process requires the addition of chemicals 2. The generation of wastewater 3. The emission of gases during the carbonizing process	[47, 48]

(continued)

Table 2 (continued)

Recycling method	Process	Advantages	Challenges	References
Production of cigarette-derived functional carbon	1. Manually stripping to remove remaining tobacco and impurities 2. Crushing CBs into a fluffy material and dried at 80 °C 3. Pyrolysis of CBs with K_2CO_3 at 800 °C 4. Washing carbonized samples several times with HCL and deionized water	1. Re-using the discarded cigarette filters 2. Activated carbon obtained from the filter had a high specific surface area 3. The final product had good ability on adsorption	1. Some part of the CBs is lost during the process of manufacturing activated carbon 2. Using the washing process leads to wastewater production 3. The process requires the use of chemicals such as HCl 4. The emission of gases during the carbonizing process	[53]
The production of Biodiesel	1. This method involves heating cigarette butts in an oxygen-limited environment, converting them into bio-oil or syngas in a fixed-bed reactor	1. No addition of chemicals to CBs 2.Energy recovery (cheap biodiesel) 3. Producing Triacetin rich-oil	1. The emission of gases during the carbonizing process 2. The process require an energy source for thermal conversion	[37]
Vector control by the chemicals extracted from CBs	1. Submerging CBs in distilled water 2. The extraction of chemical trapped in water	1. No addition of chemicals to CBs	1. No use of remaining filters after extracting the trapped materials 2. Effect of final product on organisms must be considered by further researches	[54]

5 Declaration of Competing Interest

The authors declare that they have no known competing financial interests or personal relationships that could have appeared to influence the work reported in this paper.

Acknowledgements The authors gratefully acknowledge the support provided by Mohamed V University in Rabat, Morocco.

Funding This research received no external funding.

Data Availability Data will be made available on in Supplementary data related to this article.

References

1. El Fadili, H., Ben Ali, M., Rahman, M.N., El Mahi, M., Lotfi, E.M., Louki, S.: Bioavailability and health risk of pollutants around a controlled landfill in Morocco: Synergistic effects of landfilling and intensive agriculture. Heliyon **10**, e23729 (2024). https://doi.org/10.1016/j.heliyon.2023.e23729
2. Zafeiridou, M., Hopkinson, N.S., Voulvoulis, N.: Cigarette smoking: an assessment of tobacco's global environmental footprint across its entire supply chain (2018). https://doi.org/10.1021/acs.est.8b01533
3. Mohajerani, A., Kadir, A.A., Larobina, L.: A practical proposal for solving the world's cigarette butt problem: recycling in fired clay bricks. Waste Manag. **52**, 228–244 (2016). https://doi.org/10.1016/j.wasman.2016.03.012
4. Rebischung, F., Chabot, L., Biaudet, H., Pandard, P.: Cigarette butts: a small but hazardous waste, according to European regulation. Waste Manag. **82**, 9–14 (2018). https://doi.org/10.1016/j.wasman.2018.09.038
5. El Fadili, H., Ben Ali, M., El Mahi, M., Lotfi, E.M., N. Labjar, Ibn, S., Abdelhamid, E. : Determination of properties and environmental impact due to the inclusion of cigarette fibers in mortar: a new solution to mitigate the CB pollution. Environ. Sci. Pollut. Res. (2023). https://doi.org/10.1007/s11356-023-26491-7
6. Dobaradaran, S., Soleimani, F., Akhbarizadeh, R., Schmidt, T.C.: Environmental fate of cigarette butts and their toxicity in aquatic organisms: a comprehensive systematic review. Environ. Res. **195**, 110881 (2021). https://doi.org/10.1016/j.envres.2021.110881
7. Torkashvand, J., Farzadkia, M., Reza, H., Esrafili, A.: Littered cigarette butt as a well-known hazardous waste: a comprehensive systematic review. J. Hazard. Mater. **383**, 121242 (2020). https://doi.org/10.1016/j.jhazmat.2019.121242
8. Ghasemi, A., Golbini Mofrad, M.M., Parseh, I., Hassani, G., Mohammadi, H., Hayati, R., Alinejad, N.: Cigarette butts as a super challenge in solid waste management: a review of current knowledge, Environ. Sci. Pollut. Res. **29**, 51269–51280 (2022). https://doi.org/10.1007/s11356-022-20893-9
9. Kurmus, H., Mohajerani, A.: The toxicity and valorization options of cigarette butts. Waste Manag. **104**, 104–118 (2020). https://doi.org/10.1016/j.wasman.2020.01.011
10. Kurmus, H.: enhanced fired-clay bricks incorporating cigarette butts (2021)
11. Christina, M., Araújo, B., Costa, M.F., Oceanography, C., De Oceanografia, D., Federal, U., De M, P., Via, S.N.: Review article A critical review of the issue of cigarette butt pollution in coastal environments, Environ. Res. **172**, 137–149 (2019). https://doi.org/10.1016/j.envres.2019.02.005

12. Kurmus, H., Mohajerani, A.: Energy savings, thermal conductivity, micro and macro structural analysis of fired clay bricks incorporating cigarette butts. Constr. Build. Mater. **283**, 122755 (2021). https://doi.org/10.1016/j.conbuildmat.2021.122755
13. Torkashvand, J., Godini, K., Jonidi, A., Esra, A., Farzadkia, M.: Environment assessment of littered cigarette butt in urban environment, using of new cigarette butt pollution index (CBPI) **769** (2021). https://doi.org/10.1016/j.scitotenv.2020.144864
14. Mohajerani, A., Tanriverdi, Y., Nguyen, B.T., Wong, K.K., Dissanayake, H.N., Johnson, L., Whitfield, D., Thomson, G., Alqattan, E., Rezaei, A.: Physico-mechanical properties of asphalt concrete incorporated with encapsulated cigarette butts. Constr. Build. Mater. **153**, 69–80 (2017). https://doi.org/10.1016/j.conbuildmat.2017.07.091
15. Novotny, T.E., Lum, K., Smith, E., Wang, V., Barnes, R.: Cigarettes butts and the case for an environmental policy on hazardous cigarette waste 1691–1705 (2009). https://doi.org/10.3390/ijerph6051691
16. Kadir, A.A., Mohajerani, A.: Possible utilization of cigarette butts in light-weight fired clay bricks 153–157 (2008)
17. Yousefi, M., Kermani, M., Farzadkia, M., Godini, K., Torkashvand, J.: Challenges on the recycling of cigarette butts. Environ. Sci. Pollut. Res. **28**, 30452–30458 (2021). https://doi.org/10.1007/s11356-021-14058-3
18. Moriwaki, H., Kitajima, S., Katahira, K.: Waste on the roadside, "poi-sute" waste: its distribution and elution potential of pollutants into environment. Waste Manag. **29**, 1192–1197 (2009). https://doi.org/10.1016/j.wasman.2008.08.017
19. Dobaradaran, S., Schmidt, T.C., Kaziur-Cegla, W., Jochmann, M.A.: BTEX compounds leachates from cigarette butts into water environment: a primary study. Environ. Pollut. **269**, 116185 (2021). https://doi.org/10.1016/j.envpol.2020.116185
20. Chevalier, Q., El Hadri, H., Petitjean, P., Bouhnik-Le Coz, M., Reynaud, S., Grassl, B., Gigault, J.: Nano-litter from cigarette butts: Environmental implications and urgent consideration. Chemosphere **194**, 125–130 (2018). https://doi.org/10.1016/j.chemosphere.2017.11.158
21. Dobaradaran, S., Schmidt, T.C., Lorenzo-Parodi, N., Jochmann, M.A., Nabipour, I., Raeisi, A., Stojanović, N., Mahmoodi, M.: Cigarette butts: an overlooked source of PAHs in the environment? Environ. Pollut. **249**, 932–939 (2019). https://doi.org/10.1016/j.envpol.2019.03.097
22. El Fadili, H., Ben Ali, M., Touach, N., El Mahi, M., Lotfi, E.M.: Ecotoxicological and pre-remedial risk assessment of heavy metals in municipal solid wastes dumpsite impacted soil in morocco. Environ. Nanotechnol. Monit. Manag. **17**, 100640 (2022). https://doi.org/10.1016/j.enmm.2021.100640
23. El Fadili, H., Ben Ali, M., El Mahi, M., Cooray, E., Lotfi, E.M.: A comprehensive health risk assessment and groundwater quality for irrigation and drinking purposes around municipal solid waste sanitary landfill: A case study in Morocco. Environ. Nanotechnology, Monit. Manag. **18**, 100698 (2022). https://doi.org/10.1016/j.enmm.2022.100698
24. Akhbarizadeh, R., Dobaradaran, S., Parhizgar, G., Schmidt, T.C., Mallaki, R.: Potentially toxic elements leachates from cigarette butts into different types of water: a threat for aquatic environments and ecosystems? Environ. Res. **202**, 111706 (2021). https://doi.org/10.1016/j.envres.2021.111706
25. Roder Green, A.L., Putschew, A., Nehls, T.: Littered cigarette butts as a source of nicotine in urban waters, J. Hydrol. **519**, 3466–3474 (2014). https://doi.org/10.1016/j.jhydrol.2014.05.046.
26. Torkashvand, J., Farzadkia, M., Reza, H., Esrafili, A.: Littered cigarette butt as a well-known hazardous waste: a comprehensive systematic review A: B. J. Hazard. Mater. **383**, 121242 (2020). https://doi.org/10.1016/j.jhazmat.2019.121242
27. Torkashvand, J., Farzadkia, M., Sobhi, H.R., Esrafili, A.: Littered cigarette butt as a well-known hazardous waste: a comprehensive systematic review. J. Hazard. Mater. **383**, 121242 (2020). https://doi.org/10.1016/j.jhazmat.2019.121242
28. Dobaradaran, S., Schmidt, T.C., Lorenzo-parodi, N., Telgheder, U.: Polycyclic aromatic hydrocarbons (PAHs) leachates from cigarette butts into water*. Environ. Pollut. **259**, 113916 (2020). https://doi.org/10.1016/j.envpol.2020.113916

29. Slaughter, E., Gersberg, R.M., Watanabe, K., Rudolph, J., Stransky, C., Novotny, T.E.: Toxicity of cigarette butts, and their chemical components, to marine and freshwater fish. Tob. Control. **20**, 25–29 (2011). https://doi.org/10.1136/tc.2010.040170
30. Yousefi Nasab, A., Oskoei, V., Rezanasab, M., Alinejad, N., Hosseinzadeh, A., Kashi, G.: Cigarette butt littering consequences: a study of pollution rate on beaches and urban environments, Environ. Sci. Pollut. Res. **29**, 45396–45403 (2022). https://doi.org/10.1007/s11356-022-19155-5
31. Novotny, T.E., Zhao, F.: Consumption and production waste: another externality of tobacco use. Tob. Control. **8**, 75–80 (1999)
32. Montalvão, M.F., Sampaio, L.L.G., Gomes, H.H.F., Malafaia, G.: An insight into the cytotoxicity, genotoxicity, and mutagenicity of smoked cigarette butt leachate by using Allium cepa as test system. Environ. Sci. Pollut. Res. **26**, 2013–2021 (2019). https://doi.org/10.1007/s11356-018-3731-2
33. Mathers, C.D., Loncar, D.: Projections of global mortality and burden of disease from 2002 to 2030. PLoS Med. **3**, 2011–2030 (2006). https://doi.org/10.1371/journal.pmed.0030442
34. Freire Lima, C., Amaral dos Santos Pinto, M., Brasil Choueri, R., Buruaem Moreira, L., Braga Castro, Í.: Occurrence, characterization, partition, and toxicity of cigarette butts in a highly urbanized coastal area. Waste Manag. **131**, 10–19 (2021). https://doi.org/10.1016/j.wasman.2021.05.029
35. Stigler-Granados, P., Fulton, L., Patlan, E.N., Terzyk, M., Novotny, T.E.: Global health perspectives on cigarette butts and the environment. Int. J. Environ. Res. Public Health. **16** (2019). https://doi.org/10.3390/ijerph16101858
36. Koroleva, E., Mqulwa, A.Z., Norris-Jones, S., Reed, S., Tambe, Z., Visagie, A., Jacobs, K.: Correction to: Impact of cigarette butts on bacterial community structure in soil (Environmental Science and Pollution Research, (2021), 28, 25, (33030–33040). https://doi.org/10.1007/s11356-021-13152-w). Environ. Sci. Pollut. Res. 28, 33041 (2021). https://doi.org/10.1007/s11356-021-13657-4
37. Yousef, S., Eimontas, J., Zakarauskas, K., Striūgas, N.: Pyrolysis of cigarette butts as a sustainable strategy to recover triacetin for low-cost and efficient biodiesel production. J. Anal. Appl. Pyrolysis. **175** (2023). https://doi.org/10.1016/j.jaap.2023.106167
38. Singh, A., Dayu, X., Ituen, E., Ansari, K., Quraishi, M.A., Kaya, S., Lin, Y.: Tobacco extracted from the discarded cigarettes as an inhibitor of copper and zinc corrosion in an ASTM standard D1141–98(2013) artificial seawater solution. J. Mater. Res. Technol. **9**, 5161–5173 (2020). https://doi.org/10.1016/j.jmrt.2020.03.033
39. Li, X., Zhang, J., Song, W., Chang, Z., Han, Y., Cai, R.: Cigarette Butts water extracts as corrosion inhibitor in oil field acidizing fracture using different metal chlorides intensifiers. In: Chinese material conference (2017). Springer, pp 1155–1165
40. Zhao, J., Zhang, N., Qu, C., Wu, X., Zhang, J., Zhang, X.: Cigarette butts and their application in corrosion inhibition for N80 steel at 90 °c in a hydrochloric acid solution. Ind. Eng. Chem. Res. **49**, 3986–3991 (2010). https://doi.org/10.1021/ie100168s
41. Escobar, V.G., Maderuelo-Sanz, R.: Acoustical performance of samples prepared with cigarette butts. Appl. Acoust. **125**, 166–172 (2017)
42. Maderuelo-Sanz, R., Gómez Escobar, V., Meneses-Rodríguez, J.M.: Potential use of cigarette filters as sound porous absorber. Appl. Acoust. **129**, 86–91 (2018). https://doi.org/10.1016/j.apacoust.2017.07.011
43. d'Heni Teixeira, M.B., Duarte, M.A.B., Raposo Garcez, L., Camargo Rubim, J., Hofmann Gatti, T., Suarez, P.A.Z.: Process development for cigarette butts recycling into cellulose pulp. Waste Manag. **60**, 140–150 (2017). https://doi.org/10.1016/j.wasman.2016.10.013
44. Matos, I., Bernardo, M., Fonseca, I.: Porous carbon: a versatile material for catalysis. Catal. Today **285**, 194–203 (2017). https://doi.org/10.1016/j.cattod.2017.01.039
45. Girondi, G.D., Marvila, M.M., de Azevedo, A.R.G., de Souza, C.C., Souza, D., de Brito, J., Vieira, C.M.F.: Recycling potential of powdered cigarette waste in the development of ceramic materials. J. Mater. Cycles Waste Manag. **22**, 1672–1681 (2020). https://doi.org/10.1007/s10163-020-01058-7

46. Kadir, A.A., Mohajerani, A.: Recycling cigarette butts in lightweight fired clay bricks. Proc. Inst. Civ. Eng. Constr. Mater **164**, 219–229 (2011). https://doi.org/10.1680/coma.900013
47. Xiong, Q., Bai, Q., Li, C., Lei, H., Liu, C., Shen, Y., Uyama, H.: Cost-effective, highly selective and environmentally friendly superhydrophobic absorbent from cigarette filters for oil spillage clean up. Polymers (Basel) **10**, 1101 (2018)
48. Li, S., Shi, C.L., Wang, Y., Sun, X.D., Qi, D., Wu, D., Wang, Y., Zhu, J.: New efficient selective adsorbent of tobacco specific nitrosamines derived from discarded cigarette filters. Microporous Mesoporous Mater. **284**, 393–402 (2019)
49. Ardanuy, M., Claramunt, J., García-Hortal, J.A., Barra, M.: Fiber-matrix interactions in cement mortar composites reinforced with cellulosic fibers. Cellulose **18**, 281–289 (2011). https://doi.org/10.1007/s10570-011-9493-3
50. Morales-Segura, M., Porras-Amores, C., Villoria-Sáez, P., Caballol-Bartolomé, D.: Characterization of gypsum composites containing cigarette butt waste for building applications. Sustainability **12** (2020). https://doi.org/10.3390/su12177022
51. Romero-Gómez, M.I., Silva, R.V., Costa-Pereira, M.F., Flores-Colen, I.: Thermal and mechanical performance of gypsum composites with waste cellulose acetate fibres. Constr. Build. Mater. **356**, 1–11 (2022). https://doi.org/10.1016/j.conbuildmat.2022.129308
52. Mohajerani, A., Hui, S.Q., Shen, C., Suntovski, J., Rodwell, G., Kurmus, H., Hana, M., Rahman, M.T.: Implementation of recycling cigarette butts in lightweight bricks and a proposal for ending the littering of cigarette butts in our cities. Materials (Basel) **13** (2020). https://doi.org/10.3390/ma13184023
53. Hamzah, Y., Umar, L.: Preparation of creating active carbon from cigarette filter waste using microwave-induced KOH activation. J. Phys. Conf. Ser. IOP Publishing 12027 (2017)
54. Dieng, H., Rajasaygar, S., Hassan, A., Ahmad, H., Salmah, C., Fatma, W., Satho, T., Miake, F., Fukumitsu, Y., Ramli, A., Abd, I., Enrique, R., Vargas, M., Hafiz, A., Majid, A., Abubakar, S.: Acta Tropica Turning cigarette butt waste into an alternative control tool against an insecticide-resistant mosquito vector. Acta Trop. **128**, 584–590 (2013). https://doi.org/10.1016/j.actatropica.2013.08.013

Conversion of Smart Nanomaterials to Achieve Sustainable Goals

Mavia Rashid, Muhammad Arshad, Muhammad Salman Habib, and Ehsan Ul Haq

Abstract The implementation of smart materials consisting of nanoparticles into diverse applications has become a revolutionary strategy to tackle worldwide sustainability objectives. Nanomaterials have unique properties due to high surface area and quantum confinement effect that helps in achieving Sustainable Development Goals (SDGs). Nano materials in the form of sensors, motors, films, membranes, dendrites, particles, wires, fibers, plates, etc. are used for achieving SDGs in agriculture, medical field, waste water treatment and purification, clean energy, its production, storage, pollution control, waste management, infrastructure design and development, etc.

Keywords Smart materials · Sustainable Development Goals (SDGs) · Surface area · Quantum confinement

1 Introduction

Materials consisting of at least one nano scale dimension ranging between 1 and 100 nanometers—are known as "nanomaterials". In this branch of materials science, primarily materials are manipulated and controlled at the nanoscale to provide desired features or capabilities. As these are tiny with lesser specific surface area in comparison to the bulk materials, nanomaterials can have special and unusual properties. Two main aspects are responsible for the unusual properties that are, high surface area and the second is quantum confinement [1]. High surface area is responsible for higher number of broken bonds available and the quantum confinement for the higher & discrete levels of energies similar to atoms in free space. Nanomaterials come in a variety of forms, however they can be generically categorized as nanoparticles, nanocomposites, nanotubes, nanowires, quantum dots (QD), liposomes, nanoshells, nanorobots, and nanosensors are examples of some of the nano materials.

M. Rashid · M. Arshad · M. S. Habib · E. U. Haq (✉)
Department of Metallurgical and Materials Engineering, Faculty of Chemical, Metallurgical and Polymer Engineering, University of Engineering and Technology, Lahore, Punjab, Pakistan
e-mail: amonehsan@uet.edu.pk

K. K. Kesari et al. (eds.), *Functionalized Cellulose Materials*, Engineering Materials,
https://doi.org/10.1007/978-3-031-76953-5_10

In many respects, nanomaterials are important for advancement to achieve the Sustainable Development Goals (SDGs) of the United Nations (UN). Here are some crucial elements emphasizing their significance.

2 Nano Materials for Sustainable Agriculture (UNSDG 2)

2.1 Nano Fertilizers and Their Impact on Crop Yield

Undoubtedly, the significant rise in global food production can be attributed to the increased application of inorganic fertilizers. On the other hand, several studies suggest that the increased usage of conventional fertilizers has led to significant environmental problems, such as contaminated soil, growth of unwanted micro-organisms and water bodies. Farmers incur increased costs due to the unscientific and incorrect application of higher quantities of conventional fertilizers, which also seriously contaminates the environment [2]. Conventional fertilizers release nitrate, or nitrogen and phosphorus, into the environment, which can lead to eutrophication (excessive plant and algal growth) and other problems. Major human illnesses are caused by the deterioration of groundwater caused by leached Nitrate (NO_3) and Nascent Nitrogen (N) [3]. With the aid of nano fertilizer, nanotechnology offers a developing new avenue for resolving issues brought on by conventional fertilizers formulations. Figure 1 illustrates the elements affecting the uptake, absorption, movement, and infiltration of nanoparticles (NPs) in plants crop scientists have become interested in nano fertilizers because they offer various benefits over conventional ones, including higher yields, better soil fertility, and lower pollution [4]. Because of their large surface area, nano fertilizers may store a significant number of nutrient ions, which are then released gradually (solvent exchange) in accordance with crop needs [5]. Nano fertilizers are designed with specific qualities including high solubility, stability, effectiveness, and time-oriented release to optimize their target activity through an efficient concentration of nutrient ions [6]. When compared to conventional fertilizer, nanoencapsulated phosphorus produced 14.5% greater nitrogen use efficiency (NUE) [7]. Because the surface mass ratio of particles is increased due to the reduction in size of nanomaterials, a significant amount of nutritional ions are adsorbed and desorbed gradually and consistently over an extended period of time [8]. It is crucial to assess the efficacy and dosage of NFs in order to safeguard the environment and public health [9].

2.2 Nanosensors for Precision Agriculture

To create a database system in which the data of the entire farm can be obtained is known as Satellite farming or Precision Farming. It helped farmers to remotely access

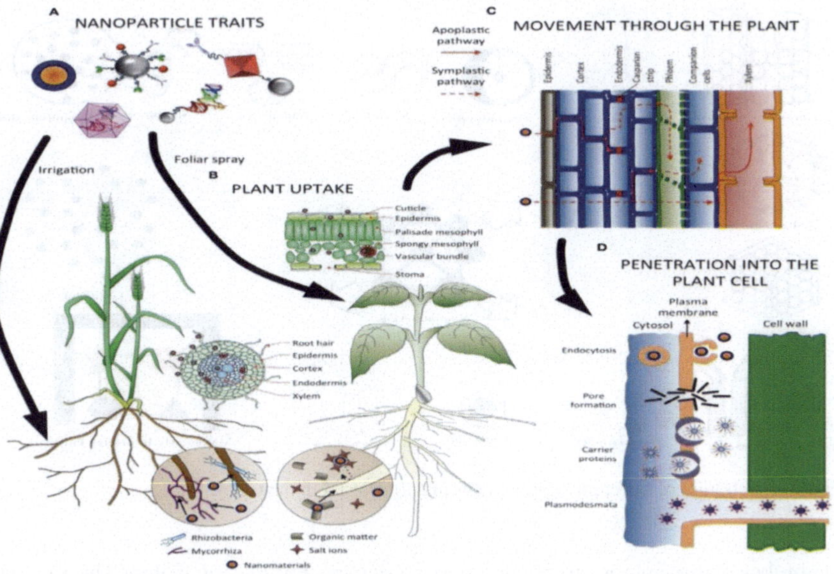

Fig. 1 Illustration of the elements affecting the uptake, absorption, movement, and infiltration of nanoparticles (NPs) in plants, including: **a** Absorption and translocation of NPs inside the flora; **b** NPs working mechanism different compounds and microorganisms; **c** NPs traveling up and down the plant via the cell wall and protoplast pathways; and **d** The process of NP uptake by inside the plant [10]

their farms. Many technologies are developed that are used in Precision Farming. Few of them are Global Navigation System, Satellite imaging, Global Positioning System, sensors, Robots, etc. [11–14]. The physical parameters processed into digital signals by a sensor. These signals are digitally observed by an instrument after processing. Nano-scale (1–100nm) sensors have an advantage over traditional sensors due to presence of nano particles having large surface area. Nano sensors are portable, quite sensitive and cost effective [15]. The working principle of nanosensors includes the conversion of physical signals given by analyte into electrical signals, which are then processed and displayed. Top down lithography, Bottom down approach, etc. is used to fabricate nanosensors [16].

Nanosensors are classified into 5 classes:

- Nanoparticles nanosensors
- Nanoprobes
- Nanowire based nanosensors
- Nanostructured nanosensors
- Nanoelectromechanical (NEMS) nanosensors [16]

Researchers have developed nanosensors that performed different functions such as detection of shortage of water by printed nanosensors on leaves, detection of contaminants like organophophates present in water, injured plants, etc. Carbon

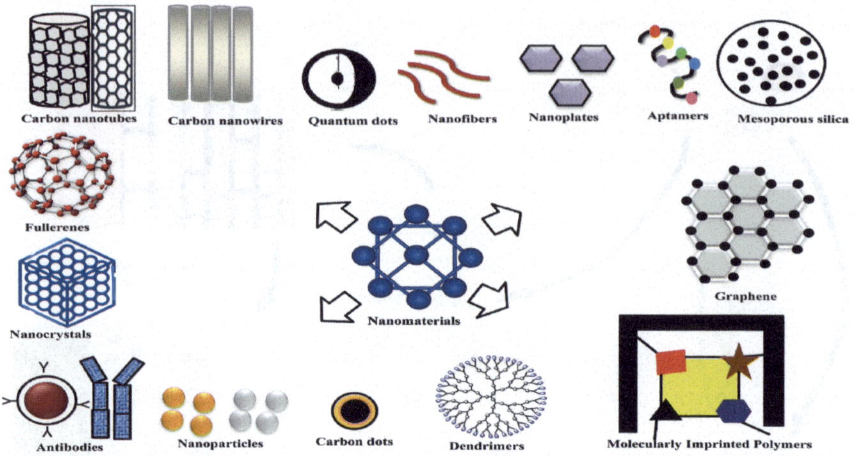

Fig. 2 Illustration of different type of Nano materials used in nanosensors

nanotubes based nanosensors are used to detect shortage of water. The opening and closing of stomata plays an important role in generating electrical signals. Electrical signals are generated when stomata are closed and these signals are detected and observed by CNT sensors. On the other hand, no electric signals generated when stomata are opened. These nanosensors even sensed gases and produce electric signals [17–19]. For advancement in Precision agriculture, new tools and nanosensors are developed (Fig. 2).

2.3 Nano Pesticides

Nano pesticides consist of nanoparticles having large surface area. It helps nano particles in controlling the attack of pest in fields with less amount of pesticide being consumed. To promote target delivery and increase effectiveness, pesticides are encapsulated. To control the quantity of Nano pesticide being supplied, farmers used Smart field system that is installed in farms [20]. Smart field system consists of wireless sensors that are connected to satellite and computer or any monitoring device. Nano pesticides are made by using carrier systems in which pesticides are dispersed. The carrier systems include Silica, Alginate, Chitosan, etc. Nano silica, ZnO based Nano pesticides, Nano-Cu are the example of nano pesticides. Nano pesticides like Nano silica absorbed in the exoskeleton of the insects and results in death. However, some nano pesticides like ZnO based nano pesticides, Nano-Cu control the growth of fungi [21, 22].

2.4 Nano Materials for Pest and Disease Management

Pesticides are used to kill pests (insects, fungi and harmful weeds, etc.). But it also has harmful impact on plants and crops which results in serious health as well as environmental issues. To avoid the harmful impact caused by the use of traditional pesticides, nanotechnology is introduced to control the release of pesticides to the target favoring pest management. Fungal Pathogens also have harmful impact on agriculture and food supply. To allow the target delivery of fungicides, the development of various types of nano formulations are done [23]. To increase the efficiency of fungicides, metal and polymer based nano fungicides, nanocomposite, and nano capsulation are developed. It increases the activity of fungicides by providing large surface area and reduces the amount of fungicide consumed due to targeted delivery.

The formulation of pesticides plays an important role in pest and disease management [24, 25]. The Nano based pesticides formulation includes Nano emulsion, Nano suspension and Nano encapsulation. Nano emulsion is the mixture of water (hydrophilic) and oil (hydrophobic) in which particle size range is between 5 and 200nm. It favors the production of hydrophobic pesticides and hydrophilic pesticides [26]. While Nano suspension or Nano colloidal suspension is a mixture of pesticide particles and liquid. The particle size is less than and equal to 100nm. This formulation is suitable for those pesticides that are water insoluble [27]. While Nano encapsulation is the coating of pesticides in nano-sized range. It protects the pesticides that are sensitive to light and prevent degradation [28, 29].

3 Nano Materials for Health and Well-Being (UNSDG 3)

3.1 Nano Medicine and Drug Delivery Systems

Nanoparticles having particle size range between 1 and 100 nm are inorganic or organic in nature. They are similar to DNA plasmids and antibodies (proteins that protect the body from unwanted substance) [30, 31]. The medical science is influenced by Nanoparticles through microfluidics, drug delivery, nano biotechnology, biosensors and microarrays [32]. The transport of drug to the target within the body is the most promising use of nano materials. It eliminates the tumorous growth without causing any harm to healthy cells [33]. To deliver drugs through BBB (Blood Barrier Brain) to the targeted site (brain tumors), modification or functionalization of nano particles has been done which gives brilliant results [34]. The advanced growth in Pharmacotherapy has been brought by introducing Nano colloidal drug carriers [35]. These drug carriers act as the most promising carrier for different categories of drugs which includes anti-hypertensive, hormones and anti-cancer drugs, etc. [36]. Most commonly used NPs used for the delivery of drug and diagnosis is Submicron colloidal particles [37, 38]. CaCO3 coated retinoic acid nanoparticle as a new drug delivery system was made to stimulate secretion of insulin from islets. These Trans coated nano particles formed aggregates on spray drying while split up in water to stimulate secretion [39].

3.2 *Diagnostic Applications of Nano Materials*

Nanotechnology has not only used for development of nano materials but also give most useful results in the diagnosis and treatment of severe diseases like cancer. Nanotechnology has been introduced in gene therapy, drug carriage, targeted therapy, molecular imaging, detection and diagnosis of diseases [40]. Nano particles like quantum dots, gold nano particles and nanoshells are used for the diagnosis of cancerous cells [41]. Nanotechnology has been introduced in Molecular diagnostics such as the development of biomarkers through nanotechnology. These biomarkers can detect the cancers more accurately in less time [42].

3.3 *Nano-Enabled Therapies for Various Diseases*

Nanotechnology has also been used for treatment of various diseases. Nano carriers used for therapy are liposomes, carbon nanotubes, dendrimers and polymeric micelles consisting of hydrophilic shell and hydrophobic core, etc. These carriers are used for treatment of cancer described in Table 1.

4 Nano Materials for Quality Education (UNSDG 4)

The fast expanding domains of nanotechnology and nanoscience are becoming more and more well-known. Traditional methods of engineering goods design, analysis, and manufacture are changed as a result of research and development in nanotechnology [52]. Nano devices and nano systems differ greatly from micro devices and microsystems, these scientists and engineers are able to conceive, analyze, and manufacture nano devices and nano systems using their math, scientific, and engineering skills. Because they are composed of atoms and molecules, nano devices and nano systems have special quantum properties and abilities that need to be taken advantage of. Therefore, advanced concepts, procedures, instruments, and technologies must to be effectively taught [53]. Learners and communicators are motivated and engaged through immersive experiences and interactive media made possible by nanomaterials. For instance, holographic and three-dimensional (3D) projection technologies made possible by nanomaterials have displayed dynamic and lifelike images and films. It promotes flexible and adoptive learning.

Table 1 Different types of Nano-carriers, their principle and name of drugs used to treat tumor cells

Nano carriers	Principle	Drugs	Reference
Liposomes	Carries drug and kept drugs that are hydrophilic in nature, to remain in the one layer liposome while drugs that are hydrophobic in nature, remain in more than one layer liposome. These drugs fuse with the cell and released in intracellular space	Doxorubicin Paclitaxel Mitomycin	43, 44
Carbon nanotubes (CNT)	Heating of CNT by absorbing light from NIR region, target tumor cells	Paclitaxel	45, 46, 47
Dendrimers	Accumulation of drug encapsulated dendrimers on tumour cell	Indometacin Oligonucleotides	48
Polymeric micelles	Capturing of hydrophobic drugs and switching of hydrophobic and hydrophilic blocks done to allow interaction between hydrophilic molecules (proteins, nucleic acid, etc.) and hydrophilic block in the core to neutralize the charged molecule	Camptothecin Paclitaxel	49, 50,51

5 Nano Materials for Clean Water and Sanitation (UNSDG 6)

Water pollution has not only detrimental effect on human health and environment but it has negative influence on financial progress and societal perspective of the respective societies. Contaminated water means that water bodies or tanks in which the unwanted substances are present that make it inappropriate for consuming and other practical uses. Many physical, chemical and mechanical methods are developed to overcome this rising problem. To improve water purification and waste water treatment methods at low cost, researchers from all over the world are working and brought new and advanced technologies [54, 55]. Nanotechnology is the most emerging field that provides a special offer to treat contaminated water by removing hazardous pollutants with high efficiency at low cost [56]. It is illustrated in Fig. 3.

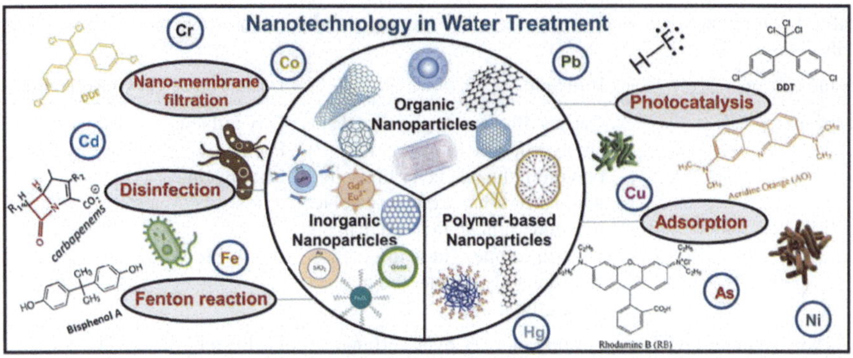

Fig. 3 Illustration of use of Nanotechnology in treatment of waste water [116]

5.1 *Nanotechnology Applications in Water Purification*

Nano materials (NM are not only used to improve traditional wastewater treatment methods but with the emergence of nanotechnology, it has also been used to introduce innovative methods for waste water treatment. As NM have dimensions in nano scale range, they possess unique properties like large surface area, quantum confinement effect, super paramagnetism, strong sorption, fast dissolution, etc. All these size dependent properties contribute in waste water treatment. High sensitivity as well as selectivity of NM towards water contaminants may be provided by high densities of low-coordinated atoms at the surface, vortex, and edges. Because of its multifunctional process and excellent efficiency, NM-based Technology for treating wastewater is less reliant on a complex instrumentation setup.

5.2 *Nano Materials for Wastewater Treatment*

Experts discovered that nanomaterial is a better alternative for purifying wastewater since it has several unique qualities including nano size, immense surface area, highly reactive, and strong solution mobility recently, when the globe was suffering significant concerns with drinking water [57], excellent mechanical properties, hydrophilicity, dispersibility, porosity, and hydrophobicity [58–60]. Various hazardous microorganisms, organic and inorganic contaminants, and some heavy metals like Pb, Mo, etc. have all been reported to be successfully eliminated by employing different nanomaterials. [61–65]. More advanced developments in nanomaterials have recently been made, including nano photocatalysts, nano motors, nano membranes, and nano sorbents (which have a high sorption capacity and are useful for a variety of water treatment techniques) described in Table 2. Certain imprinted

Table 2 Nano materials used for waste water treatment with examples; their working mechanism, pro & cons

Nano materials	Examples	Working mechanism	Pros	Cons	Reference
Nano photocatalyst	Metal oxide nano photocatalyst CdS/TiO2 composite	Photocatalysis: nano photocatalyst absorb energy from sunlight (hf > band gap) and the couple of electron–hole are produced. This pair goes to the surface of catalyst where pollutant molecules are present. Redox reaction occur results in Degradation and Mineralization of pollutant favoring formation of carbon dioxide and water	Less toxic, cheap, chemically and physically stable, easily approachable, environmental friendly and outstanding capacity to absorb light with nano size range between 1 and 100 nm	Toxicity during water treatment and difficulty in recovery of catalysts	66–71
Nano motors	PEDOT/Pt bilayer nanomotor Pt coating Activated carbon-based motors Au NPs/TiO2/Pt nanomotor	Conversion of energy taken from different resources into other forces followed by different mechanism depending upon nature of pollutant (adsorption in case of hydrophobic pollutant and toxic metal while adsorption and degradation for organic pollutants). It includes energetic mixing which favors waste water treatment	No toxicity, more efficient towards waste water treatment	Life cycle of nano motors reduced in case of treatment waste water having high viscosity	72,73
Nano membranes	Nano fiber membrane Nano composite membrane Aquaporin-based membranes	Filtration (nano infiltration, ultra infiltration and reverse osmosis)	Easy to handle, High permeability favoring infiltration	Less efficiency, reliability, selectivity and stability. High operational and maintenance cost	74–77
Nano sorbents	Graphenes oxide Ag/polyaniline Ag/carbon C/TiO2 Nano iron oxide Zeolite	Filtration followed by sorption	Large surface area, superior adsorption rate at short time	Less stability	78–80

polymers can also be used to treat contaminated water. To put it briefly, the investigation of nanomaterial applications in water treatment is thought to evaluate favorable aspects [65].

5.3 Nano Sensors for Water Quality Monitoring

To ensure that all society members have access to clean water, quantitative measurement of water pollution is essential. The Environmental Protection Agency (EPA) and World Health Organization (WHO) are two regulatory agencies that set the highest permitted concentrations of water pollutants that are safe to consume because of their harmful effects [81]. As an alternative, before disposing of industrial effluents, they must be checked to see if the pollution level surpasses the federal regulations for wastewater discharge [82] them into the water's top. Either surface water or ground water contains the contaminants, depending on how soluble they are in water. Common NMs, including metal-oxide NPs, noble metals, Carbon nanotubes, Quantum dots, etc., are employed to produce nanosensors using a variety of ideas and techniques including field-effect transistors, SERS, and ultraviolet–visible spectrum analysis, chemiluminescence, and electrochemiluminescence, fluorescence, photochemical, and electrochemical techniques, among others [83].

6 Nano Materials for Affordable and Clean Energy (UNSDG 7)

Making the shift from an economy dependent on fossil fuels to a low-carbon footprints, it is necessary to address the conversion of renewable energy sources and their storage. Because of their incredibly tiny building blocks and low volume to interface ratio, nanostructured materials provide a wealth of tools for scientists looking to increase the effectiveness of energy conversion or a material's ability as well as energy density.

The following are a few illustrations of advanced tools made possible by Nano science:

(i) Rapid photo excited charge separation and collection, preventing recombination problems
(ii) Fasten ion and atom movement along nano crystallite interfaces
(iii) High photocatalytic activity owing to high specific surface area
(iv) Improved light harvesting because of the low reflectivity at the surfaces of nanostructured material

6.1 Nano Materials in Solar Energy Harassing

The utilization of the particle nature of light in conventional photovoltaic cells, where absorbed photons generate electron–hole pairs, and the collective electromagnetic nature of light, where sunlight is captured by antennas and rectified, are two of the proposed mechanisms for converting solar energy to electricity. Studying the three generations of PV, which are distinguished by the trade-off between cost reduction and efficiency enhancements, is the most effective approach to monitor technological advancement. [84]. Solar cells with semiconducting p–n junctions, such as single crystal and poly-grain Si, are included in the first generation of PV. Through the introduction of thin-film technologies, second generation PV lowered costs. Amorphous Si, thin film Si, CdTe, CuInSe2, and dye-sensitized photochemical cells are a few examples. The third generation of PV that is now in use should result in higher efficiency. High efficiency multi-gap tandem cells, multiple exciton generation solar cells, intermediate band solar cells (IBSCs), hot carrier converters, thermophotovoltaics/thermophonic, etc. are some examples of proposed enhanced solar cells. Additionally, it has been suggested that sunlight may be collected by (nano)antennas and corrected by a diode by taking use of the light's wave form [84–87]. The fundamental components of modern solar cells are engineered nanomaterials. Semiconducting nanocrystals may be used in multiple exciton production solar cells, where a single, high-energy photon may excite a semiconductor nanocrystal and produce a few electron–hole pairs. [84, 88–95], or nanomaterials for the construction of optical antennas that support modes with incredibly short wavelengths and have typical dimensions of roughly 100–1000 nm and robust electromagnetic field confinement at the nanoscale [84, 87, 96]. The field of engineered nanomaterials has expanded, and concentrating solely on technological advancements is now insufficient. Improved fundamental knowledge of the electrical, optical, and transport properties of nanomaterials is crucial for future technological advancements, as is figuring out how best to move technology from academic labs to commercial settings. Recent developments in computational nanoscience and cutting-edge research may make such discoveries possible. Figure 4 Illustrates the use of nano wires in rectifier and antenna system.

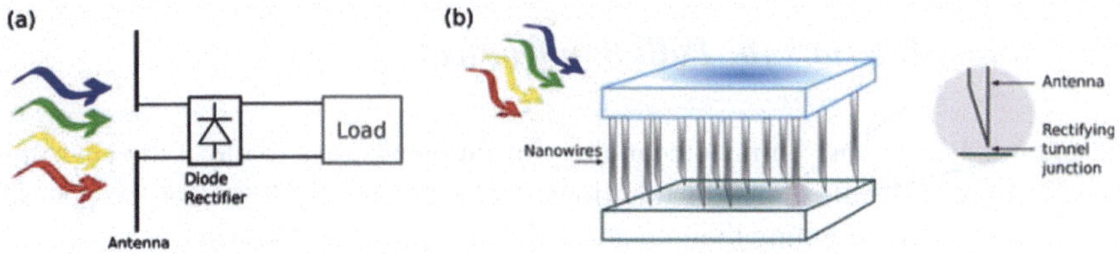

Fig 4 **a** Load and Rectifier & antenna system block diagram; **b** representation of rectifier and antenna system where the system utilizing nano-wires which act as antennas and its surface and tip form rectifying tunneling junctions [87]

6.2 Nanostructured Materials for Energy Storage

We may be able to create energy storage gadgets having highest energy-densities possible while retaining the strong power, rapid recharging, and extended life cycle of an electrochemical capacitor thanks to nanomaterials and hybrid nanomaterials. Following are the reasons showing why nano structured materials are preferred for energy storage:

(1) Charging times of batteries may be reduced through low diffusion rate or draw a significant current in lesser period of time (raise efficiency and the power).
(2) Utilization of silicon in Li-ion battery anodes is another good example, where nanoparticles can tolerate cycling but can also experience considerable swelling which may cause failure and low cycle lives. Consequently, we may utilize materials that have a higher energy storage capacity or prolong the life of materials that are already in use.
(3) Because multivalent ion transport is slow in bulk materials, the introduction of nanomaterials may empower feasible magnesium- or Al-ion rechargeable batteries, able to retain charge significantly to higher degree than Li-ion batteries which in current use.

The formation of "black silicon," an anti-reflective nanostructured material, and other nanoscale structures with a broad surface area and a record high aspect ratio on Si are crucial for battery anodes that are Si-based and applications for solar systems, where the capacity to reduce surface reflection leads to improved solar harvesting [86]. This way an efficient etching technique may be originated which could be used for mass production. Because of their tendency to absorb light and transform it into extraordinarily active carriers of charge and thermal energy, plasmonic nanoparticles are gaining attention in the field of photocatalysis (Fig. 5).

7 Nano Materials for Sustainable Cities and Communities (UNSDG 11)

7.1 Nano Materials for Pollution Control

Research is being done on nanotechnology to find new ways to clean the environment and boost the efficiency of traditional technologies. By lowering the release of contaminants or stopping their creation, this technique is also being investigated as a means of decreasing pollution. Environmental improvement has been designated as one of the eight cross-cutting areas of nanotechnology by the US National Nanotechnology Initiative [97]. The main environmental concern nowadays is the management and elimination of pollutants in land, sediments, and water due to recent developments. Nanoparticles size and large surface area give them several benefits. Because these nanoparticles are easily carried by water [98], they can be utilized

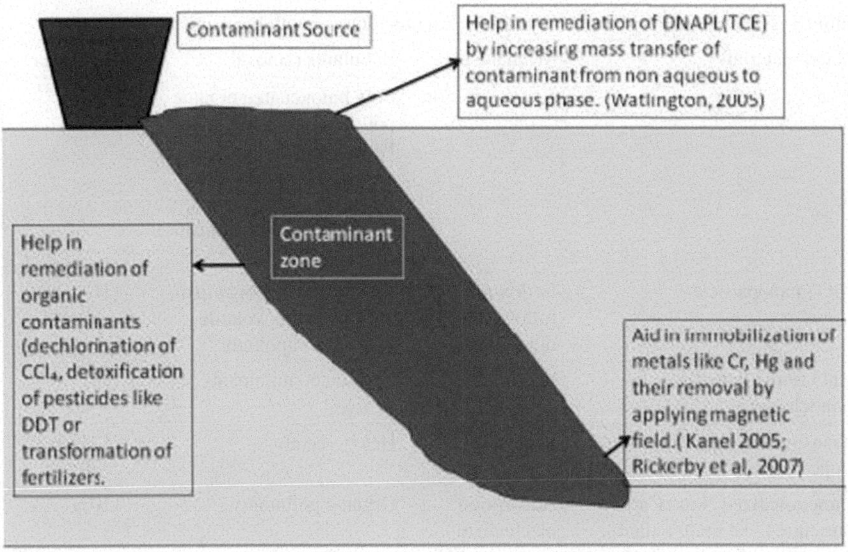

Fig. 5 Schematic diagram demonstrating the degradation of different contaminants like inorganic, organic and petrochemical compounds by using nanoparticles of iron [106]

to handle poisons below the surface without the need for removal. These particles may be delivered into incredibly tiny locations because to their small size, and they can even stay active for extended periods of time [97]. To prevent pollutants from having an adverse effect on the environment, pollution must be continuously sensed or monitored. Either by concentrating on the binding between the recognition element and pollutant or by enhancing the electronic interface and transduction to the sensor layer, research is aimed at enhancing the degree of selectivity, specificity, and environmental monitoring sensors' sensitivity. Nanoparticle-based sensors are used to detect biological organisms, inorganic pollutants, or organic toxins. Gold nanocatalysts, a metal oxide nanocatalyst exhibit encouraging outcomes in terms of stopping or minimizing pollution at its source. This is accomplished by low-light autocatalysts, pollution control applications like air cleansing or hydrogen stream purification for fuel cells. Nanotechnology has made a large variety of materials available that can be used for various spills, including metal oxides, carbon nanoparticles, bimetallic particles, dendrimers, liposomes and nanoscale zeolites as described in Table 3. Figure 6 demonstrates the degradation of different contaminants by using nanoparticles of iron.

Table 3 Different nano materials used for pollutants removal with mechanism

Nano materials	Mechanism	Pollutants removal	References
Iron	Redox reaction	Poly halogenated organic pollutants, Heavy metals, Petrochemical compounds, organic solvents and Pesticides, fertilizers (sub-surface contaminants), etc	[99]
TiO2 nanoparticle	Redox reaction followed by UV radiations	Dyes and microorganism, Nitrous oxide, volatile organic compounds	[100]
ZnO semiconductor nanocluster	Photocatalysis	Organic contaminants	[99]
Transition metal (Vanadium) doped Titania	Photocatalysis	Heavy metals	[101]
Functionalized Nano Carbon structures like multiwalled nanotubes (A concoction of sulphuric acid and nitric acid)	Adsorption mechanism	Organic pollutants	[102]
Fe nanoparticles coated with ferromagnetic carbon	Carbon shell physical adsorption mechanism	Cr (VI)	[103]
Gold nanocatalyst	Redox reaction followed by UV radiations	Carbon monoxide	[104]
Nanosized zeolites	Redox reaction	Hydrocarbon like toluene	[105]

7.2 *Smart Nanomaterials for Infrastructure Design and Development*

Products based on nanotechnology have the potential to increase construction's overall competitiveness. The following areas have seen the most application of nanotechnology in the building industry: (1) more robust and lightweight structural composites; (2) coatings that required low maintenance; (3) enhanced properties of cementitious materials; (4) nano-sensors related to construction; and (5) lowered fire retardant and insulation transfer heat rate. Currently, there are fewer nanomaterials being used in construction, primarily due to the following reasons: low knowledge about which nanomaterials are suitable for construction and how they behave; a lack of precise guidelines for the planning and implementation of construction components that make use of nanoparticles; a decrease in the availability of nanoproducts; a shortage of comprehensive data regarding their composition; expensive; and a lack of awareness about the hazards to health posed by nanoparticles. Researchers studying the concrete industry have benefited greatly from nanotechnology. In reality, it has

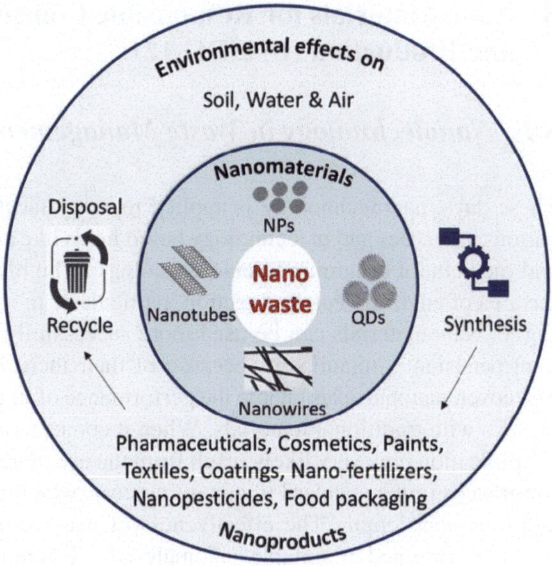

Fig. 6 Processing of nanomaterials (NP, nanowires, quantum dots, and nanotubes) and how they are utilized make a variety of Nano-products that end up in Nano waste and this have a deleterious print on the environment [10]

sparked the creation of novel additives, nanocomposites, cements, and concretes. Studies have shown that the addition of nanoparticles, such as concrete pore fillers, would increase the durability of concrete through chemical and physical interactions.

Researchers have created reinforced concrete by including fibers that range in length from nano to micrometer and are composed of steel, carbon, or polymers. Concrete's mechanical qualities can be enhanced by adding nano-silica (SiO2), which increases the amount of crushed micro and nanostructures. In the real world, nano sized silica prolongs the life of concrete by lowering water penetration and the amount of calcium in the water required to soak the cement. Concrete that can mend itself is another example of nanotechnology's success. This kind of concrete is still the subject of ongoing study. The microcapsules that are inserted into the cracked concrete break and allow the healing agent to enter the affected area by capillary action. The aforementioned substance makes contact with the concrete's internal catalyzer, releasing a polymer that covers the crack's surface [107–109].

8 Nano Materials for Responsible Consumption and Production (UNSDG 12)

8.1 Nanotechnology in Waste Management

These days, nanotechnology is applied to many facets of daily life and is used in various industries and in technology based fields like electronics, power, food safety, and medicine. Furthermore, nanotechnology is highly innovative advancements in the area of environmental protection, particularly in managing waste represented in Fig. 6. Nanomaterials can be used more successfully than traditional chemicals to treat persistent contaminants because of their increased reactivity and selectivity. Moreover, nanoparticles enable the performance of impossible chemical reactions to achieve with traditional materials. When it comes to application efficacy, treatment simplification may very likely result from the use of nanoparticles. This would mean fewer steps in the standard treatment process, which could reduce the energy, cost, and treatment length. The effectiveness of nanotechnology in managing waste is mostly determined by suitable nanomaterials' selection for certain contaminants and the surrounding environment. Yet, many have been made in research laboratories up to this point, very few nanomaterials have eventually been manufactured at an industrial scale. At the moment, metal and carbon-based nanomaterials are the most often utilized types, primarily iron with zero-valent (nZVI) being the most frequently used in industry based on environmental protection [110, 111].

8.2 Sustainable Nanomaterial Production Processes

It is essential to create environmentally friendly processes for producing these nanomaterials. Stability requirements are crucial in order to avoid the accumulation issues in NPs that lower their interfacial free energy and specific surface area, thereby decreasing their intended reactivity profile. This necessitates the additional use of surfactants or polyelectrolytes, which increases waste generation [112, 113].

9 Nano Materials for Climate Action (UNSDG 13)

9.1 Nanotechnology in Capturing of Carbon and Its Storage for Greenhouse Gas Reduction

The impact of greenhouse gases (GHGs) has demonstrated that the presence of carbon products in large quantity in environment has an effect of global warming. It has potentially leading to air pollution and catastrophic weather effects. This continuous burning of fossil fuels is done due to rise in demand for energy and thus leading to the development of affordable and energy-efficient carbon dioxide technologies. The design and development of natural reservoir systems, such as components for CO2 storage, is the focus of current efforts. However, it is necessary to separate CO2 from the waste gasses before its treatment. The gases produced from chimneys are known as flue gases and they contain water, nitrogen gas, dust particles and greenhouse gas (CO2). So the extraction of greenhouse gas from these gases is the basic step in the capturing of CO2. Solvent absorption is typically needed for this process. For carbon dioxide capture to be both economical and energy-efficient, better methods are required. Over time, developments have also spurred study into the straight conversion of greenhouse gas into valuable items, as an alternative to simple sequestration. The use of nanotechnology can significantly lower worldwide CO2 emissions [114]. A few notable benefits of nanotechnology include its ability to produce materials with unique physiochemical characteristics and a low active volume to surface area ratio [115]. There are four categories for nanomaterials: nanoporous, nanocrystalline, nanocomposites, and nanohollow structure materials. Given that different nanomaterials are used as precursors in the fabrication process, nanomembranes exhibit a higher sorption rate than conventional sorbents. However, the gas that is intended for adsorption affects the permeability and selectivity of the membrane. These nanomaterials include mesoporous silica nanoparticles, carbon nano adsorbents, nano-sized zeolites, and metal–organic framework (MOF) containing embedded nanoparticles. Alkanes, CO2, and water are captured using pores with a size range of 4–10 nm [114]. Because of their high adsorption capacity, which ranges from 4 to 300 mg of CO2/g under mild conditions, they are employed to remove CO2 from the air [115, 116].

10 Conclusion

To sum up, transforming smart Nano materials is a revolutionary way to accomplish sustainable objectives in a variety of industries. The creation of cutting-edge sensors, medication delivery systems, effective energy storage solutions, and other technologies is facilitated by the special qualities of nanomaterials. The incorporation of nanotechnology into the design and use of materials has demonstrated great promise in addressing urgent environmental, financial, and social issues. Research and development procedures must be conducted carefully because of the possible negative effects that nanomaterials may have on the environment, as well as ethical and safety issues related to their use. It is imperative to establish strong regulatory frameworks and risk assessment methods to guarantee the secure and long-term incorporation of intelligent nanomaterials into diverse applications.

Challenges and Ethical Considerations—Environmental impact of nano materials—Safety and ethical concerns in nano material applications—Regulatory frameworks for nano materials.

Future Perspectives and Conclusions—Emerging trends in nano materials for UNSDGs—Potential breakthroughs and innovations—Summary and concluding remarks.

This chapter would explore the diverse applications of nano materials in addressing the various UN Sustainable Development Goals, highlighting their potential to contribute to a more sustainable and equitable future. Each section would delve into specific examples, current research, and future prospects related to the utilization nano-materials in achieving respective SDGs.

References

1. Downing, C. A., Portnoi, M. E.: Quantum confinement in Dirac-like nanostructures, pp 1–5 (2023). https://doi.org/10.1016/B978-0-323-90800-9.00074-3
2. Guo, M., Liu, M., Hu, Z., Zhan, F., Wu, L.: Preparation and properties of a slow release NP compound fertilizer with superabsorbent and moisture preservation. J. Appl. Polym. Sci. **96**, 2132–2138 (2005). https://doi.org/10.1002/app.21140
3. Xia, Y., et al.: Recent advances in control technologies for non-point source pollution with nitrogen and phosphorous from agricultural runoff: current practices and future prospects. Appl. Biol. Chem. **63**(1), 8 (2020). https://doi.org/10.1186/s13765-020-0493-6
4. Shang, Y., Hasn, M.K., Ahammed, G.J., Li, M., Yin, H., Zhou, J.: Applications of nanotechnology in plant growth and crop protection: a review. Molecules **24**, 2558 (2019). https://doi.org/10.3390/molecules24142558
5. Subramanian, K. S., Manikandan, A., Thirunavukkarasu, M., Rahale, C. S.: Nano-fertilizers for Balanced Crop Nutrition BT—Nanotechnologies in Food and Agriculture. In: Rai, M., Ribeiro, C., Mattoso, L., Duran, N., Eds. Cham: Springer International Publishing, pp. 69–80 (2015)
6. Torney, F., Trewyn, B.G., Lin, V.S.-Y., Wang, K.: Mesoporous silica nanoparticles deliver DNA and chemicals into plants. Nat. Nanotechnol. **2**(5), 295–300 (2007). https://doi.org/10.1038/nnano.2007.108
7. Hegab, R., Kotp, Y.H., Eissa, D.: Using nanotechnology for enhancing phosphorus fertilizer use efficiency of peanut bean grown in sandy soils. J. Adv. Pharm. Educ. Res. **8**, 59–67 (2018)
8. Dhaliwal, S.S., Naresh, R.K., Mandal, A., Singh, R., Dhaliwal, M.K.: Dynamics and transformations of micronutrients in agricultural soils as influenced by organic matter build-up: a review. Environ. Sustain. Indicators **1–2**, 100007 (2019). ISSN 2665-9727, https://doi.org/10.1016/j.indic.2019.100007
9. Deepak, V.K., Sapan, P., Kalyan, K.S.: Effects of nanoparticles on seed germination, growth, phytotoxicity and crop improvement. Agric. Rev. **42**(1), 1–11 (2020). https://doi.org/10.18805/ag.R-1964
10. P´erez-De-Luque, A.: Interaction of nanomaterials with plants: what do we need for real applications in agriculture? Front. Environ. Sci. **5**, 12 (2017). https://doi.org/10.3389/fenvs.2017.00012
11. The State of Site-Specific Management for Agriculture. (1997). https://doi.org/10.2134/1997.stateofsitespecific Pierce, F. J., & Sadler, E. J. (1997). The State of Site-Specific Management for Agriculture. (F. J. Pierce & E. J. Sadler, Eds.) The State of Site-Specific Management for Agriculture. American Society of Agronomy, Crop Science Society of America, Soil Science Society of America. Definition and interpretation of potential management zones in Australia B. M. Whelan and Alex McBratney
12. Reyns, P., Missotten, B., Ramon, H., Baerdemaeker, J.D.: A review of combine sensors for precision farming. Precision Agric. **3**(2), 169–182 (2002). https://doi.org/10.1023/a:1013823603735

13. Mcbratney, A., Whelan, B., Ancev, T., Bouma, J.: Future directions of precision agriculture. Precision Agric. **6**(1), 7–23 (2005). https://doi.org/10.1007/s11119-005-0681-8
14. Admin. (2018, June 15). Important tools to succeed in precision farming. Retrieved May 22, 2020, from https://precisionagricultu.re/important-toolsto-succeed-in-precision-farming/
15. Chen, K.Y., Janz, K.F., Zhu, W., Brychta, R.J. (2012) Redefining the roles of sensors in objective physical activity monitoring. Med Sci Sports Exercise 44. https://doi.org/10.1249/mss.0b013e3182399bc8
16. Nanosensors applications in agriculture and food industry Enisa Omanović-Mikličanina , Mirjana Maksimovićb Donohoe, J., Helmich, R., Mack, & National Institute of Food and Agriculture. (2019, November 15). New Nanosensor Detects Microscopic Contaminants in Water. Retrieved May 22, 2020, from https://www.usda.gov/media/blog/2019/11/15/new-nanosensor-detects-microscopiccontaminants-water
17. Nanosensors applied to plant leaves warn of water shortage. (2017, November 08). Retrieved May 22, 2020, from https://www.nanowerk.com/nanotechnologynews/newsid=48578.php
18. Nanosensor can alert a smartphone when plants are stressed. (2020, April 15). Retrieved May 22, 2020, from http://www.sciencedaily.com/releases/2020/04/200415133512.html
19. Guan, H., Chi, D., Yu, J., Li, X.: A novel photodegradable insecticide: preparation, characterization and properties evaluation of nano-Imidacloprid. Pestic. Biochem. Physiol. **92**(2), 83–91 (2008). https://doi.org/10.1016/j.pestbp.2008.06.008
20. Shang, Y., Hasan, M.K., Ahammed, G.J., Li, M., Yin, H., Zhou, J.: Applications of nanotechnology in plant growth and crop protection: a review. Molecules **24**(14), 2558 (2019)
21. Giannousi, K., Avramidis, I., DendrinouSamara, C.: Synthesis, characterization and evaluation of copper based nanoparticles as agrochemicals against *Phytophthora infestans*. RSC Adv. **3**(44), 21743 (2013). https://doi.org/10.1039/c3ra42118j
22. Horbach, R., Navarro-Quesada, A.R., Knogge, W., Deising, H.B.: When and how to kill a plant cell: infection strategies of plant pathogenic fungi. J. Plant Physiol. **168**, 51–62 (2011)
23. Surendra, K.D.: The new integrated pest management paradigm for the modern age. J. Integr. Pest. Manag. **10**(1), 1–9 (2019)
24. Huang, W., Wang, C., Duan, H., Bi, Y., Wu, D., Du, J., Yu, H.: Synergistic antifungal effect of biosynthesized silver nanoparticles combined with fungicides. Int. J. Agric. Biol. **20**, 1225–1229 (2018)
25. Khan, M.R., Rizvi, T.F.: Nanotechnology: scope and application in plant disease management. Plant Pathol. J. **13**, 214–231 (2014)
26. Patel, V.R., Agrawal, Y.K.: Nanosuspension: an approach to enhance solubility of drugs. J. Adv. Pharm. Technol. Res. **2**, 81–87 (2011)
27. Kah, M., Beulke, S., Tiede, K., Hofmann, T.: Nanopesticides: state of knowledge, environmental fate, and exposure modeling. Crit. Rev. Environ. Sci. Technol. (2013). https://doi.org/10.1080/10643389.2012.671750
28. Nuruzzaman, M., Rahman, M.M., Liu, Y., Naidu, R.: Nanoencapsulation, nano-guard for pesticides: a new window for safe application. J. Agric. Food Chem. **64**, 1447–1483 (2016). https://doi.org/10.1021/acs.jafc.5b05214
29. Whitesides, G.M.: The 'right' size in nanobiotechnology. Nat. Biotech. **21**, 1161–1165 (2003)
30. Lowe, C.R.: Nanobiotechnology: the fabrication and applications of chemical and biological nanostructures. Curr. Opin. Chem. Biol. **10**, 428–434 (2000)
31. Arayne, M.S., Sultana, N., Qureshi, F.: Review: nanoparticles in delivery of cardiovascular drugs. Pak. J. Pharm. Sci. **20**, 340–348 (2007)
32. Wang, L., Zhao, W., Tan, W.: Bioconjugated silica nanoparticles: development and applications. Nano Res. **1**, 99–115 (2008)
33. Nazarov, G.V., Galan, S.E., Nazarova, E.V., Karkishchenko, N.N., Muradov, M.M., Stepanov, V.A.: Nanosized forms of drugs (a review). Pharm. Chem. J. **43**, 163–170 (2009)
34. Arayne, M.S., Sultana, N., Sabah, N.: Fabrication of solid nanoparticles for drug delivery. Pak. J. Pharm. Sci. **20**, 251–259 (2007)
35. Bala, I., Hariharan, S., Kumar, M.N.: PLGA nanoparticles in drug delivery: the state of the art. Crit. Rev. Ther. Drug Carrier Syst. **21**, 387–422 (2004)

36. Gelperina, S., Kisich, K., Iseman, M.D., Heifets, L.: The potential advantages of nanoparticle drug delivery systems in chemotherapy of tuberculosis. Am. J. Respir. Crit. Care Med. **172**, 1487–1490 (2005)
37. Vauthier, C., Labarre, D., Ponchel, G.: Design aspects of poly(alkylcyanoacrylate) nanoparticles for drug delivery. J. Drug Target. **15**, 641 (2007)
38. Yamaguchi, Y., Igarashi, R.: Nanotechnology for therapy of type 2 diabetes. Nihon Rinsho **64**, 295–300 (2006)
39. Hu, J., Huang, W., Huang, S., ZhuGe, Q., Jin, K., Zhao, Y.:. Magnetically active Fe3O4 nanorods loaded with tissue plasminogen activator for enhanced thrombolysis. Nano Res. 9:2652–61 (2016) [Google Scholar]
40. Huang, L., Hu, J., Huang, S., Wang, B., Siaw-Debrah, F., Nyanzu, M.: Nanomaterial applications for neurological diseases and central nervous system injury. Prog Neurobiol. (2017) [PubMed]
41. Tran, S., Degiovanni, P.J., Piel, B., Rai, P.J.C., Medicine, T.: Cancer nanomedicine: a review of recent success in drug delivery. **6**, 44 (2017). [PMC free article] [PubMed]
42. Bozzuto, G., Molinari, A.: Liposomes as nanomedical devices. Int. J. Nanomed. **10**, 975–999 (2015)
43. Chou, H., Lin, H., Liu, J.M.: A tale of the two PEGylated liposomal doxorubicins. Oncol. Targets. Ther. **8**, 1719–1720 (2015)
44. Brennan, M.E., Coleman, J.N., Drury, A., Lahr, B., Kobayashi, T., Blau, W.J.: Nonlinear photoluminescence from van Hove singularities in multiwalled carbon nanotubes. Optics Lett. **28**, 266–8 (2003). [Google Scholar]
45. Kam, N.W., O'Connell, M., Wisdom, J.A., Dai, H.: Carbon nanotubes as multifunctional biological transporters and near-infrared agents for selective cancer cell destruction. In: Proceedings of the National Academy of Sciences of the United States of America, **102**, 11600–5 (2005). [PMC free article] [Google Scholar]
46. Burlaka, A., Lukin, S., Prylutska, S., Remeniak, O., Prylutskyy, Y., Shuba, M., et al.: Hyperthermic effect of multi-walled carbon nanotubes stimulated with near infrared irradiation for anticancer therapy: in vitro studies. Exp. Oncol. **32**, 48–50 (2010)
47. Cheng, Y., Xu, Z., Ma, M., Xu, T.: Dendrimers as drug carriers: applications in different routes of drug administration. J. Pharm. Sci. **97**, 123–143 (2008)
48. Duncan, R.: The dawning era of polymer therapeutics. Nature reviews Drug Discov. **2**, 347–60 (2003) [Google Scholar]
49. Miyata, K., Christie, R.J., Kataoka, K.J.R., Polymers, F.: Polymeric micelles for nano-scale drug delivery **71**, 227–234 (2011)
50. Blanco, E., Bey, E.A., Khemtong, C., Yang, S.G., Setti-Guthi, J., Chen, H., et al.: Beta-lapachone micellar nanotherapeutics for non-small cell lung cancer therapy. Can. Res. **70**, 3896–3904 (2010)
51. http://www.actionbioscience.org/education/uddin_chowdhury.html, May 2008
52. Lyshevski, S.E., et al.: New Nano Science, Engineering and Technology course at the RIT. In: Proc. ASEE Conf. Engineering on the Edge: Engineering in the New Century, Binghamton, NY, pp E.5.1–E.5.6 (2005)
53. Umar, K., Haque, M.M., Mir, N.A., Muneer, M.: Titanium dioxide-Mediated photocatalyzed mineralization of Two Selected organic pollutants in aqueous suspensions. J. Adv. Oxid. Technol. **16**, 252–260 (2013)
54. Umar, K., Ibrahim, M.N.M., Ahmad, A., Rafatullah, M.: Synthesis of Mn-Doped TiO2 by novel route and photocatalytic mineralization/intermediate studies of organic pollutants. Res. Chem. Intermediat. **45**, 2927–2945 (2019)
55. Baruah, S., Khan, M.N., Dutta, J.: Perspectives and applications of nanotechnology in water treatment. Environ. Chem. Lett. **14**, 1–14 (2016)
56. Wu, Y., Pang, H., Liu, Y., Wang, X., Yu, S., Fu, D., Chen, J., Wang, X.: Environmental remediation of heavy metal ions by novel-Nanomaterials: A review. Environ. Pollut. **246**, 608–620 (2019)

57. Daer, S., Kharraz, J., Giwa, A., Hasan, S.W.: Recent applications of nanomaterials in water desalination: a critical review and future opportunities. Desalination **367**, 37–48 (2015)
58. Yaqoob, A.A., Ibrahim, M.N.M.: A review article of nanoparticles; synthetic approaches and wastewater treatment methods. Int. Res. J. Eng. Technol. **6**, 1–7 (2019)
59. Tang, W.W., Zeng, G.M., Gong, J.L.: Impact of humic/fulvic acid on the removal of heavy metals from aqueous solutions using nanomaterials: a review. Sci. Total. Environ. **468**, 1014–1027 (2014)
60. Mir, N.A., Haque, M.M., Khan, A., Umar, K., Muneer, M., Vijayalakshmi, S.: Semiconductor mediated photocatalysed reaction of two selected organic compounds in aqueous suspensions of Titanium dioxide. J. Adv. Oxid. Technol. **15**, 380–391 (2012). Water **12**, 495 (2020) 21 of 30
61. Umar, K.: Water contamination by organic-pollutants: TiO2 photocatalysis. In: Oves, M., Khan, M.Z., Ismail, I.M.I., Eds.: Modern age environmental problem and remediation. Springer Nature: Basel, Switzerland, pp. 95–109 (2018)
62. Kalhapure, R.S., Sonawane, S.J., Sikwal, D.R.: Solid lipid nanoparticles of clotrimazole silver complex: an efficient nano antibacterial against Staphylococcus aureus and MRSA. Colloid Surf. B **136**, 651–658 (2015)
63. Fang, X., Li, J., Li, X., Pan, S., Zhang, X., Sun, X., Han, J.S.W., Wang, L.: Internal pore decoration with polydopamine nanoparticle on polymeric ultrafiltration membrane for enhanced heavy metal removal. Chem. Eng. **314**, 38–49 (2017)
64. Sekoai, P.T., Ouma, C.N.M., Du Preez, S.P., Modisha, P., Engelbrecht, N., Bessarabov, D.G., Ghimire, A.: Application of nanoparticles in biofuels: an overview. Fuel **237**, 380–397 (2019)
65. Sherman, J. Nanoparticulate Titanium Dioxide Coatings, and Processes for the Production and Use Thereof. U.S. Patent No, 6653356B2, 25 November 2003.
66. Ali, I., Ghamdi, K.A., Wadaani, F.T.A.: Advances in iridium nano catalyst preparation, characterization and applications. J. Mol. Liq. **280**, 274–284 (2019)
67. Bhanvase, B.A., Shende, T.P., Sonawane, S.H.: A review on grapheme-TiO2 and doped grapheme-TiO2 nanocomposite photocatalyst for water and wastewater treatment. Environ. Technol. Rev. **6**, 1–14 (2017)
68. Tahir, M.B., Kiran, H., Iqbal, T.: The detoxification of heavy metals from aqueous environment using nano-photocatalysis approach: a review. Environ. Sci. Pollut. Res. **26**, 10515–10528 (2019)
69. Ciambelli, P., La Guardia, G., Vitale, L.: Nanotechnology for green materials and processes. Stud. Surf. Sci. Catal. **179**, 97–116 (2019)
70. Rajabi, H.R., Shahrezaei, F., Farsi, M.: Zinc sulfide quantum dots as powerful and efficient nanophotocatalysts for the removal of industrial pollutant. J. Mater. Sci. Mater. Electron. **27**, 9297–9305 (2016)
71. Jurado-Sánchez, B., Wang, J.: Micromotors for environmental applications: a review. Environ. Sci. Nano **5**, 1530–1544 (2018)
72. Eskandarloo, H., Kierulf, A., Abbaspourrad, A.: Nano-and micromotors for cleaning polluted waters: focused review on pollutant removal mechanisms. Nanoscale **9**, 13850–13863 (2017)
73. Hirata, K., Watanabe, H., Kubo, W.: Nanomembranes as a substrate for ultra-thin lightweight devices. Thin Solid Film. **676**, 8–11 (2019)
74. Rashidi, H.R., Sulaiman, N.M.N., Hashim, N.A., Hassan, C.R.C., Ramli, M.R.: Synthetic reactive dye wastewater treatment by using nano-Membrane filtration. Desalin. Water Treat. **55**, 86–95 (2015)
75. Bello˘n, T., Polezhaev, P., Vobecká, L., Slouka, Z.: Fouling of a heterogeneous anion-Exchange membrane and single anion-Exchange resin particle by ssdna manifests differently. J. Membr. Sci. **572**, 619–631 (2019)
76. Naeem, F., Naeem, S., Zhao, Z., Shu, G.Q., Zhang, J., Mei, Y., Huang, G.S.: Atomic layer deposition synthesized ZnO nanomembranes: a facile route towards stable supercapacitor electrode for high capacitance. J. Power Source **451**, 227740 (2020)
77. Yu, L., Ruan, S., Xu, X., Zou, R., Hu, J.: Review One-Dimensional nanomaterial-Assembled macroscopic membranes for water treatment. Nano Today **17**, 79–95 (2017)

78. Khajeh, M., Laurent, S., Dastafkan, K.: Nanoadsorbents: Classification, preparation, and applications (with emphasis on aqueous media). Chem. Rev. **113**, 7728–7768 (2013)
79. Zhang, Y.H., Hu, C.Z., Liu, F., Yuan, Y., Wu, H., Li, A.: Effects of ionic strength on removal of toxic pollutants from aqueous media with multifarious adsorbents: a review. Sci. Total. Environ. **646**, 265–279 (2019)
80. US-EPA, Risk assessment, management and communication of drinking water contamination, Washington, DC (1989)
81. US-EPA, Final 2014 Effluent Guidelines Program Plan and 2014 Annual Effluent Guidelines Review Report, Report EPA-HQ-OW-2014-0170 (2014)
82. Su, S., Wu, W., Gao, J., Lu, J., Fan, C.: J. Mater. Chem. **22**, 18101–18110 (2012)
83. Conibeer, G.: Third-generation photovoltaics. Mater. Today **10**, 42–50 (2007)
84. Bailey, R.L.: A proposed new concept for a solar-energy converter. J. Eng. Power **94**, 73–77 (1972)
85. Corkish, R., Green, M.A., Puzzer, T.: Solar energy collection by antennas. Sol Energy J. **73**, 395–401 (2002)
86. Miskovsky, N.M., Cutler, P.H., Mayer, A., Weiss, B.L., Willis, B., Sullivan, T.E., Lerner, P.B.: Nanoscale devices for rectification of high frequency radiation from the infrared through the visible: a new approach. J. Nanotechnol. **2012**, 512379 (2012)
87. Nozik, A.: Quantum dot solar cells. Physica E **14**, 115–120 (2002)
88. Ellingson, R.J., Beard, M.C., Johnson, J.C., Yu, P., Micic, O.I., Nozik, A.J., Shabaev, A., Efros, A.L.: Highly efficient multiple exciton generation in colloidal PbSe and PbS quantum dots. Nano Lett. **5**, 865–871 (2005)
89. Nozik, A.J.: Multiple exciton generation in semiconductor quantum dots. Chem. Phys. Lett. **457**, 3–11 (2008)
90. McGuire, J.A., Joo, J., Pietryga, J.M., SchallerKlimov, R.D.: New aspects of carrier multiplication in semiconductor nanocrystals. Acc. Chem. Res. **41**, 1810–1819 (2008)
91. Beard, M.C., Ellingson, R.J.: Multiple exciton generation in semiconductor nanocrystals: toward efficient solar energy conversion *Laser Photon.* Rev. **2**, 377–399 (2008)
92. Nair, G., Geyer, S.M., Chang, L.Y., Bawendi, M.G.: Carrier multiplication yields in PbS and PbSe nanocrystals measured by transient photoluminescnece. Phys. Rev. B **78**, 125325 (2008)
93. Ji, M., Park, S., Connor, S.T., Mokari, T., Cui, Y., Gaffney, K.J.: Efficient multiple exciton generation observed in colloidal PbSe quantum dots with temporally and spectrally resolved intraband excitation. Nano Lett. **9**, 1217–1222 (2009)
94. Semonin, O.E., Luther, J.M., Choi, S., Chen, H.-Y, Gao, J., Nozik, A.J., Beard, M.C.: Peak external photocurrent quantum efficiency exceeding 100% via MEG in a quantum dot solar cell. Science **334**, 1530–3 (2011)
95. Yifat, Y., Iluz, Z., Eitan, M., Friedler, I., Hanein, Y., Boag, A., Scheuer, J.: Quantifying the radiation efficiency of nano antennas. Appl. Phys. Lett. **100**, 111113 (2012)
96. Tratnyek, P.G., Johnson, R.L.: Nanotechnologies for environmental cleanup. Nano Today **1**, 44–48 (2006). https://doi.org/10.1016/S1748-0132(06)70048-2
97. Karn, B., Kuiken, T., Otto, M.: Nanotechnology and in situ remediation: a review of the benefits and potential risks. Environ. Health Perspect **117**, 1823–1831 (2009). https://doi.org/10.1289/ehp.0900793
98. Zhang, W.X.: Nanoscale iron particles for environmental remediation: an overview. J. Nanopart. Res. **5**, 323–332 (2003). https://doi.org/10.1023/A:1025520116015
99. Kamat, P.V., Meisel, D.: Nanoscience opportunities in environmental remediation. C R Chimie **6**, 999–1007 (2003). https://doi.org/10.1016/j.crci.2003.06.005
100. Mirghani, M.S.: Vanadium doped titania nanoparticles for photocatalytic removal of heavy metals from aqueous solutions. J. Exp. Nanosci. **16**(1), 51–61 (2021). https://doi.org/10.1080/17458080.2021.1886277
101. Hoa, L.T.M. (2018) Characterization of multi-walled carbon nanotubes functionalized by a mixture of HNO3/H2SO4. Diamond Relat. Mater. 89, pp. 43–51 (2018). ISSN 0925-9635, https://doi.org/10.1016/j.diamond.2018.08.008

102. Zhang, D., Wei, S., Kaila, C., Su, X., Wu, J., Karki, A.B., Guo, Z.: Carbon stabilized iron nanoparticles for environmental remediation. Nanoscale **2**, 917–919 (2010). https://doi.org/10.1039/C0NR00065E
103. Parkin, I.P., Palgrave, R.G.: Self-cleaning coatings. J. Mater. Chem. **15**, 1689 (2005). https://doi.org/10.1039/B412803F
104. Panov, A.G., Larson, R.G., Totah, N.I., Larsen, S.C., Grassian, V.H.: Photooxidation of toluene and *p*-xylene in cation-exchanged zeolites X, Y, ZSM-5, and beta: the role of zeolite physicochemical properties in product yield and selectivity. J. Phys. Chem. B **104**, 5706–5714 (2000). https://doi.org/10.1021/jp000831r
105. Mehndiratta, P., et al.: Environmental pollution and nanotechnology. Environ. Pollut. **2**(2), 49–58 (2013). https://doi.org/10.5539/ep.v2n2p49
106. Saurav: Application of nanotechnology in building material. Int. J. Eng. Res. Appl. **2**(5), 1077–1082 (2012)
107. Karthick Ganesh, V.: Nanotechnology in civil engineering. Eur. Sci. J. **8**(27), 96–109
108. Ge, Z.: Application of nanotechnology and nanomaterials in construction. In: First International conference on Construction in Developing Countries, August 4–5 (2008)
109. Dermatas, D., Mpouras, T., Papassiopi, N. et al. (in press) Adsorption of groundwater pollutants by iron nanomaterials. In: Litter, M., Quici, N., Meichtry, M., et al. (eds) Iron nanomaterials for water and soil treatment. Pan Stanford Publishing Pte. Ltd., Singapore
110. Dermatas, D., Mpouras, T., Panagiotakis, I.: 'Application of Nanotechnology for waste management: Challenges and limitations. Waste Manag Res SAGE Publications Ltd STM **36**(3), 197–199 (2018). https://doi.org/10.1177/0734242X18758820
111. Virkutyte, J., Varma, R.S.: Green Synthesis of Nanomaterials; Environmental Aspects in "Sustainable Nanotechnology and the Environ ment: Advances and Achievements. In: Shamim, N., Ed.; ACS Symposium Series, Vol. 1124. American Chemical Society: Washington, DC, 2013; Chapter 2, pp 11−39
112. Sustainable preparation of metal nanoparticles-methods and applications. Luque, R., Varma, R. S., Eds.; RSC Green Chemistry # 19; Royal Society Chemistry: Thomas Graham House, Cambridge, United Kingdom (2013)
113. Kumar, R., Mangalapuri, R., Hossein Ahmadi, M., et al.: The role of nanotechnology on post-combustion CO2 absorption in process industries. Int. J. Low-Carbon Technol. **15**(3), 361–367 (2020)
114. Alonso, A., Moral-Vico, J., Markeb, A.A., Busquets-Fité, M., Komilis, D., Puntes, V., Sánchez, A., Font, X.: Critical review of existing nanomaterial adsorbents to capture carbon dioxide and methane. Sci. Total. Environ. **595**, 51–62 (2017)
115. Ashley, M., Magiera, C., Ramidi, P., Blackburn, G., Scott, T.G., Gupta, R., Wilson, K., Ghosh, A., Biswas, A.: Nanomaterials and processes for carbon capture and conversion into useful by products for a sustainable energy future. Greenhouse Gases Sci. Tech. **2**(6), 419–444 (2012)
116. Ajith, M.P., Aswathi, M., Priyadarshini, E., Rajamani, P.: Recent innovations of nanotechnology in water treatment: A comprehensive review. Biores. Technol. **342**, 126000, (2021), ISSN 0960-8524. https://doi.org/10.1016/j.biortech.2021.126000

Smart Sustainable Materials: A Blueprint for a Better Tomorrow

Harshita Thakur, Simrandeep Kaur, Abhijeet Singh, Gajendra B. Singh, and Gaurav Mudgal

Abstract This chapter explores the potential of sustainable materials to address pressing environmental challenges and drive sustainable development. It highlights the diverse applications of sustainable materials across various sectors, including healthcare, fashion, technology, construction, and transportation. These applications are revolutionary and is greatly beneficial for humans by integrating sustainable materials into our daily lives, we can reduce our environmental footprint. It emphasizes the importance of education, awareness, and collaboration among stakeholders to accelerate the adoption of sustainable materials and transition towards a more sustainable society. As the use of conventional materials is utilizing high energy and cost, we require a robust strategy to mitigate these challenges. Smart-Sustainable materials are such strategies that targets wide variety of fields. This cutting-edge technology provides numerous benefits that can lead to a sustainable future. The transition to a sustainable future requires a multifaceted approach that addresses not only the technical aspects of sustainable materials but also the social, economic, and ethical implications. By understanding the broader context and actively working towards a more sustainable society, we can create a future where human prosperity and environmental well-being coexist harmoniously.

Keywords Sustainability · Smart · Nano · SDG · Ecosystem · Economic · Environment · Biodegradable

Harshita Thakur, Simrandeep Kaur and Abhijeet Singh have equal contribution.

H. Thakur · S. Kaur · A. Singh · G. B. Singh · G. Mudgal (✉)
University Institute of BioTechnology, Chandigarh University, Mohali, Punjab 140413, India
e-mail: gauravmdgl@gmail.com

G. B. Singh
e-mail: gajendrabiotech@gmail.com

G. Mudgal
Center for Waste Management and Renewable Energy, Saveetha Dental College and Hospitals, Saveetha Institute of Medical and Technical Sciences, Saveetha University, Chennai 600077, India

K. K. Kesari et al. (eds.), *Functionalized Cellulose Materials*, Engineering Materials,
https://doi.org/10.1007/978-3-031-76953-5_11

1 Introduction

In the era of smart sustainable materials there is a possibility of a world where buildings could adjust to the weather, packaging dissolves harmlessly in air. This chapter describe this transformative concept and explore its potential to revolutionise our lives. For centuries humans have relied on traditional materials like steel, concrete and plastic. These served well, but often at a cost. Their production consumes vast resources and energy, often leading behind a trail of pollution. Like, a concrete block in a building -strong and durable, but susceptible to cracking and prone to poor insulation. During hot summers, these structures trap heat, requiring significant energy expenditure for air-conditioning. Cracks left unattended can worsen over time, compromising the structural integrity and necessitating repairs or even demolition. This is where smart sustainable materials enter the scene. These are not the average bricks or mortar; these are engineered innovations that combine traditional materials with cutting edge technology. Now, if that same concrete block is infused with microscopic fibres that can self-heal minor cracks. This 'smart 'concrete can extend the lifespan of structures, reducing the need for frequent repairs and demolitions. The beauty of smart sustainable materials lies in their ability to address environmental and economic concerns simultaneously. We begin exploring a gist of how these can benefit in real time. Researchers are developing concrete infused with bacteria called Bacillus. These bacteria lie dormant until exposed to water and a specific nutrient cocktail. When a crack appears, water seeps in, activating the bacteria. As they grow, they deposit calcium carbonate, essentially filling the crack and mimicking the natural healing process of bone. Pilot programs conducted in the Netherlands have shown promising results with 'bio-concrete 'demonstrating the ability to selfheal minor cracks. This technology has the potential to revolutionise infrastructure maintenance, reducing costs, extending the lifespan of roads and bridges, minimising environmental impact by lowering the need for demolition and reconstruction [1]. There exists windows that automatically adjust to the outside temperature, keeping your home cool in the summer and warm in the winter. A thin layer of electrochemical material within the window pane changes its transparency when an electric current is applied. By adjusting the amount of sunlight and heat entering a building, electrochemical windows can significantly reduce energy consumption used for heating and air conditioning. Companies like Sage Glass are already commercially producing these 'smart windows. This technology can lead to substantial energy savings for homeowner and building managers, not only reducing utility Bills but also lowering a building 'a carbon footprint [2]. Other than this, mycelium, the root network of fungi, is being harnessed to create sustainable alternatives to traditional plastics. This fast-growing network of threadlike structures can be cultivated using agricultural waste and moulded into various shapes. The resulting "fungal foams" are not only strong and lightweight, but also decompose naturally after use. Companies like *Eco active design* are pioneering this technology, creating packaging materials, furniture components and even potential building blocks from mycelium. This shift towards biodegradable materials can significantly reduce plastic pollution and waste, creating more suitable for our planet [3].

The convergence of materials science and sustainability has precipitated a paradigm shift in technological innovation. Traditional materials, often characterized by their resource-intensive production and limited-service life, have proven inadequate in addressing the escalating environmental challenges of our era. This has necessitated a transition towards materials that not only fulfill our evolving needs but also minimize their ecological footprint. Smart sustainable materials, a confluence of cutting-edge technology and eco-conscious design, offer a promising solution to this imperative.

The journey towards sustainable materials has been marked by a historical evolution. While the industrial revolution ushered in an era of unprecedented technological advancement, it also laid the groundwork for environmental degradation. The unsustainable extraction and consumption of natural resources, coupled with the release of harmful pollutants, led to a growing awareness of the deleterious consequences of human activities on the planet. This realization catalyzed a shift in scientific and industrial thinking, paving the way for the development of materials that could mitigate these adverse effects.

Despite significant progress in sustainable materials research, numerous challenges persist. The increasing global population, coupled with rising standards of living, has placed an unprecedented burden on natural resources. Climate change, driven by anthropogenic greenhouse gas emissions [4], poses a formidable threat to ecosystems and human health. Furthermore, the disposal of traditional materials, often characterized by their limited recyclability and durability, contributes to waste accumulation and environmental pollution. Sustainable materials, characterized by their ability to adapt to changing environmental conditions and optimize resource utilization, offer a potential solution to these pressing challenges. These materials incorporate advanced functionalities, such as self-healing properties, energy harvesting capabilities, and biodegradability, to enhance their sustainability and performance. By integrating intelligent design principles and innovative technologies, smart sustainable materials can contribute to a more circular economy, reduce waste generation, and minimize environmental impact.

The development and adoption of smart sustainable materials require a multidisciplinary approach that involves collaboration among scientists, engineers, policymakers, and industry leaders. Continued research and innovation are essential to address the technical and economic barriers that hinder the widespread implementation of these materials. By investing in sustainable materials research and development, we can create a future where technological advancement and environmental stewardship coexist harmoniously [5].

Pollution acts as an extensive covering over the ecosystems, from darkened skies resulting from emissions of fossil fuel combustion, including CO2, SO2, NOx, and particulate matter, which helps destroy the quality of air and is a threat to human health. This also includes industrial effluent replete with heavy metals and noxious chemicals that poison waterways, kill aquatic life, and at the same time contaminate most drinking water sources. Chemical runoffs and haphazard disposal of wastes sow seeds for soil contamination, ruin agricultural lands, and endanger food security and further he toxins produced in the clothing and textile industries which ultimately harm environmental balance [6]. The tangent of climate change: CO2 and methane, emitted from the energy-consuming industry, raise global temperatures and destabilise weather patterns by boosting the greenhouse effect. Thirst for non-renewable sources of energy escalates climate volatility, moving in a cycle of environmental peril. The heartbeat of biodiversity weakens: broadening urban jungles and industrial juggernauts choke natural habitats, with some estimates suggesting thousands of species perish globally while their delicate ecosystems are thrown completely off balance. Overfishing, indiscriminate harvesting, strip ecosystems bare, rendering once vibrant habitats as stark and bare as possible, bereft of biodiversity. The proliferation of waste has been a testament to excess: it heaps up in mountains, overwhelming landfills and despoiling a landscape marred by noxious refuse. Indomitable, ubiquitous plastics choke rivers and seas, entrap marine life, and cast a long shadow over environmental stewardship.

Water is an increasingly threatened lifeline, under mounting pressures. The industrial juggernaut and thirsty agribusinesses guzzle dry water reserves; in arid regions of the developing world, it is the communities pitted against one another. Chemical contaminants seep into aquifers and rivers, and yesterday's pristine water sources become today's wellsprings of peril. Widespread deforestation and intensive agriculture strip away the protective cover of the earth and open up avenues for rampant soil erosion and desertification. An incessant chase after high yields leeches' soil of its nutrients to the extent of turning once-fertile lands into arid stretches devoid of vitality. Under toxic emissions and industrial by-products that poison the very air one breathes, the fuel for respiratory diseases, and a long shadow falls on public health. The workers, toiling under hazardous conditions, bear the brunt of chemical exposures and occupational hazards due to which their health is thus sacrificed at the altar of progress in industry.

Henc amidst the rich tapestry of 21st-century challenges today, it becomes essential that smart and sustainable materials align with an entangled web of environmental exigencies, economic imperatives, regulatory frameworks, and transformative strides in technological fields. These materials are exemplary cases of cutting -edge innovation in the discipline of material science and have huge transformative potentials across various sectors—something that makes them indisputably indispensable in forging a resilient, regenerative for our sustainable future (Fig. 1).

2 Revolutionising Healthcare with a Green Touch

The healthcare industry, while dedicated to human wellbeing, often faces criticism for its environmental footprint. From single-use plastics in medical equipment to hazardous chemicals in drug production, the industry generates a significant amount of waste. However, a revolution is brewing, fuelled by smart sustainable materials

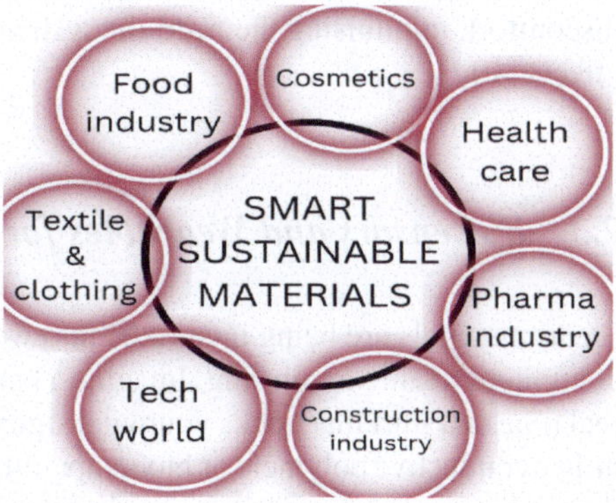

Fig. 1 Above are numerous fields that have been benefited by these smart sustainable materials

that offer a beneficial for both patients and the planet. Here's how these materials are transforming healthcare.

2.1 Biocompatible Implant and Tissue Engineering

What if the implant seamlessly integrates into the body and mimics like natural tissue? This vision is becoming a reality with biocompatible materials derived from natural sources like collagen and hydroxyapatite. These materials not only offer superior biocompatibility, reducing rejection risks but are also often biodegradable eliminating the need for future surgeries to remove implants [7]. For example, researchers at the Mayo Clinic are developing injectable scaffolds made from silk fibroin, a protein derived from silkworms. These scaffolds act as a temporary support structure, encouraging the growth of new cartilage and promoting natural healing in damaged knees. This eliminates the need for traditional metal or plastic implants, offering a less invasive and more sustainable solution [8].

2.2 Targeted Drug Delivery with Smart Polymers

Smart polymers are revolutionising drug delivery with their ability to release medication in a controlled and targeted manner. Imagine a pill that releases its contents only when it reaches the specific area of the body requiring treatment. These smart polymers can be designed to respond to specific triggers like changes in pH or temperature, ensuring the drug reaches its target size with minimal side effects [9]. Researchers at the University of California, San Francisco are developing dissolvable microneedles made from hyaluronic acid, naturally occurring sugar molecules in the body. These microneedles can deliver pain medication directly to the source of discomfort, minimising the need for systemic medications and their associated side effects [10].

2.3 *Biosensors and Wearables for Real-Time Monitoring:*

Smart materials are being integrated into wearable devices and biosensors to enable continuous health monitoring. Imagine a bandage that can monitor wound healing in real time or a shirt that tracks your vital signs throughout the day. These devices often utilise conductive polymers or biocompatible hydrogels to collect and transmit health data, allowing for early detection of potential problems and personalised treatments [11]. A group from the University of Manchester are developing smart wound dressings embedded with sensors that can monitor wound temperature, moisture levels and bacterial growth. This real time data allows healthcare professionals to optimise treatment plans for diabetic foot ulcer, a common and potentially debilitating complication [12].

2.4 *Biodegradable Medical Packaging and Devices*

The mountains of single -use plastics generated by healthcare facilities provide a significant environmental challenge. Sustainable materials like polylactic acid (PLA), derived from corn starch are paving the way from biodegradable alternatives. Imagine syringes, catheters, and packaging that decomposes after use reducing waste and environmental impact [13]. Companies like *Bio structures* are pioneering the use of PLA in sutures. These biodegradable sutures offer comparable strength and functionality to traditional sutures but decompose naturally after the wound heals eliminating the need for suture removal and offering a more sustainable option [14].

The development of sustainable materials in healthcare is rapidly evolving field. As research continues, we can expect more innovative applications to emerge. These advancements hold a potential to improve patients' outcomes, reduce healthcare costs and lessen the environmental impact of the industry. By embracing sustainable materials, the healthcare industry can create a healthier future for both patients and their planet (Fig. 2).

3 A Natural Glow: Up for the Cosmetic Industry

The world of cosmetics is all about enhancing beauty, but traditional practices often come at a cost to the environment. From ingredients derived from unsustainable sources to excessive packaging waste, the industry's environmental footprint is significant. However, a revolution is underway, fielded by smart sustainable materials that

Fig. 2 Sustainable medicine: leading the charge for green healthcare

promise a greener approach to beauty. Here's how these materials are transforming the cosmetic industry.

3.1 Biodegradable Packaging

Mountains of plastic packaging clog landfills and pollute ecosystems. What if lipstick in tubes is made from recycled paper or makeup compacts crafted from mushroom mycelium. Sustainable materials like bamboo, recycled paper, and even fungi-based bioplastics are offering ecofriendly alternatives to traditional plastic packaging. For instance, companies like *Lush* are pioneering the use of ocean waste plastic for their cosmetic packaging. This initiative not only reduces reliance on virgin plastic but also helps clean up our oceans [15].

3.2 Natural and Organic Ingredients

Consumers are increasingly seeking cosmetics formulated with nature and organic ingredients. Sustainable materials like plant-based oils, butter and waxes offer gentle and effective alternatives synthetic chemicals. Mica, a mineral commonly used in cosmetics for shimmer and glitter, can be unethically sourced. Companies like

Hourglass Cosmetics are developing alternatives made from plant-based cellulose, offering a cruelty -free and sustainable solution [16].

3.3 Biodegradable and Refillable Products

Single use makeup wipes and disposable applicators contribute significantly to waste. Sustainable materials like plant based cellulose and reusable applicators made from bamboo or recycled materials offer a greener alternative. Additionally, refillable compacts and containers encourage responsible consumption by allowing consumers to reuse packaging [17].

3.4 Microfiber Free Formula

Synthetic microfibers from facial scrubs and cleansers and up in our waterways harming marine life. Sustainable materials like natural exfoliating scrubs made from ground pumice or fruit pits offer eco-friendly alternatives. Companies like *Lush* utilise finely ground olive stones in their facial scrubs, providing effective e exfoliation without the environmental impacts of synthetic microfibers. The development of sustainable materials in cosmetics is a rapidly evolving field. As research continues, we can expect even more innovative applications to emerge. These advancements hold the potential to create a more sustainable beauty industry, cater to environmentally conscious consumers and offer safe and effective cosmetic products. By embracing sustainable materials, the cosmetic industry can redefine beauty while safeguarding the environment for future generations [18].

3.5 Bio Based Bottles from Algae

Microalgae, a tiny aquatic organism, is emerging as a sustainable source of bio plastic. Companies like *Albèa* are developing cosmetic bottles made from algae oil, promising a renewable and biodegradable alternative to traditional plastic [19].

3.6 Compostable Packaging from Wheat Straw

Wheat straw, a by -product of wheat production, offers a sustainable source of packing material. *Love beauty and planet* utilises wheat straw for their cosmetic tubes demonstrating the potential of agriculture waste to become valuable resources [20].

3.7 *Reusable Silicone Moulds for Makeup Palettes*

Instead of disposable plastic pallets, some brands like *KNC beauty* are exploring reusable silicone moulds. These can be filled with individual eyeshadow or blush pans minimising waste and encouraging a more customizable makeup experience. Companies are finding innovative ways to minimise waste by utilising ingredients derived from food waste. Brands like *Future cosmetics* incorporate apple and orange peels into their lipstick, offering vibrant and natural colours while reducing food waste [21] (Fig. 3).

4 Revolutionising the Food Industry

The food industry struggles with the challenges of balancing consumers' needs with environmental responsibility. Traditional practices often rely on excessive packaging, resource -intensive production methods and chemical preservatives, creating a significant ecological footprint. However, a wave of innovation is transforming the food landscape -sustainable materials are revolutionising the way we produce, package and consume food. Here's a deep dive into why these materials is crucial for the future of food. The first and foremost important is to maintain an extended shelf life of food. Certain sustainable materials offer improved preservation, such as, biodegradable films with controlled permeability. These films are designed to regulate the exchange of gases like oxygen and carbon dioxide. Companies like *Appel Sciences*

Fig. 3 Sustainable Shine: The Natural Revolution in Cosmetics

offer plant-based coatings for fruits and vegetables that extend shelf life by regulating respiration rates [22]. Next importance should purpose for reduced reliance on chemical preservatives. Some sustainable materials have inherent antimicrobial properties. Packaging infused with essential oils—essential oils like thyme or oregano have natural antimicrobial properties. Companies like *Green Wrap* offer food Wraps infused with essential oils to help control the growth of harmful bacteria on food surfaces potentially reducing the need for harmful chemicals [23]. Next requirement is for enhanced traceability and Safety. Smart packaging with embedded sensors can monitor food temperature during storage and transport. Companies like *Geotab* offer smart labels with temperature sensors that transmit real time data on food temperature throughout the supply chain, this data can help ensure food safety prevent outbreaks of foodborne illnesses and minimise spoilage [24]. Citrus growers in Florida are partnering up with companies like *Citrus Peel Power* to transform a waste product—orange peels into a valuable resource. These peels are processed into a slow-release fertiliser, rich in nutrients that nourishes citrus trees and reduces reliance on chemicals and fertilisers. Next requisite is smart irrigation systems. Such system would utilize sensors embedded in the soil to precisely deliver water where and when crop needs. Materials like peanut shell sensors optimise water usage in agriculture, a critical resource facing increasing pressure due to climate change and population growth. This can lead to significant water saving particularly in drought -prone regions. It promotes water conservation, a crucial step towards a more sustainable future. It also help farmers improve crop yields and resource utilisation, leading to greater food security [25] (Fig. 4).

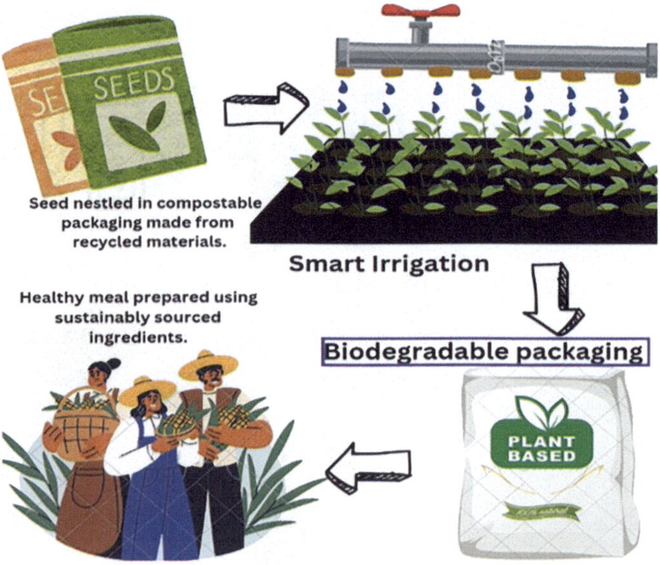

Fig. 4 Freshness redefined: revolutionizing food safety and sustainability

5 Textile and Clothing Industry

The textile and clothing industry, while a significant contributor to global GDP, faces a harsh reality. Its traditional practices leave a hefty environmental footprint. From excessive water usage in cotton production to the microfiber pollution caused by synthetic fibres. The need for change is undeniable. Here is how the sustainable materials provide a breakthrough:

5.1 Traditional Materials and Their Drawbacks

As a natural fibre crop, cotton cultivation demands vast amounts of water. Studies published in the Journal of cleaner Production (2020) reveal that conventional cotton production requires an average of 2,700 L of water per kilogram of fibre. This intensive water usage can strain resources, particularly in arid regions. Similarly, polyester a dominant synthetic fibre is derived from petroleum. Research published in Environmental science and Technology (2015) indicates that scientific fibre production contributes significantly to greenhouse gas emissions and micro plastic pollution. Micro plastics released during washing synthetic garments contaminate waterways and pose a threat to marine ecosystem [26].

5.2 Solution: Sustainable Materials

These offer a paradigm shift, addressing the environmental concerns associated with traditional fabrics.

Organic cotton—cultivated using organic farming practices, organic cotton minimises water consumption and reliance on synthetic pesticides and fertilisers. A 2017 study in the Journal of Sustainable Agriculture found that organic cotton production uses 71% less water and 62% less energy compared to conventional cotton [27]

Cellulose based fibres—Derived from wood pulp or other plant sources, these fibres offer a renewable and biodegradable alternative. Researchers found that that cellulose fibres possess good moisture management properties, making them suitable for comfortable and breathable clothing [26]

Mycelium based materials—The mycelium network of fungi can be cultivated into a leather like material. Various studies highlight the low environmental impact of mycelium production compared to animal leather. Mycelium requires minimal water and land and is naturally biodegradable.

Biomimicry—Scientists are exploring biomimetic materials that mimic natural structures for enhanced performance. For example, research published in Material science and Engineering details the development of super hydrophobic textiles

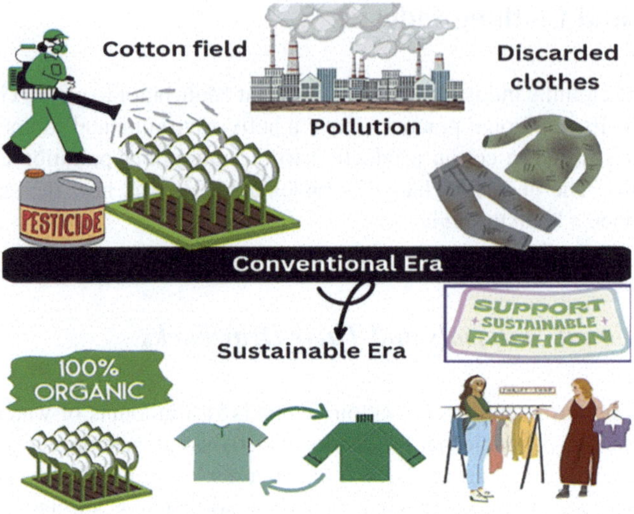

Fig. 5 **Fashion Meets Function:** Smart Sustainable Materials in Textiles

inspired by lotus leaves. These textiles repel water and dirt potentially reducing the need for frequent washing and associated water usage [28]

Self-Healing Materials—Researchers are developing polymers that can repair small tears or abrasions. A 2019 study in Nature showcased a self-healing polymer composite that could revolutionise garment longevity and reduce textile waste [12].

The transition to a sustainable textile industry requires a collaborative effort. The industry is moving towards closed-loop systems where used textiles are recycled into new fibres, minimising the reliance on traditional fibre. Advancements in material science aim to fully develop biodegradable garments that decompose naturally at the end of their lifespan. As consumers become increasingly informed about the environmental impacts of clothing the demand for sustainable fashion will continue to rise. By embracing sustainable materials and fostering innovation the textile and clothing industry can move toward a future with fashion aligns with environmental responsibility ensuring a healthier planet f0r our generation to come [29] (Fig. 5).

6 Pharmaceutical and Biomedical Avenues

The horizons of the biomedical and pharmaceutical industries are rapidly expanding with help of latest advanced green technologies the use of which has been rendered more appropriate with the young generation of smart sustainable materials—nanomaterials with characteristics defined at atomic and molecular levels. Some cutting-edge examples that showcase the transformative potential of this technology.

6.1 Self-Healing Hydrogels for Drug Delivery: Self-Healing Hydrogels for Drug Delivery

Researchers are working on self-healing hydrogels made from biodegradable polymers and nanofiller components, nano clays in this case. Hydrogels of this type can be loaded with drugs and administered at the site of the desired action. After that, the nano clay particles come in contact within the network of hydrogel which leads to the formation of dynamic covalent bonds and the ability to self-heal if the hydrogel is slightly damaged. This also enhances the life of this drug delivery system while reducing emissions and possible side effects [30].

6.2 Stimuli-Responsive Nanoparticles for Targeted Therapy: Stimuli-Responsive Nanoparticles for Targeted Therapy

Traditional chemotherapy is normally married to the systematic delivery of drugs whereby several side effects are realised. Scientists are working on what are called "smart" nanoparticles, that can deliver drugs only when certain conditions are met like the pH or temperature of the local environment. These nanoparticles can be synthesised from biocompatible polymers and the exterior part of these nanoparticles can be modified with targeting agents which will bind to receptors on the diseased cells. Upon these nanoparticles, reaching at the target site the pH conditions (e. g., acidic tumour) are altered and the drug is released directly to the cancerous cells and not harming the normal cells [31].

6.3 Biomimetic Nanorobots for Targeted Drug Delivery and Surgery

As discussed earlier, the totally new vision of nanotechnology in medicine is presented in the formation of biomimetic nanorobots, although this idea is still in development. These milli/micro-scale robots mimicking biological ones (e.g. bacteria) will contain on-board navigation tools, on-board sensors and control abilities along with their environment. Consider the situation in which biomimetic insta-Nanorobots are implanted into the affected zone of the body and move according to the cells' map towards a tumour, where it provides drugs to the affected cells or performs micro-surgery [32] (Fig. 6).

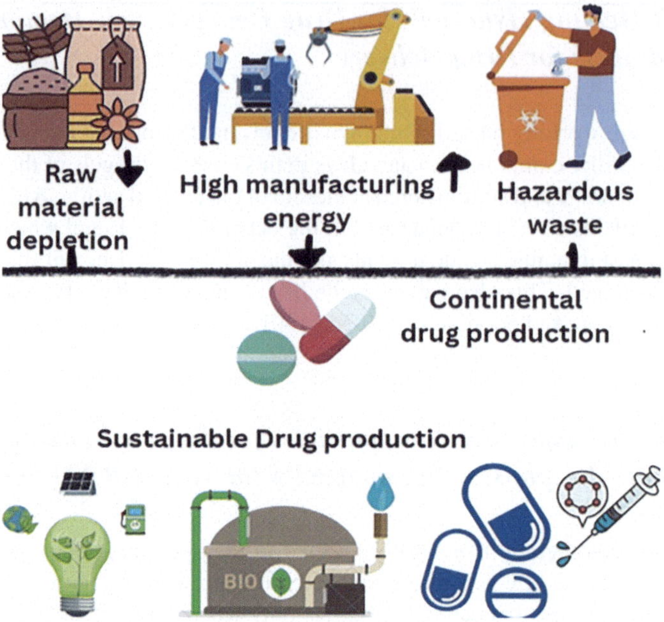

Fig. 6 **Smart biomaterials:** pioneering eco-friendly medical breakthroughs

7 Sustainable Materials: Revolutionising the Tech World

The tech industry is a force of innovation but it also has a big problem—its footprint. Traditional components are mined, hazardous and energy hungry. But a change is coming. Sustainable materials are changing how we design, build and use tech and it's good for the environment, safety and the bottom line.

7.1 Environmental Benefits

Reduced Resource Depletion: The demand for technology means pressure on finite resources. Sustainable materials offer multiple solutions. By using post-consumer and post-industrial recycled materials like metals from old devices, we reduce the need for virgin resource extraction. Life Cycle Assessments (LCAs) by the National Bureau of Standards and Technology (NBS) in 2021 show that using recycled aluminium in laptops can reduce environmental impact by a lot compared to virgin aluminium [33]. Renewable resources like bamboo or algae are eco-friendly alternatives to traditional plastics. Research publicahed in in Green Chemistry (2020) by Y. Wang et al. looks into bio-based polymers from algae for electronic device casings. These materials are biodegradable and have lower environmental impact than conventional plastics. Further, energy hungry manufacturing processes for tech components is a big contributor to greenhouse gas emissions. Sustainable materials can help into this. Energy Efficient Manufacturing: Processes that use renewable energy like solar or wind power, like *Dell* do in their manufacturing facilities, reduce the carbon footprint throughout the product life cycle. Other than this, compostable or recycled packaging for devices, like *Apple's iPhone* packaging made from recycled paper, reduces waste and energy consumption during production. A 2019 study in the Journal of Industrial Ecology shows the environmental benefits of switching to recycled paper packaging for electronic devices [34]. The growing mountain of e-waste lumps is a big environmental problem. Sustainable solutions offer a way forward into this. First is with durable and Long-Lasting Materials. Devices made with materials like scratch resistant sapphire glass displays last longer, reducing the need for frequent replacements and e-waste generation. A 2017 study in journal Resources, Conservation and Recycling by A. Bakker et al. looks at design for longevity in electronic products and the importance of durable materials. Further, materials designed for easy disassembly and recycling at end of life, such as modular components, can make a big difference in e-waste management. The European Union's Restriction of Hazardous Substances Directive (RoHS) encourages the use of recyclable materials and bans hazardous substances in devices [35].

7.2 Safety and Health Benefits

Some traditional materials contain lead or mercury. Sustainable alternatives have safer options. The health risks associated with the long-term usage of devices could be reduced by using nontoxic, compatible materials with human skin; for example, some bio-based polymers. A work published in Biomaterials in 2016 by P. Bajaj et al. presents research into the synthesis of biocompatible polymers for wearable electronic devices [36]. Conflict-Free Materials are a new concept. Source sustainably to ensure materials are not sourced through unethical means, some of which are associated with their mining. The CFSI is an industry-driven initiative to promote responsible sourcing of minerals used in electronic devices [37].

7.3 Economic Benefits

Sustainable materials can indeed maximise resource efficiency. Incorporating Life Cycle Costing (LCC) techniques throughout the entire product life cycle can recognize opportunities to cut down on waste and promote resource efficiency. The Society of Environmental and Resource Economics encourages the use of LCC in many sectors, including technology, through an initiative led by *Cost Savings*. While some of these sustainable materials are costlier from a short-term point of view, their advantages in the long run offset this. Manufacturing with renewable resources could

Fig. 7 Eco-engineered: smart materials powering tech evolution

provide cost-effective and sustainable production processes in the long run. A 2020 study by Environmental Science & Technology presents how there is an economic benefit associated with shifting to renewable sources of energy for manufacturing. There is a higher demand by consumers for green products. Brands that begin using them develop a differentiation in the marketplace (Fig. 7).

8 Construction Industry and Sustainable Materials

Construction is a globally significant industry in the development process and poses some of the most critical challenges through its dependence on conventional materials, engendering enormous environmental impacts. The landscape is changing due to innovation in this area. Within the construction industry specifically, sustainable materials are increasingly turning out to be innocuous alternatives that offer a wide array of benefits across the environmental, economic, and social spheres [38]. Here is a detailed look at the advantages, along with examples of real-world successes.

8.1 Environmental Benefits

Traditional materials, like steel and concrete, have a very high embodied energy. This is now being taken care of with sustainable substitutes. Mass Timber: Engineered wood products, like cross-laminated timber (CLT), require less embodied energy compared with concrete and steel. According to a 2019 study in Construction and Building Materials by A. Sartori et al., there is a comparison of the embodied energy of CLT with concrete and steel in different building applications. Recycled Materials: Reuse of existing material, like reclaimed wood and demolition debris, reduces the need for virgin resources and related energy consumption. According to J. Crawford, in an article about Resources, Conservation, and Recycling, 2018, in the case of construction, using recycled concrete aggregates in construction helps in saving the environment by assessing the life cycle [39]. Improved Energy Efficiency: Use of sustainable materials in energy-efficient buildings is possible because of the many reasons. Cellulose insulation or recycled cork, for example, offers better thermal insulation compared to traditional ones. A study published by P. del Coz Díaz et al., in 2020 in the journal Energy and Buildings [3], reviews different insulation materials in terms of their thermal performance and shows that cellulose insulation greatly reduces the energy consumption of the building [40]. Window advanced coatings or integrated blinds have potential for reducing either heat gain or loss, thus increasing building energy performance. Applied Energy 2017 by W. Li et al. evaluates the potential of changeable window technologies to save energy; their contribution in terms of a decrease in building use of energy is shown [41]. Traditional or conventional construction generates a tremendous amount of C&D waste. Sustainable materials also offer ways to reduce the same, including: i) Prefabricated Construction: Prefabrication of building components at off-site places minimises the generation of waste at a construction site. A 2019 study published in Journal of Cleaner Production on the role of prefabrication in reducing construction waste by M. Khan et al. explores these benefits [42]; and ii) Modular Construction: For modular buildings constructed from prefabricated units, the units can be dismantled and the material reprocessed at the end of their service life, significantly reducing wastes. Research by S. Aye et al. published in 2017 by Sustainable Cities and Society analysed environmental sustainability-related benefits of modular construction techniques against conventional methods.

8.2 Economic Benefits

Sustainable materials can enhance resource efficiency. Life Cycle Cost Analysis: LCCA methodologies practised across the life cycle can calculate Reduction Opportunities of material use and optimise resources. The National Institute of Standards and Technology, NIST, promotes the use of LCCA in construction processes to make effective decisions. Reduced Operational Costs: This is one area where sustainable materials can play their part in reducing.

The improvement in thermal insulation properties, due to the use of sustainable materials, shall help reduce energy for heating and cooling of buildings. A 2018 article by M. A. Santamouris et al. in Building and Environment explains the interrelation between building materials and energy efficiency, emphasising the potential cost savings by making sustainable choices [43]. The demand for sustainable construction is growing. Green Building Rating Systems: Systems, like that of LEED—Leadership in Energy and Environmental Design—encourage the use of sustainable materials, hence creating an increasing market for such innovative solutions. The U.S. Green Building Council administers LEED certification, which promotes the cause of sustainability in construction [44]. Improved IAQ— Some green materials, particularly natural fibres or low-emitting material varieties, can contribute to the enhancement of a healthier indoor environment. Sustainable materials with reduced VOC emissions can minimise indoor air pollution and improve occupant health. A study on the influence of building materials in indoor air quality was published in Indoor Air in 2016 by D. W. Li et al., stressing that low-VOC materials are very important [45]. Sustainable materials can result in safer construction.

The transformative power of sustainable materials does not stop here. The list of examples keeps growing with such things as automotive parts made from recycled plastics and sustainable building materials, which enable minimal consumption of energy. Only further research and more aggressive development can predict more runaway breakthroughs in the coming years.

Mass adoption of sustainable materials presents a strong opportunity to decouple growth from environmental degradation. This innovation can create a world where industries thrive but at the same time preserve the planet for subsequent generations. Here are the key takeaways:

1. Environmental Benefits: Sustainable Materials Offer Substantial Environmental Benefits through Reduced Resource Depletion, Lower Carbon Footprints, and Minimise Waste Generation across Multiple Industries.
2. Viability in Economics: Because of this, such materials can help in optimising the use of resources for reduced resource use, reduced costs, and increased life cycle economics. Additionally, a growing market demand exists for these kinds of products, hence opening up opportunities in new markets.
3. Societal Impacts: Sustainable materials promote public health through safer health practices, reduced toxicity, and resource efficiency in the food industry.

A clean revolution in sustainable materials would, therefore, obviously not be about technological improvement but a paradigm shift. It is the way our relationship with the environment has been envisioned, one that is underpinned by a belief in a future where human ingenuity and innovation coexist harmoniously with the health of the planet. Collaborative work from all sectors and industries will go a long way into tapping the potential of sustainable materials for creating a better future for all (Fig. 8).

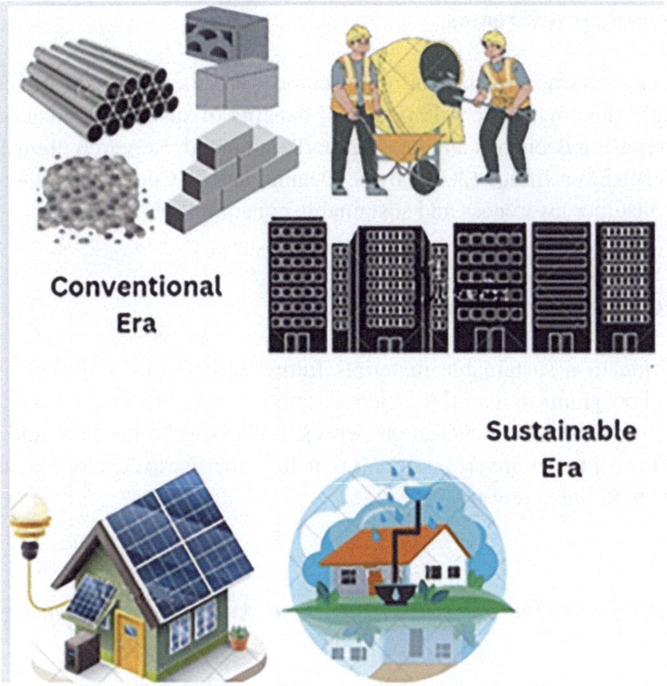

Fig. 8 Eco-empowered construction: building a sustainable legacy

9 Education and Awareness: The Cornerstone of a Sustainable Material Future

The burgeoning field of sustainable materials holds immense promise for environmental and economic well-being. However, for this potential to be fully realised, a well-informed public and a skilled workforce are essential. Here, we explore the crucial role of education and awareness in fostering a sustainable materials future.

9.1 Knowledge Gaps and the Need for Education

9.1.1 Scientific Literacy

A fundamental understanding of the environmental impact of conventional materials and the benefits of sustainable alternatives is crucial. Educational initiatives can bridge this knowledge gap by incorporating life cycle assessments (LCAs) and environmental impact studies into relevant curriculums across various disciplines, from engineering to material science.

9.1.2 Consumer Awareness

Empowering consumers to make informed choices is critical. Educational campaigns can highlight the environmental and social benefits of sustainable materials, influencing purchasing decisions and driving market demand. Research published in the Journal of Business Ethics (2020) by S. Ozanne and T. Cuddihy explores the link between consumer awareness and sustainable consumption practices.

9.1.3 Workforce Development

The transition to a sustainable materials future necessitates a skilled workforce. Educational programs in materials science, engineering, and design should integrate knowledge of sustainable material properties, processing techniques, and life cycle analysis. This equips future professionals with the expertise to develop and implement sustainable material solutions.

9.2 Plans for Effective Education and Awareness Campaigns

Interdisciplinary Approach: Scientists, engineers, educators, and communication experts should work together. Picture a team of experts developing engaging online learning modules about sustainable materials, replete with interactive quizzes and real-world examples. Digital Learning Platforms could create appealing learning experiences with wider audiences through online platforms or other interactive tools. Consider for a moment an app where consumers can scan product barcodes to learn about the materials used and their impact on the environment. Case studies and real-world examples can help illustrate the success stories of sustainable materials, inspiring and motivating stakeholders to adapt to change. Take the case of companies like Patagonia, using recycled raw material in their attires, or Dell, in some of its products, using ocean plastic waste. These examples go on to prove that the use of sustainable materials is both viable and is increasingly being practised in various industries [46]. Government policies and incentives can very strongly fast-track any progress. Just imagine the tax breaks available to companies that can make use of sustainable materials, or subsidies for research and development in this field. They motivate innovation to make these sustainable materials more accessible.

9.3 Current Scenario and Emerging Initiatives

International Awareness: The urgency of sustainability has acquired international awareness. One of the many targets categorised within the United Nations' SDGs pertains to responsible consumption and production, underscoring the need for sustainable materials as described under SDG12. Educational Initiatives: The universities/research institutions have, of late, offered courses and research programs related to sustainable materials. For example, MIT launched the Materials Research Laboratory Sustainability Initiative to enhance research and education in that area [47]. Leading companies collaborate with higher educational institutions to finally develop curricula in sustainable materials and drive innovation. An example is how BASF collaborates with the New Plastics Economy initiative of the Ellen MacArthur Foundation to encourage responsible plastic use and developing sustainable alternatives.

10 Conclusion

Our earth, a rich tape of life and abundant with species, is facing an evident fact. The very discoveries that have driven our advancement, the concrete, steel and plastic that form our cities and industries, are seen as a danger to its future. However, amidst this challenge, a beacon of hope emerges: a journey through sustainable materials. This chapter is not simply an analysis of these materials it is an initiative to join a crucial quest in prospecting a future of development and sustainability. The topics include the role of sustainability materials in various sectors of the economy. In healthcare, products such as biodegradable stents and collagen scaffolds [48] to introduce biocompatibility into medical practice by affording minimally invasive alternatives while assisting tissue regeneration to form the basis of a future wherein the medical science is complementary to the patient's welfare. The fashion industry that has been accused of polluting our environment is undergoing a revival through the fashion of organic cotton and recycled material. Picture elastic bands that are both stylish and do not use or require excessive amounts of water in their making or require the use of dangerous chemicals. This revolution is not exclusive to the fashion industry; it is applicable in food industry packaging as well Biodegradable packaging and plant-based coatings reduce waste and increase the shelf life of food products thus reducing wastage in the food chain [49]. Technology which is at the heart of innovation is also aligning itself to this sustainable future. Used parts are being incorporated in the gadgets, and green power sources are powering the product manufacturing. This means that, as a result, electronic devices are more environmentally friendly and troubling for the climate [50]. A similar fashion, the pharma business is also looking forward to sustainable packaging and delivery system solutions that will make practices more sustainable in general throughout the life cycle of a drug. In construction, sustainable utilising materials and pre-fabricated parts are transforming the structures of buildings and they do not waste much of the earth's resources. These are only few applications about smart about sustainable materials it encompasses wide range. There are many more aspects like a system of transportation in which cars require self-healing composites making them almost free of maintenance and with greatly increased durability or the concepts that allow buildings to adapt to the external environment and control the temperature within their structure using materials inspired by living organisms. Such opportunities, which earlier could be discussed only in the framework of sci-fi movies, today are quite tangible due to constant advancements achieved in the course of research and new upcoming technologies. To bring these materials at a large scale it requires people to unite which will focus more on the education process and making people aware of the problem. Provider awareness and patient awareness are two different things in a layman's sense. Lively discussions for self-sufficiency and production of other items through recycled and eco-friendly resources with a feeling of the ability to make a change could lead to widespread awareness. The adoption of the ideas of cooperation between institutions, industry giants, and policy makers, which will plant the seeds of development and knowledge sharing. In this way, through educating the population and developing skilled workers, people will be able to learn how to create sustainable materials and improve the world by turning this idea into practice.

There is no time like the present to do something. The world is at a crucial juncture where great decisions have to be made with regard to the future of the earth. It is high time to welcome sustainable materiality not only as a requirement but as a potentiality. Therefore, let's take advantage of them to establish a world where creativity and care for nature. will be the central values. This is not just a chapter in a book; it's a rallying cry; an opportunity to embrace the concept of sustainable materials and to be part of the generation that forms the future as interesting as the material that it will be made of. This odyssey invites action and calls for a new story to be written—sustainable is no longer the exception but the rule.

References:

1. a, M.B., b, C.H.-B.; ; a, A.A.; ; d, B.L.F.; ; c, I.J.-R.; ; et al. Advancements in bacteria based self-healing concrete and the promise of modelling **358** (2022). https://doi.org/10.1016/j.conbuildmat.2022.129412
2. Brzezicki, M.: A systematic review of the most recent concepts in smart windows technologies with a focus on electrochromics **13**, 9604 (2021)
3. a, S.M.; ; a, M.W.U.; ; b, M.U.-I.; ; a, O.M.A.; ; a, G.Y. Synthesis and applications of fungal mycelium-based advanced functional materials. **2021**, *6*, https://doi.org/10.1016/j.jobab.2021.01.001.
4. Wang, J., Chen, Y., Liao, W., He, G., Tett, S.F.B., Yan, Z., Zhai, P., Feng, J., Ma, W., Huang, C., et al.: Anthropogenic emissions and urbanization increase risk of compound hot extremes in cities. Nat. Clim. Chang. **11**, 1084–1089 (2021). https://doi.org/10.1038/s41558-021-01196-2
5. Mohanty, A.K., Wu, F., Mincheva, R., Hakkarainen, M., Raquez, J.-M., Mielewski, D.F., Narayan, R., Netravali, A.N., Misra, M.: Sustainable polymers. Nat Rev Method Primers **2**, 46 (2022). https://doi.org/10.1038/s43586-022-00124-8
6. Muthu, B.S.S.: Assessing the environmental impact of textiles and the clothing. Matthew deans Elsevier (2020)
7. Han, W.B, S.M.Y., Rajaram, K., Hwang, S.-W.: Materials and fabrication strategies for biocompatible and biodegradable conductive polymer composites toward bio-integrated electronic systems (2022). https://doi.org/10.1002/adsu.202100075
8. Teo, A.K.K., N.H.J.N.: Ho, B.K.: Innovations in bio-engineering and cell-based approaches to address immunological challenges in islet transplantation.. Frontiers 15 (2024). https://doi.org/10.3389/fimmu.2024.1375177

9. Van Gheluwe, L., Chourpa, I., Gaigne, C., Munnier, E.: Polymer-based smart drug delivery systems for skin application and demonstration of stimuli-responsiveness **13**, 1285 (2021)
10. Jin, J.H.J.S.G.: Microneedle for transdermal drug delivery: current trends and fabrication. J. Pharm. Invest. 51 (2021). https://doi.org/10.1007/s40005-021-00512-4
11. Hong, Y., Lin, Z., Yang, Y., Jiang, T., Shang, J., Luo, Z.: Biocompatible conductive hydrogels: applications in the field of biomedicine **23**, 4578 (2022)
12. Manu Goyal, N.D.R., Rajbhandari, S., Ahmad, N., Wang, C., Yap, M.H.: Recognition of ischaemia and infection in diabetic foot ulcers: dataset and techniques. Elsevier 117 (**2020**). https://doi.org/10.1016/j.compbiomed.2020.103616
13. Lili Sha, Z.C., Chen, Z., Zhang, A., Yang, Z.: Polylactic Acid Based Nanocomposites: Promising Safe and Biodegradable. Mater Biomed. Field (2016). https://doi.org/10.1155/2016/6869154
14. Alhulaybi, Z.A.J.C.: Fabrication and characterization of poly (lactic acid)-based biopolymer for surgical sutures, 7, 98 (2023). https://doi.org/10.3390/chemengineering7050098
15. De, S.K., Kawdia, P., Gupta, D., Pragya, N.J.M.o.E.Q.A.I.J. :Packaging plastic waste management in the cosmetic industry 34, 820–842 (2023). https://doi.org/10.1108/MEQ-03-2022-0053
16. Quelhas, A., A.C.T.: Mimicking natural-colored photonic structures with cellulose-based materials (2023). https://doi.org/10.3390/cryst13071010
17. Morganti, P., Morganti, G., Gagliardini, A., Lohani, A.: From cosmetics to innovative cosmeceuticals—non-woven tissues as new biodegradable carriers (2021) 8, 65
18. Wering, D.v.: Plastic: the hidden beauty ingredient (2022)
19. Ruibin Wang, Y.F., Li, D., Li, K., Yan, Y.: Towards the sustainable production of biomass-derived materials with smart functionality: a tutorial review. Green Chem. (2024). https://doi.org/10.1039/D4GC01771D
20. The ultimate guide to wheat straw plastic (2024)
21. Giles, C.: Cosmetics and sustainability (2020)
22. Pusik, L., Pusik, V., Bondarenko, V., Didukh, N., Semenenko, I.: Effectiveness of protective edible coatings to extend shelf lives of fresh fruits and vegetables (2024). https://repo.btu.kharkov.ua//handle/123456789/49080
23. Perumal, A.B., Huang, L. , Nambiar, R.B. , He, Y. , Li, X., Sellamuthu, P.S.J.F.c: Application of essential oils in packaging films for the preservation of fruits and vegetables: a review **375**, 131810 (2022). https://doi.org/10.1016/j.foodchem.2021.131810
24. Hanie Yousefi, H.-M.S., Imani, S.M., Alkhaldi, K., Filipe, C.D.M., Dida, T.F.: Intelligent food packaging: a review of smart sensing technologies for monitoring food quality. ACS Publication (2019). https://doi.org/10.1021/acssensors.9b00440
25. Casadei, S., Peppoloni, F., Ventura, F., Teodorescu, R., Dunea, D., Petrescu, N.: Application of smart irrigation systems for water conservation in Italian farms. Environ. Sci. Pollut. Res. **28**, 26488–26499 (2021). https://doi.org/10.1007/s11356-021-12524-6
26. Sait, S.T., Sørensen, L., Kubowicz, S., Vike-Jonas, K., Gonzalez, S.V., Asimakopoulos, A.G., Booth, A.M.J.E.P.: Microplastic fibres from synthetic textiles: Environmental degradation and additive chemical content **268**, 115745 (2021). https://doi.org/10.1016/j.envpol.2020.115745
27. Jans, Y., von Bloh, W., Schaphoff, S., Müller, C.: Global cotton production under climate change—Implications for yield and water consumption. Hydrol. Earth Syst. Sci. (2021). https://doi.org/10.5194/hess-25-2027-2021
28. Suresh Kumar, N., Padma Suvarna, R., Chandra Babu Naidu, K., Banerjee, P., Ratnamala, A., Manjunatha, H. A.: review on biological and biomimetic materials and their applications. Appl. Phys. A 126, 445 (2020). https://doi.org/10.1007/s00339-020-03633-z
29. Daria, M., Krzysztof, L., Jakub, M.: Characteristics of biodegradable textiles used in environmental engineering: a comprehensive review **268** (2020). https://doi.org/10.1016/j.jclepro.2020.122129s
30. Lin, S.H., A.P.-H.H., Hsu, S.-H..: Injectable, micellar chitosan self-healing hydrogel for asynchronous dual-drug delivery to treat stroke rats (2023). https://doi.org/10.1002/adfm.202303853

31. Minkoa, J.M.: Multifunctional and stimuli-responsive nanocarriers for targeted therapeutic delivery. Expert Opin. Drug Deliv. (2020). https://doi.org/10.1080/17425247.2021.1828339
32. Hu, M.G.X., Chen, X., Mao, W., Qian, X., Yuan, W.-E.: Micro/nanorobot: a promising targeted drug delivery system. Pharmaceutics **2020** (2020). https://doi.org/10.3390/pharmaceutics12070665
33. Habibi, Y., Lucia, L.A., Rojas, O.J.J.C.r.: Cellulose nanocrystals: chemistry, self-assembly, and applications. 110, 3479–3500 (2010). https://doi.org/10.1021/cr900339w
34. Usman, M., Radulescu, M.J.S.o.T.T.E.: Examining the role of nuclear and renewable energy in reducing carbon footprint: does the role of technological innovation really create some difference? 841, 156662 (2022). https://doi.org/10.1016/j.scitotenv.2022.156662
35. Recyled Products Available online: https://www.apple.com/in/environment/#:~:text=Recycled%20and%20renewable%20materials%20often,end%20our%20reliance%20on%20mining. Accessed on
36. Bajaj, P., Harris, J.F., Huang, J.-H., Nath, P., Iyer, R.: Advances and challenges in recapitulating human pulmonary systems: at the cusp of biology and materials. ACS Biomater. Sci. Eng. **2**, 473–488 (2016). https://doi.org/10.1021/acsbiomaterials.5b00480
37. Kenry, Y.J., Lim, C.: Emerging flexible and wearable physical sensing platforms for healthcare and biomedical applications. Emerging flexible and wearable physical sensing platforms for healthcare and biomedical applications. Microsyst. Nanoeng. 2 (2016). https://doi.org/10.1038/micronano.2016.43
38. Sartori, T., Calmon, J.L.J.J.o.B.E: Analysis of the impacts of retrofit actions on the life cycle energy consumption of typical neighbourhood dwellings 21, 158–172 (**2019**). https://doi.org/10.1016/j.jobe.2018.10.009
39. Natt Makul, R.F., Zeyad, A.M., Amran, M.: Use of recycled concrete aggregates in production of green cement-based concrete composites: a review. Research Gate 2021. https://doi.org/10.3390/cryst11030232
40. Aditya, L., T.M.I.M., Behzad Rismanchi. A review on insulation materials for energy conservation in buildings. Research Gate (2017). https://doi.org/10.1016/j.rser.2017.02.034
41. Gao, Y., Zheng, Q., Jonsson, J.C., Lubner, S., Curcija, C., Fernandes, L., Kaur, S., Kohler, C.J.A.E.: Parametric study of solid–solid translucent phase change materials in building windows **301**, 117467 (2021). https://doi.org/10.1016/j.apenergy.2021.117467
42. Kamyar Kabirifar, M.M.M., Cynthia Changxin Wang, Vivian WY Tam.: Construction and demolition waste management contributing factors coupled with reduce, reuse, and recycle strategies for effective waste management: a review. J. Clean. Prod. (2020). https://doi.org/10.1016/j.jclepro.2020.121265
43. Nikolaos Papadakis, D.A.K.: A review of energy efficiency interventions in public buildings. https://doi.org/10.3390/en16176329
44. LEED (2024)
45. Mohammad Arar, C.J., Ahmed Alqasimi, N.: Investigating the influence of the building material on the indoor air quality in apartment in Dubai. Frontiers (2022). https://doi.org/10.3389/fbuil.2021.804216
46. 1, P.R.a., S.M.a.c., S.P.b.; E.G.; Sharma, S.K.: Recent advances in the sustainable design and applications of biodegradable polymers (2021). https://doi.org/10.1016/j.biortech.2021.124739
47. Johnson, W.L., S.A.K., White, G.S., Herzberger, J.K.L., Heyliger, P.a.P.R.: Material matter (2016). https://doi.org/10.1063/1.4940511
48. S.R.V.R., T.R.W., A.W.N.: Biodegradable stent use for congenital heart disease. **61** (2021). https://doi.org/10.1016/j.ppedcard.2021.101349
49. M.H., S.J., M.F., F.G.e, S.A. et al.: Plant protein-based food packaging films; recent advances in fabrication, characterization, and applications **120** (2022). https://doi.org/10.1016/j.tifs.2022.01.013
50. Marcelo Pilotto Cenci, T.S., Munchen, D.D., Dartora, P.C., Veit, H.M., Bernardes, A.M., Dias, P.R.: Eco-friendly electronics—a comprehensive review (2021). https://doi.org/10.1002/admt.202001263

Game Changer: How Cellulose-Based Bioleather is Transforming the Market

Gaurav Mudgal, Rashi, Jaspreet Kaur, Jaspreet kaur, Sanjana Singh, Sanjoy K. Dhar, Ramasamy Vasanthrekha, Abhijeet Singh, Harshita Thakur, Simrandeep Kaur, Vani Sharma, Sanjhi Paliwal, Rohit Kumar Mishra, Barathi Seetharaman, Subhashini Sivaji, Lyudmila Asyakina, and Victor Atuchin

Abstract Leather has long been a cherished material in human civilization, adorning our garments and furnishings with its supple embrace. However, the traditional leather industry now faces a reckoning as the world awakens to the pressing need for environmental stewardship. The toll of livestock farming, the pollution of tanning processes, and the ethical concerns surrounding animal welfare have cast a shadow over this once-hallowed craft. Amidst this dilemma, cellulosic leathers emerge as a promising solution. These innovative materials are derived from the most abundant biopolymer on Earth—cellulose. By harnessing the power of this renewable resource, the fashion industry can now envision a future where style and sustainability coexist in perfect harmony. This comprehensive compilation delves deep into the potential

Gaurav Mudgal, Sanjoy K. Dhar, Rashi, Jaspreet Kaur, and Jaspreet kaur have equally contributed to this article

G. Mudgal (✉) · Rashi · J. Kaur · S. Singh · S. K. Dhar · A. Singh · H. Thakur · S. Kaur · V. Sharma
University Institute of Biotechnology, Chandigarh University, Mohali, Punjab 140413, India
e-mail: gauravmdgl@gmail.com

Rashi
e-mail: rashi221133@gmail.com

S. Singh
e-mail: sanju8433rw@gmail.com

G. Mudgal
Center for Waste Management and Renewable Energy, Saveetha Dental College and Hospitals, Saveetha Institute of Medical and Technical Sciences, Saveetha University, Chennai 600077, India

J. kaur
Department of Human Genetics, Punjabi University, Patiala, Punjab NH64, 147002, India

R. Vasanthrekha · B. Seetharaman · S. Sivaji
Department of Biotechnology, Faculty of Engineering and Technology, SRM Institute of Science and Technology, Kattankulathur, Chengalpattu Dt, Tamil Nadu 603203, India
e-mail: barathis@srmist.edu.in

S. Paliwal
Centre for Research Impact and Outcome (CRIO), Chitkara University, Punjab 140401, India

K. K. Kesari et al. (eds.), *Functionalized Cellulose Materials*, Engineering Materials,
https://doi.org/10.1007/978-3-031-76953-5_12

of cellulosic leather, tracing its historical context and the environmental challenges that have spurred its development. Through a meticulous examination of the latest advancements in production techniques, readers will discover how these plant-based materials are redefining the very essence of luxury and utility. Yet, the journey to a fully sustainable future is not without its obstacles. From the need for scalable manufacturing processes to the complexities of integrating cellulosic leather into established supply chains, the path ahead demands the collective efforts of visionaries, innovators, and policymakers. By addressing these challenges head-on, the fashion industry can unlock the true potential of cellulosic leathers, ushering in an era of ethical style and environmental stewardship. As the curtain rises on this new chapter in the story of leather, the possibilities are as boundless as the fields that nurture the very fibers that will shape our future. With each stitch, each design, the tapestry of sustainable fashion unfurls, inviting us to embrace a world where the beauty of our garments is matched only by the care we show for our precious planet.

Keywords Cellulose · Leather · Health · Ethics · Animal · Environment · SDG · Resources · Microbial · Vegan

R. K. Mishra
Centre of Science and Society, Faculty of Science, Central University of Allahabad, Prayagraj 211002, India
e-mail: rohit_ernet@yahoo.co.in

L. Asyakina
Laboratory for Phytoremediation of Technogenically Disturbed Ecosystems, Kemerovo State University, Krasnaya Street, 6, Kemerovo 650000, Russia

V. Atuchin
Laboratory of Optical Materials and Structures, Institute of Semiconductor Physics, SB RAS, Novosibirsk 630090, Russia

Research and Development Department, Kemerovo State University, Kemerovo 650000, Russia

Department of Industrial Machinery Design, Novosibirsk State Technical University, Novosibirsk 630073, Russia

R&D Center "Advanced Electronic Technologies", Tomsk State University, Tomsk 634034, Russia

S. K. Dhar
Banophool Adibashi Green Heart College, Plot-4, Block-A, Ahsanullah Road-1, Section-Mirpur-13, Dhaka 1216, Bangladesh

1 Introduction

Early humans used animal hides for basic protection from the elements, a practice not dissimilar to how other animals might use their environment for shelter. However, our intellectual capacity allowed us to refine this process, developing tanning techniques that transformed perishable skins into durable, versatile leather. This innovation exemplifies how human ingenuity has taken a natural resource and dramatically expanded its utility and longevity. This accelerated development is evident in the dramatic transformation of our lifestyles, which have evolved to provide increasing levels of comfort and access to resources. The leather industry itself illustrates this transformation. What began as a simple means of using animal byproducts has grown into a global industry, with leather being used not just for clothing and shelter, but for luxury goods, automotive interiors, and even specialized industrial applications. This evolution mirrors our broader societal shift from basic survival needs to complex, multi-faceted industries that cater to both necessities and desires. Our intellectual and technological adaptations have allowed us to modify our environment and way of life to an extent unparalleled in the natural world. In the case of leather, we've developed synthetic alternatives, pioneered eco-friendly tanning processes, and explored the biofabrication of leather-like materials. These advancements demonstrate how our problem-solving capabilities extend beyond mere utilization of natural resources to the creation of entirely new materials and processes. However, this rapid progress also brings new challenges, particularly in terms of sustainability and our relationship with the natural world. The leather industry, for instance, grapples with issues of animal welfare, the environmental impact of tanning processes, and the sustainability of synthetic alternatives. These challenges underscore the need for a more holistic approach to innovation, one that considers long-term environmental and ethical implications alongside immediate human needs and desires.

The fashion industry's reliance on animal-based materials, such as leather, has raised significant environmental and ethical concerns. Traditional leather production processes often involve harmful chemicals and contribute to water pollution, deforestation, and greenhouse gas emissions. As a result, there is a growing demand for sustainable alternatives, and vegan leather has emerged as a promising solution. A recent article explores the potential of plant-based substrates and food waste as raw materials for producing vegan leather [1]. By utilizing these sustainable resources, it is possible to create leather alternatives that are both environmentally friendly and have comparable physical properties to traditional leather. Examples of plant-based leather include mushroom leather, bacterial cellulose, and materials derived from pineapple and jellyfish. As research and development in this area continue, we can expect to see a wider range of sustainable and biodegradable leather alternatives entering the fashion market.

This chapter delves into the exciting world of bioleather, exploring its potential as a game-changer in the sustainable material landscape. We will begin by examining the environmental shortcomings of traditional leather and introduce cellulose, a readily available and renewable resource, as the foundation for bioleather production. Here, we will explore the diverse plant sources with high cellulose content, like fast-growing grasses and agricultural byproducts, that hold immense promise for sustainable bioleather production. Following this, we will embark on a journey through the fascinating process of transforming cellulose into a leather-like material. We will dissect the various stages involved, from the selection and pre-treatment of plant materials to the extraction and modification of cellulose, shaping, finishing, and ultimately, the creation of bioleather with desired properties. The chapter then dives into the exciting potential of bioleather. We will explore its functional characteristics, comparing its strength, breathability, and versatility to traditional leather. We will delve into the diverse applications of bioleather, showcasing its potential use in apparel, footwear, accessories, and even beyond fashion, such as in furniture and automotive interiors. Crucially, we will unveil the environmental benefits of bioleather, highlighting its reduced water footprint, lower reliance on harsh chemicals, and potential for biodegradability. We will explore how bioleather production can contribute to a more circular economy, utilizing waste streams from existing agriculture. Finally, we will peer into the future of bioleather, exploring ongoing research and development efforts. This includes advancements in bioconversion techniques, the potential of using transgenics for improved cellulose production, and the continuous push for scalability and cost-effectiveness to make bioleather a fully viable and accessible sustainable alternative. Throughout this chapter, we will be guided by the vision of a more sustainable future for the fashion industry. By harnessing the power of nature's innovation—cellulose—bioleather offers a compelling solution to bridge the gap between style and sustainability.

However, it's important to note that cellulosic leathers are not without challenges. Questions remain about their durability compared to traditional leather, the energy requirements for their production, and the potential for large-scale adoption by industries accustomed to traditional materials. Moreover, the emergence of cellulosic leathers raises complex socio-economic questions. While they offer environmental benefits, a widespread shift to these materials could disrupt traditional leather industries, potentially affecting livelihoods in regions where leather production is a significant economic activity.

The development of cellulosic leathers exemplifies how human innovation continues to evolve in response to new challenges. It represents a potential solution that balances our desire for leather-like materials with growing environmental and ethical concerns. However, like many technological advancements, it also introduces new complexities that we must navigate. As we move forward, the key will be to approach these innovations holistically, considering not just their immediate benefits, but also their long-term implications for our environment, economy, and society. The story of leather—from animal hides to cellulosic alternatives—serves as a powerful reminder of our capacity for innovation and the ongoing need to balance progress with sustainability.

2 Challenges with Traditional Leather and Its Emerging Cellulosic Correctives

While a significant contributor to global economies, the fashion and textile industry faces a pressing challenge—its environmental impact. Countries like India, Brazil, Russia, and China contribute a total of more than 45% of the world's leather export market worth over $74 billion in 2023 [2]. Traditional leather production has been a mainstay material for centuries. Leather production is one of humanity's oldest industries, historically providing an essential material for various applications. Before the advent of modern materials like glass, paper, and plastics, leather was indispensable. However, the traditional leather tanning process was environmentally harmful, relying on a combination of chemistry, biotechnology, and trial-and-error methods. This involved using substances like blood, dung, and eggs, resulting in significant pollution of air, water, and land.

At the outset, the leather industry has long been associated with environmental degradation. It thus also shapes a major opportunity to fulfill jobs for millions of people. Leather has been a coveted material, prized for its strength, durability, and timeless elegance. From the iconic armor of ancient warriors to the luxurious handbags of today, leather has weaved itself into the fabric of human history and fashion. Its traditional production methods, however, cast a long shadow on our environment. Multiple factors amount to this long shadow from the leather business. Firstly, as an industrial production, it has been a heftily resource-intensive process. Water is a critical resource in leather production, serving as the primary medium for processing raw hides and skins into finished leather. The industry is highly water-intensive, consuming approximately 40–45 L per kilogram of raw material. The tanning process relies heavily on about 130 different kinds of chemicals including harsh constituents like chromium, formaldehyde, and various dyes [3]. Raw wastewater generated from leather processing is characterized by its poor biodegradability and high salt content [4]. This complex composition presents significant challenges in wastewater treatment and disposal [5]. These chemicals can pollute waterways if not properly disposed of, harming aquatic life and potentially influencing human health in surrounding communities. Leather tanning effluent is categorized as one of the ten most polluting industrial wastewaters by several environmental regulatory agencies [6]. This severe classification underscores the significant environmental impact of the leather industry. The leather manufacturing process is divided into four primary stages: beamhouse, tanning, post-tanning, and finishing. While each stage contributes to wastewater generation, the post-tanning process is particularly significant, accounting for approximately 38% of the overall wastewater produced in the industry [7]. Consequently, the leather industry faces increasing pressure to develop and implement cleaner, more sustainable wastewater treatment technologies [8–10]. This imperative is driven by the need to protect both the environment and the industry's economic viability.

Besides this, the leather industry can contribute to deforestation as some sources rely on clearing land for animal grazing or cultivating plants used in tanning processes. This loss of forests disrupts ecosystems, reduces biodiversity, and contributes to greenhouse gas emissions. Bovine leather production is intricately linked to deforestation, particularly in regions like Brazil, a major exporter of this material [11−13]. From the ethical viewpoint, the leather industry's reliance on extensive cattle ranching contributes significantly to the clearing of forests for grazing land. This complex relationship between leather production and deforestation poses challenges for ensuring sustainable and ethical supply chains. A large portion of leather comes from animals raised on factory farms. These animals might be subjected to cramped living conditions, limited access to natural behaviors, and potential overuse of antibiotics [14]. The very essence of traditional leather production raises ethical concerns for those who oppose animal cruelty. The act of slaughtering animals for their hides goes against the values of many consumers who seek cruelty-free alternatives.

In addition to these cues, the leather business is also faced with the issue of limited Sustainability. This is included with a global landfill Burden. Traditional leather is not readily biodegradable, contributing to landfill waste. Although some leather products can last for years, they eventually end up in landfills. Next is the issue with high cost. Due to the complex processing methods and reliance on animal hides, leather products are often more expensive than alternatives, limiting accessibility for some consumers. All these culprits are highlighted in Fig. 1.

Hence, despite its rich history and undeniable functionality, traditional leather production comes with significant environmental and ethical burdens. As the world strives for a more sustainable future, leather emerges as a promising alternative, offering a chance to retain the beauty and utility of leather without environmental and ethical drawbacks. This is where the emerging field of cellulosic leather materials enters the picture, offering a promising avenue that aligns with our need for more sustainable and ethical alternatives. Cellulosic leathers, derived from plant- and/or microbial-based sources such as agricultural waste, wood pulp, or even lab-grown cellulose, represent a convergence of our technological capabilities and our growing environmental consciousness. While plants are the primary source of cellulose, microorganisms also produce this versatile polymer, known as bacterial cellulose or biocellulose (BC) [15–19]. Renowned for its low cost, eco-friendly nature, and adaptability, biocellulose has garnered significant research interest in recent years, driving exploration of its diverse applications. Although sharing similar chemical formula with plant cellulose, BC exhibits distinct structural properties. While plant cellulose fibers are relatively thick (13–22 μm) with moderate crystallinity (44–65%) [20, 21], BC boasts a nanofibrillar structure (10–100 nm) and exceptionally high crystallinity (around 90%). These structural differences endow BC with superior tensile strength (~70 Newtons) and hydrophilicity due to abundant hydroxyl groups [22, 23]. BC thus offers a promising alternative to traditional textile materials. Its unique properties, including high tensile strength, biodegradability, and moldability, make it a desirable material for the fashion industry. The ability to produce BC in various shapes and sizes through fermentation, coupled with its compatibility with natural dyes, positions it as a sustainable and eco-friendly textile option. By utilizing

Fig. 1 Compelling view showing the multiple burdens from production and wasted conventional leather

agro-industrial waste as substrates for BC production, the industry can further reduce its environmental impact and enhance the material's economic viability (Fig. 2).

These new materials offer several potential advantages over traditional leather:

1. Sustainability: Cellulosic leathers can be produced from renewable resources, potentially reducing the environmental impact associated with livestock farming.
2. Ethical production: As a plant-based alternative, cellulosic leather addresses animal welfare concerns associated with traditional leather.
3. Customization: The production process of cellulosic leather allows for greater control over properties like texture, durability, and color, potentially surpassing the limitations of natural hides.
4. Consistency: Unlike animal hides, which can vary in quality and appearance, cellulosic leathers can be produced with consistent characteristics.
5. Reduced water usage and chemical pollution: Many cellulosic leather production processes require less water and fewer harmful chemicals compared to traditional tanning methods.
6. Circular economy potential: Some cellulosic leathers are designed to be biodegradable or recyclable, aligning with circular economy principles.

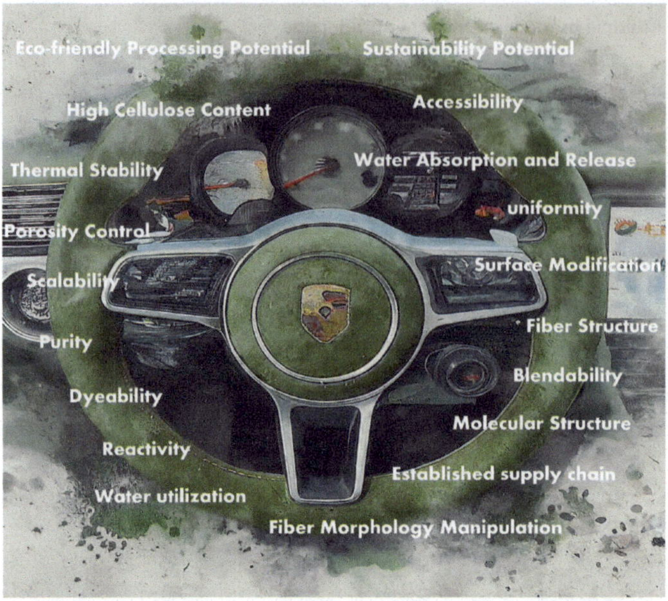

Fig. 2 Various decisive factors to be optimised, searched and researched for more efficient Bioleather production requisite

3 Cellulose: A Promising Material for Bioleather

Cellulose stands as one of the most abundant and renewable organic polymers on Earth, making it an ideal candidate for sustainable material development. This remarkable biopolymer is the primary structural component of plant cell walls, constituting approximately 33% of all plant matter. Its ubiquity in the natural world is staggering - from towering trees in ancient forests to the smallest blades of grass, cellulose forms the backbone of the plant kingdom. The renewability of cellulose is rooted in the continuous cycle of plant growth, driven by photosynthesis, which effectively transforms atmospheric carbon dioxide into complex organic structures. This process not only replenishes cellulose sources but also contributes to carbon sequestration, aligning with global efforts to mitigate climate change.

4 Cellulosic Sources

Cellulose can be derived from a vast array of sources (Table 1) beyond traditional forestry, including agricultural residues, fast-growing plants like bamboo, and even certain bacteria. The diversity of cellulose sources ensures a steady and adaptable supply chain, resistant to localized disruptions. Additionally, ongoing research into optimizing plant growth and cellulose extraction techniques promises to further enhance the efficiency and sustainability of cellulose production. This abundance and renewability, coupled with cellulose's biodegradability and versatile chemical structure, position it as a cornerstone material for the development of sustainable technologies across various industries, from textiles and packaging to advanced materials and biomedical applications. Let us discuss the prevalent existing and new sources of cellulose for use in bioleather materials.

4.1 Plant-Based Sources

4.1.1 Cotton as a Versatile Cellulosic Source

Several plants are particularly rich in cellulose content, making them valuable sources for cellulose extraction and utilization in various industries. One of the purest natural sources of cellulose is cotton, which contains up to 90–95% cellulose, the highest among natural fibers. This high purity reduces the need for extensive purification steps, streamlining the production process. Cotton fibers have a unique twisted ribbon-like structure, providing high surface area. This structure facilitates better chemical reactions and modifications during the bioleather production process. A high degree of polymerization and crystallinity contributes to the strength and durability of the resulting bioleather material. The cellulose in cotton is more accessible than wood or other plant sources, allowing for easier chemical treatments and modifications. Unlike wood or other plant sources, cotton fibers naturally contain minimal lignin or hemicellulose. This natural purity reduces the need for harsh chemical treatments to remove non-cellulosic components. Other than these attributes, cotton fibers are relatively uniform in structure and composition leading to more consistent bioleather products. The hydroxyl groups in cotton cellulose are readily available for chemical modifications, allowing for easier functionalization to impart leather-like properties. Given these, another intriguing factor is the prevalently established supply chains both nationally and internationally, thus ensuring a steady and reliable source of raw material for bioleather production. Processing technologies are well-developed and can be adapted allowing for easier scaling up of production processes. Organic and sustainably grown cotton can be used, enhancing the eco-friendly profile of the bioleather. Above all, cotton is biodegradable, contributing to the end-of-life sustainability of the product, and because waste from the textile industry can be repurposed for bioleather production, cotton-based bioleather manufacture may add a circular economy aspect to the process. Cotton's hydrophilic nature can be modulated to create bioleather with controlled moisture management properties. This can lead to more comfortable and breathable bioleather. Also, that this material readily accepts a wide range of dyes, that it allows for easy coloration of the bioleather to mimic various leather types. A thermal stability further allows for processing at higher temperatures if needed, without degradation. Being resistant to many chemicals, cotton allows for a range of chemical treatments in the bioleather production

Table 1 Compelling sources for bioleather material

Property/attribute	Cotton	Wood (Softwood)	Wood (Hardwood)	Bamboo	Flax	Hemp	Jute	Agave	Eucalyptus
Cellulose content (%)	90–95%	40–50%	40–55%	40–50%	60–80%	70–80%	60–65%	60–65%	40–50%
Fiber structure	Twisted ribbon-like	Long fibers	Shorter fibers	Long fibers	Long, strong fibers	Long, strong fibers	Long fibers	Long, coarse fibers	Short fibers
Purity	Very high	Moderate	Moderate	Moderate	High	High	Moderate	Moderate	Moderate
Uniformity	High	Moderate	Moderate	High	High	High	Moderate	Moderate	High
Accessibility of cellulose	High	Moderate	Moderate	Moderate	High	High	Moderate	Moderate	Moderate
Established supply chain	Very High	Very High	Very High	Moderate	High	Moderate	High	Low	High
Scalability	High	High	High	High	Moderate	Moderate	Moderate	Moderate	High
Eco-friendly processing Potential	High	Moderate	Moderate	High	High	High	High	High	Moderate
Reactivity	High	Moderate	Moderate	High	High	High	Moderate	Moderate	Moderate
Water absorption	High	Moderate	Moderate	High	Moderate	Moderate	High	Low	Moderate
Dyeability	High	Moderate	Moderate	High	High	High	Moderate	Moderate	High
Thermal stability	Good	Good	Good	Good	Good	Good	Good	Good	Good
Chemical Resistance	Good	Good	Good	Good	Good	Good	Moderate	Good	Good

(continued)

Table 1 (continued)

Property/attribute	Cotton	Wood (Softwood)	Wood (Hardwood)	Bamboo	Flax	Hemp	Jute	Agave	Eucalyptus
Sustainability Potential	Moderate	High	High	Very High	High	Very High	High	High	High
Waste utilization potential	High	High	High	Moderate	Moderate	Moderate	Moderate	High	High
Competition with food/textile	High	Low	Low	Low	Moderate	Low	Low	Low	Low
Growth rate	Moderate	Slow	Slow	Very Fast	Fast	Fast	Fast	Slow	Fast

process. These fibers can be easily manipulated (e.g., mercerization) to alter their properties, aiding in the tailoring of bio leather's characteristics to specific applications. Further, this cellulose can be easily blended with other materials catering to the creation of composite bioleathers with enhanced properties. The structure of cotton allows for control over the porosity of the final bioleather. This can be crucial for mimicking the breathability of natural leather. Easy applicability of enzyme treatments allow cotton processing with environmentally friendly methods compared to traditional leather. This aligns with the sustainability goals of bioleather production. However, it's important to note that the use of cotton for bioleather should be balanced with considerations of sustainable agriculture practices and potential competition with food crops or textile production. The ideal approach might involve utilizing cotton waste or developing specialized cotton varieties specifically for bioleather production.

4.1.2 Hardwood and Softwood Cellulose

Cellulose, derived from both softwoods (pine, spruce, fir) and hardwoods (oak, maple, birch), offers a promising avenue for innovation in the leather industry. While traditionally used in paper production and other applications, cellulose's abundance, renewability, and biodegradability make it an attractive alternative or complement to traditional leather materials. Softwood is particularly valued for its long cellulose fibers, which contribute to the strength and durability of the final material. The fibers from hardwood, on the other hand, offer different structural properties, allowing for customization of texture and flexibility in the resulting bioloeather. The production of bioloeather from hardwood and softwood involves breaking down the cellulose into smaller molecules, which are then reconstituted into a fibrous network. This network is combined with natural or synthetic binders to create a flexible, leather-like material. Recent advancements in biotechnological processes have optimized the conversion of wood-based cellulose into high-quality bioloeather, reducing the environmental impact and increasing the material's viability for commercial use. The versatility of hardwood and softwood-based bioloeather makes it an attractive material across various industries. In the fashion industry, it is used for making clothing, bags, and shoes, offering a sustainable alternative to animal leather without compromising on style or quality. In the automotive sector, bioloeather is being explored for use in car interiors, providing a luxurious yet eco-friendly option for upholstery. The furniture industry is also adopting bioloeather for chairs, sofas, and other upholstered items, capitalizing on its durability and aesthetic appeal. Additionally, the use of by-products from the forestry industry, such as sawdust and wood chips, further enhances the sustainability of the material by promoting waste valorization. Ongoing research is focused on improving the properties and production processes of hardwood and softwood-based bioloeather. Scientists are exploring the use of genetically modified microbes to increase the efficiency of cellulose breakdown and to produce bioloeather with enhanced properties. Studies are also investigating the combination of wood-based cellulose with other biopolymers or nanomaterials to create hybrid materials with superior performance characteristics.

The potential for customization in terms of texture, strength, and biodegradability is driving interest in hardwood and softwood-based bioloeather. As research progresses, we can expect to see new innovations that will expand the range of applications for this sustainable material.

4.1.3 Bamboo

Bamboo contains approximately 40–50% cellulose, with the added benefit of rapid growth. Bamboo cellulose presents a versatile and sustainable alternative to traditional cellulose sources. It exists in various forms, including nanocrystals, microcrystals, and hydrogels [24]. Compared to conventional cellulose from wood, bamboo cellulose offers advantages such as reduced resource demands, and a lower environmental impact. These factors contribute to its increasing popularity as a sustainable material across diverse industries. Bamboo fibers offer a compelling combination of strength, elasticity, and antimicrobial properties, making them a promising material for diverse applications [25]. These natural attributes stem from the presence of bioactive compounds like flavonoids, phenols, and lignin. While traditional extraction methods often rely on harsh chemicals or energy-intensive processes, the growing emphasis on sustainability has driven the search for eco-friendlier alternatives to unlock the full potential of bamboo fibers. A recent work aimed to optimize the production of non-woven fabric from waste bamboo fibers for potential medical applications [26]. By experimenting with different fiber sizes, fermentation times, and blending ratios with polyester, the study seeks to develop antibacterial, moldable, and high-performance fabric. The research includes assessing the physical and microbial properties of the resulting fabric, with a focus on its potential to inhibit bacterial growth. Ultimately, the goal is to create a practical and beneficial product for elderly care. Similarly in another study, the research aimed to enhance the properties of polyurethane imitation wood (PUIW) by incorporating cellulose nanofibers (CNFs) derived from bamboo [27]. By optimizing fiber size through fermentation and blending with polyester, the study sought to create a high-performance, antibacterial non-woven fabric suitable for medical applications. The research focused on improving the mechanical properties, thermal stability, and antimicrobial characteristics of the resulting material.

4.1.4 Flax

Flax, a versatile plant belonging to the Linaceae family, thrives in cooler climates. Renowned for its rapid growth cycle, flax matures within approximately 100 days, requiring minimal fertilizer and pesticide inputs compared to other crops like cotton [28]. This low environmental impact, coupled with its abundance, positions flax as a sustainable and attractive raw material for various industries. The fibers contain about 60–80% cellulose. Flax has been cultivated for centuries for its fiber and oil. Beyond textiles, flax fibers are increasingly recognized as a valuable reinforcement material for composites due to their superior mechanical properties [29]. The potential for developing biodegradable composites using flax-based resins is a growing area of interest, driven by the need for sustainable alternatives to traditional materials. Recent advancements in biotechnology have also opened doors for the development of transgenic flax varieties with enhanced properties, expanding the applications of this remarkable plant. In fact, apart from being used as a cellulosic bioleather source, Flax is also used as an additive in the chemical treatment of leather for inducing elasticity and softening [30]. This **f**atliquoring is a crucial step in leather processing that involves impregnating the leather with oils and lubricants to enhance its flexibility, softness, and durability. This process takes place after tanning and involves immersing the leather in a water-based solution containing the necessary chemicals. By coating the individual leather fibers, fatliquoring prevents them from sticking together during drying, resulting in a more supple and pliable final product. Many stores now readily offer such leather products developed from flax source [31–33].

4.1.5 Hemp

Hemp, a versatile and sustainable crop, has gained renewed attention in recent years. Its potential applications extend beyond traditional uses like textiles and paper to encompass a wide range of industries. This comprehensive review by Crini and coworkers explores the various uses of hemp, from its application in construction and insulation to its role in food, cosmetics, and pharmaceuticals [34]. With its sustainable production methods and numerous benefits, hemp offers a promising solution for addressing global challenges related to resource depletion and environmental impact. While hemp has been used for centuries, its prominence declined in the early twentieth century. However, growing concerns about resource depletion and environmental impact have led to a resurgence of interest in hemp fiber production. A recent article by Gedik and Avinc, provides an overview of hemp plant cultivation, fiber extraction, and processing methods [35]. It discusses the unique properties of hemp fiber, including its strength, durability, and sustainability. Additionally, they explored the various applications of hemp fiber in textile production, from yarn spinning to fabric weaving and finishing. Finally, it examines the economic potential of hemp fiber and its contribution to a more sustainable textile industry [35]. Hemp contains 70–80% cellulose in its fibers. The current thrust is to develop a sustainable and biodegradable leather alternative using a multilayered cellulosic material. A recent research explored the potential of a multilayered cellulosic material (MCM) as a sustainable alternative to traditional leather in footwear production [36]. By combining nonwoven cellulosic fiber mats with denim and hemp fabrics, the researchers created a biodegradable material that closely resembles the properties of calf and pigskin leather in terms of heat loss and break force. However, MCM exhibited superior breathability and evaporative properties, making it a promising option for sustainable footwear design [36]. The use of hemp fabric in MCM further contributes to its environmental sustainability, as hemp is a rapidly renewable and low-impact crop. Another study presents a preliminary life cycle assessment (LCA) comparing hemp leather to bovine leather, evaluating their environmental impacts [37]. The findings demonstrate that hemp leather offers a more sustainable alternative to traditional leather, with significantly lower greenhouse gas emissions, acidification potential, and eutrophication potential. While hemp leather requires slightly more water during manufacturing, its overall environmental footprint is substantially smaller than bovine leather. According to the researchers, further research into optimizing the manufacturing process can potentially reduce water consumption and enhance the sustainability of hemp leather [37].

4.1.6 Jute

Approximately 60–65% cellulose content. Jute, a natural fiber derived from the jute plant, offers a sustainable and biodegradable alternative to traditional leather. Its strength, durability, and resistance to abrasion make it a suitable material for various applications, including textiles, packaging, and composite materials. In recent years, researchers have explored the potential of jute as a base material for bioleather. Jute-based bioleather can be produced through a variety of processes, including chemical treatment, mechanical processing, or a combination of both. By incorporating jute fibers into a biopolymer matrix, it is possible to create materials with properties similar to traditional leather, such as flexibility, durability, and breathability. The use of jute in bioleather production offers several environmental benefits. Firstly, it benefits from offering a reduced carbon footprint. Jute is a rapidly renewable crop with a lower carbon footprint compared to animal-derived leather. Secondly, it offers reduced water consumption. The production of jute requires less water compared to traditional leather tanning processes. Then another advantage is biodegradability. Jute-based bioleather is biodegradable, reducing waste and environmental pollution. As research and development in this area continue, jute-based bioleather is poised to become a more viable and sustainable alternative to traditional leather products.

These plants offer diverse sources of cellulose, each with its own unique properties and potential applications. The choice of cellulose source often depends on factors such as local availability, growth rate, ease of extraction, and the specific requirements of the intended application. Agricultural residues like sugarcane bagasse, corn stalks, and rice straw are particularly interesting from a sustainability perspective, as they allow for cellulose extraction from what would otherwise be waste products. For example, a recent work introduces a novel, environmentally friendly leather alternative crafted from agro-waste components. By combining waste maple leaves, apple fruit pulp, kombucha biomass cellulose, biodegradable polyesters, and plasticizers, a group from the Centre of Polymer Systems, University Institute, Tomas Bata University in Zlin, Czech Republic have successfully developed biocomposite materials with comparable physical and mechanical properties to traditional leather

[38]. These biocomposites exhibit porosity, breathability, and significant mechanical strength, making them promising candidates for applications in leather accessories like handbags and shoe soles. This study contributes to the growing field of sustainable materials and offers a viable solution for reducing the leather industry's environmental impact.

4.2 Microbial Bioleather: A Frontier in Sustainable Materials

Microbial bioleather, a burgeoning field of research, explores the potential of microorganisms to produce leather-like materials with enhanced sustainability. This innovative approach offers a promising alternative to traditional animal-based leather, addressing environmental impact, animal welfare, and resource consumption concerns. A variety of microorganisms, including bacteria, fungi, and yeast, have been investigated for their ability to produce bioleather. For example, a bacteria, *Geobacillus thermodenitrificans,* produces a cellulose-based material with leather-like properties. A fungus, *Schizophyllum commune*, can be cultivated to produce a mycelium-based material with a leather-like texture. Another bacterium, *Komagataeibacter xylinus*, produces bacterial cellulose, a strong, flexible, and biodegradable material that can be used to create leather alternatives. Researchers have made significant strides in developing microbial bioleather. The first key area is material optimization. Scientists are exploring methods to enhance the mechanical properties, durability, and aesthetic appeal of microbial bioleather. Then comes optimizing scale-up production. Efforts are underway to develop scalable and cost-effective production processes for microbial bioleather. Then comes, functionalization. Many researchers are investigating ways to incorporate additional functionalities into microbial bioleather, such as antimicrobial properties, UV resistance, or enhanced breathability. Finally, one needs to emphasize how to find commercial translation of the researched bioleather outcomes. Well, many companies are beginning to introduce microbial bioleather products into the market, targeting various applications such as fashion accessories, footwear, and automotive interiors. Despite the promising advancements, several challenges remain in the development and commercialization of microbial bioleather. For example, scaling up production and reducing costs are essential for widespread adoption. Ensuring consistent quality and performance of microbial bioleather can be challenging. Similarly, building consumer awareness and acceptance of microbial bioleather as a viable alternative to traditional leather is crucial. Future research should focus on addressing these challenges and exploring new applications for microbial bioleather, such as medical devices, packaging materials, and filtration membranes. Overall, microbial bioleather represents a promising avenue for sustainable materials development. As research progresses and production costs decrease, we can expect to see growing adoption of microbial bioleather in various industries, contributing to a more sustainable and environmentally friendly future.

5 Conclusions

The development of cellulosic leathers represents a significant step towards a more sustainable and ethical fashion industry. By harnessing the power of renewable cellulose, these materials offer a viable alternative to traditional animal-based leather, addressing concerns related to environmental impact and animal welfare. While challenges such as durability and cost-competitiveness remain, continued research and development can pave the way for widespread adoption. Cellulosic leathers align with several Sustainable Development Goals, contributing to responsible consumption, innovation, climate action, and the protection of terrestrial ecosystems. By embracing this innovative material, the fashion industry can play a vital role in driving sustainable development and creating a more environmentally responsible future.

References

1. Pavani, C. et al.: in From Waste to Wealth. (eds. R.K. Arya, G.D. Verros, O.P. Verma & C.M. Hussain) pp 951–964. Springer Nature Singapore, Singapore (2024)
2. geeksforgeeks.org (2024)
3. Dandira, V., Mugwindiri, K., Chikuku, T.J.: A cleaner production exercise of a leather manufacturing company: a zimbabwean experience. Int. J. Scientif. Technol. Res. **1**, 1–5 (2012)
4. Hansen, É., de Aquim, P.M., Gutterres, M.: Environmental assessment of water, chemicals and effluents in leather post-tanning process: a review. Environ. Impact Assess. Rev. **89**, 106597 (2021)
5. Lofrano, G., Meriç, S., Zengin, G.E., Orhon, D.: Chemical and biological treatment technologies for leather tannery chemicals and wastewaters: a review. Sci. Total Environ. **461–462**, 265–281 (2013)
6. Kumar, V., Majumdar, C., Roy, P.: Effects of endocrine disrupting chemicals from leather industry effluents on male reproductive system. J. Steroid Biochem. Mol. Biol. **111**, 208–216 (2008)
7. Cassano, A., Molinari, R., Romano, M., Drioli, E.: Treatment of aqueous effluents of the leather industry by membrane processes: a review. J. Membr. Sci. **181**, 111–126 (2001)
8. Sundarapandiyan, S., Chandrasekar, R., Ramanaiah, B., Krishnan, S., Saravanan, P.: Electrochemical oxidation and reuse of tannery saline wastewater. J. Hazard. Mater. **180**, 197–203 (2010)
9. Krishnamoorthy, G., Sadulla, S., Sehgal, P.K., Mandal, A.B.: Green chemistry approaches to leather tanning process for making chrome-free leather by unnatural amino acids. J. Hazard. Mater. **215–216**, 173–182 (2012)
10. Suresh, V., Kanthimathi, M., Thanikaivelan, P., Raghava Rao, J., Unni Nair, B.: An improved product-process for cleaner chrome tanning in leather processing. J. Clean. Prod. **9**, 483–491 (2001)
11. Mammadova, A., Masiero, M., Pettenella, D.: Embedded deforestation: the case study of the Brazilian—Italian Bovine leather trade **11**, 472 (2020)

12. Mammadova, A., Behagel, J., Masiero, M., Pettenella, D.: Deforestation as a systemic risk: the case of Brazilian Bovine leather **13**, 233 (2022)
13. Mammadova, A., Behagel, J., Masiero, M.: Making deforestation risk visible. Discourses on bovine leather supply chain in Brazil. Geoforum **112**, 85–95 (2020)
14. Plannthin, D.-K.: In: S.S. Muthu & M.A. Gardetti (eds.) Green fashion: Volume 2, pp. 49–122. Springer Singapore, Singapore (2016)
15. Costa, A.F.d.S. et al.: Dyeing of bacterial cellulose films using plant-based natural dyes. Int. J. Biol. Macromol. **121**, 580–587 (2019)
16. Albuquerque, R.M. et al.: Production of a bacterial cellulose/poly(3-hydroxybutyrate) blend activated with clove essential oil for food packaging. **29**, 259–270 (2021)
17. Amorim, J.D.P., et al.: Obtainment of bacterial cellulose with added propolis extract for cosmetic applications **62**, 565–575 (2022)
18. Qasim, U., et al.: Renewable cellulosic nanocomposites for food packaging to avoid fossil fuel plastic pollution: a review. Environ. Chem. Lett. **19**, 613–641 (2021)
19. de Medeiros, A.D.L.M., et al.: Biocellulose for treatment of wastewaters generated by energy consuming industries: a review **14**, 5066 (2021)
20. Ververis, C., Georghiou, K., Christodoulakis, N., Santas, P., Santas, R.: Fiber dimensions, lignin and cellulose content of various plant materials and their suitability for paper production. Ind. Crops Prod. **19**, 245–254 (2004)
21. de Amorim, J.D.P., et al.: Plant and bacterial nanocellulose: production, properties and applications in medicine, food, cosmetics, electronics and engineering. A Review. Environ. Chem. Lett. **18**, 851–869 (2020)
22. Ng, F.M.C., Wang, P.W.: Natural self-grown fashion from bacterial cellulose: a paradigm shift design approach in fashion creation. Des. J. **19**, 837–855 (2016)
23. Gelin, K., et al.: Characterization of water in bacterial cellulose using dielectric spectroscopy and electron microscopy. Polymer **48**, 7623–7631 (2007)
24. Rashid, M.M., Abir, N., Hossain, M.A., Iqbal, M.I. & Siddique, A.B.: In: ArunRamnath, R., Sanjay, M.R., Siengchin, S., Fiore, V. (eds.) Cellulose fibre reinforced composites, pp. 23–48 Woodhead Publishing (2023)
25. Jain, S., Kumar, R., Jindal, U.C.: Mechanical behaviour of bamboo and bamboo composite. J. Mater. Sci. **27**, 4598–4604 (1992)
26. Chonsakorn, S. et al.: A novel non-woven fabric from bamboo fiber in medical lifestyle products **10** (2024)
27. Liu, Y., et al.: Bamboo-based cellulose nanofibers as reinforcement for polyurethane imitation wood. Ind. Crops Prod. **210**, 118177 (2024)
28. Duchemin, B., et al.: Ultrastructure of cellulose crystallites in flax textile fibres. Cellulose **19**, 1837–1854 (2012)
29. Wróbel-Kwiatkowska, M., et al.: New biocomposites based on bioplastic flax fibers and biodegradable polymers. Biotechnol. Prog. **28**, 1336–1346 (2012)
30. Żarłok, J., Śmiechowski, K., Mucha, K., Tęcza, A.: Research on application of flax and soya oil for leather fatliquoring. J. Clean. Prod. **65**, 583–589 (2014)
31. Meesho.com (2024)
32. Etsy.com (2024)
33. https://monkeesoffairfax.com (2024)
34. Crini, G., Lichtfouse, E., Chanet, G., Morin-Crini, N.: In: Crini, G., Lichtfouse, E. (eds.) Sustainable agriculture reviews 42: hemp production and applications, pp 37–87. Springer International Publishing, Cham (2020)
35. Gedik, G., Avinc, O.: In: Belwal, T., Belwal, N.C. (eds.) Revolutionizing the potential of hemp and its products in changing the global economy, pp 69–95. Springer International Publishing, Cham (2022)
36. Nam, C., Lee, Y.-A.: Multilayered cellulosic material as a leather alternative in the footwear industry **37**, 20–34 (2019)

37. Hultkrantz, M.: In: Mechanical engineering, Vol. Bachelor of Science in Engineering—(KTH ROYAL INSTITUTE OF TECHNOLOGY, SCHOOL OF ARCHITECTURE AND THE BUILT ENVIRONMENT (2018)
38. Saha, N., Ngwabebhoh, F.A., Nguyen, H.T., Saha, P.: Environmentally friendly and animal free leather: Fabrication and characterization. AIP Conference Proceedings **2289** (2020)

Correction to: Comparative Analysis of Nanomaterials and Artificial Intelligence for Sustainable Nutrient Management in Soil

Madhu Bala and Virat Khanna

Correction to:
Chapter "Comparative Analysis of Nanomaterials and Artificial Intelligence for Sustainable Nutrient Management in Soil" in: K. K. Kesari et al. (eds.), *Functionalized Cellulose Materials*, Engineering Materials, https://doi.org/10.1007/978-3-031-76953-5_6

In the original version of this book, both the authors were incorrectly affiliated with both Lovely Professional University and Chandigarh University, which has been updated with the following belated corrections: Madhu Bala, affiliated solely with Lovely Professional University (LPU) and Virat Khanna is affiliated solely with Chandigarh University (CU). The book has been updated with the changes.

The updated version of this chapter can be found at
https://doi.org/10.1007/978-3-031-76953-5_6

K. K. Kesari et al. (eds.), *Functionalized Cellulose Materials*, Engineering Materials,
https://doi.org/10.1007/978-3-031-76953-5_13

The manufacturer's authorised representative in the EU is Springer Nature Customer Service Centre GmbH, Europaplatz 3, 69115 Heidelberg, Germany. If you have any concerns regarding our products, please contact ProductSafety@springernature.com

Printed and bound by CPI Group (UK) Ltd, Croydon, CR0 4YY
30/09/2026
02225600-0007